74 Springer Series in Solid-State Sciences

Edited by Peter Fulde

Springer Series in Solid-State Sciences

Editors: M. Cardona P. Fulde K. von Klitzing H.-J. Queisser

40 **Semiconductor Physics** An Introduction
3rd Edition By K. Seeger

41 **The LMTO Method**
Muffin-Tin Orbitals and Electronic Structure
By H. L. Skriver

42 **Crystal Optics with Spatial Dispersion, and Excitons**
By V. M. Agranovich and V. L. Ginzburg

43 **Resonant Nonlinear Interactions of Light with Matter**
By V. S. Butylkin, A. E. Kaplan, Yu. G. Khronopulo, and E. I. Yakubovich

44 **Elastic Media with Microstructure II**
Three-Dimensional Models
By I. A. Kunin

45 **Electronic Properties of Doped Semiconductors**
By B. I. Shklovskii and A. L. Efros

46 **Topological Disorder in Condensed Matter**
Editors: F. Yonezawa and T. Ninomiya

47 **Statics and Dynamics of Nonlinear Systems**
Editors: G. Benedek, H. Bilz, and R. Zeyher

48 **Magnetic Phase Transitions**
Editors: M. Ausloos and R. J. Elliott

49 **Organic Molecular Aggregates,** Electronic Excitation and Interaction Processes
Editors: P. Reineker, H. Haken, and H. C. Wolf

50 **Multiple Diffraction of X-Rays in Crystals**
By Shih-Lin Chang

51 **Phonon Scattering in Condensed Matter**
Editors: W. Eisenmenger, K. Laßmann, and S. Döttinger

52 **Superconductivity in Magnetic and Exotic Materials**
Editors: T. Matsubara and A. Kotani

53 **Two-Dimensional Systems, Heterostructures, and Superlattices**
Editors: G. Bauer, F. Kuchar, and H. Heinrich

54 **Magnetic Excitations and Fluctuations**
Editors: S. Lovesey, U. Balucani, F. Borsa, and V. Tognetti

55 **The Theory of Magnetism II**
Thermodynamics and Statistical Mechanics
By D. C. Mattis

56 **Spin Fluctuations in Itinerant Electron Magnetism** By T. Moriya

57 **Polycrystalline Semiconductors,**
Physical Properties and Applications
Editor: G. Harbeke

58 **The Recursion Method and Its Applications**
Editors: D. Pettifor and D. Weaire

59 **Dynamical Processes and Ordering on Solid Surfaces**
Editors: A. Yoshimori and M. Tsukada

60 **Excitonic Processes in Solids**
By M. Ueta, H. Kanzaki, K. Kobayashi, Y. Toyozawa, and E. Hanamura

61 **Localization, Interaction, and Transport Phenomena**
Editors: B. Kramer, G. Bergmann, and Y. Bruynseraede

62 **Theory of Heavy Fermions and Valence Fluctuations**
Editors: T. Kasuya and T. Saso

63 **Electronic Properties of Polymers and Related Compounds**
Editors: H. Kuzmany, M. Mehring, and S. Roth

64 **Symmetries in Physics:** Group Theory Applied to Physical Problems
By W. Ludwig and C. Falter

65 **Phonons: Theory and Experiments II**
Experiments and Interpretation of Experimental Results By P. Brüesch

66 **Phonons: Theory and Experiments III**
Phenomena Related to Phonons
By P. Brüesch

67 **Two-Dimensional Systems: Physics and New Devices**
Editors: G. Bauer, F. Kuchar, and H. Heinrich

68 **Phonon Scattering in Condensed Matter V**
Editors: A. C. Anderson and J. P. Wolfe

69 **Nonlinearity in Condensed Matter**
Editors: A. R. Bishop, D. K. Campbell, P. Kumar and S. E. Trullinger

70 **From Hamiltonians to Phase Diagrams**
The Electronic and Statistical-Mechanical Theory of sp-Bonded Metals and Alloys
By J. Hafner

71 **High Magnetic Fields in Semiconductor Physics**
Editor: G. Landwehr

72 **One-Dimensional Conductors**
By S. Kagoshima, T. Sambongi, and H. Nagasawa

73 **Quantum Solid-State Physics**
Editors: S. V. Vonsovsky and M. I. Katsnelson

74 **Quantum Monte Carlo Methods** in Equilibrium and Nonequilibrium Systems
Editor: M. Suzuki

Volumes 1–39 are listed on the back inside cover

Quantum Monte Carlo Methods in Equilibrium and Nonequilibrium Systems

Proceedings of the Ninth Taniguchi International Symposium, Susono, Japan November 14–18, 1986

Editor: M. Suzuki

With 156 Figures

Springer-Verlag Berlin Heidelberg New York London Paris Tokyo

Professor Masuo Suzuki

Faculty of Science, Department of Physics, University of Tokyo,
7-3-1 Hongo, Bunkyo-ku, Tokyo 113, Japan

Series Editors:

Professor Dr., Dres. h. c. Manuel Cardona
Professor Dr., Dr. h. c. Peter Fulde
Professor Dr. Klaus von Klitzing
Professor Dr. Hans-Joachim Queisser

Max-Planck-Institut für Festkörperforschung, Heisenbergstrasse 1
D-7000 Stuttgart 80, Fed. Rep. of Germany

ISBN-13:978-3-642-83156-0 e-ISBN-13:978-3-642-83154-6
DOI: 10.1007/978-3-642-83154-6

Library of Congress Cataloging-in-Publication Data. Taniguchi International Symposium on the Theory of Condensed Matter (9th : 1986 : Susono-shi, Japan) Quantum Monte Carlo methods in equilibrium and nonequilibrium systems. (Springer series in solid-state sciences ; 74) Includes index. 1. Monte Carlo method–Congresses. 2. Quantum theory–Congresses. I. Suzuki, M. (Masuo), 1937-. II. Title. III. Series.
QC174.85.M64T36 1986 530.1'33 87-16461

Softcover reprint of the hardcover 1st edition 1987

2153/3150-543210

Foreword

Speech by Toyosaburo Taniguchi

Dr. Kubo, Chairman, Distinguished Guests, and Friends,

I am very happy, pleased and honored to be here this evening with so many distinguished guests, friends, and scholars from within this country and from different parts of the world. The Taniguchi Foundation wishes to extend a warm and sincere welcome to the many participants of the Ninth International Symposium on the Theory of Condensed Matter, which series was inaugurated eight years ago through the strenuous efforts of Dr. Ryogo Kubo, who is gracing us today with his presence.

We are deeply indebted to Dr. Kubo, Dr. Suzuki, and their associates, who have spent an enormous amount of time and effort to make this particular symposium possible. We are convinced that the foundation should not be considered as what makes our symposium a success. The success is entirely due, I feel, to the continuous efforts of the Organizing Committee and of all those who have lent their support to this program. In this sense, your words of praise about the symposium, if any, should be directed to all of them.

So far, I have met in person a total of 62 participants in this Division from 12 countries: Argentina, Belgium, Canada, Denmark, the Federal Republic of Germany, France, Ireland, Israel, Rumania, Switzerland, the United Kingdom, and the United States of America, with 133 participants from Japan. Those friends I have been privileged to make, I shall always treasure.

Whenever I meet with the participants in our symposia, both young and old, I am deeply impressed by the unselfish and sincere dedication they display in pursuing their vocation. To those younger people who are made welcome by world-famous scholars as their friends, we offer our hopes that even after climbing the ladder of academic fame within, say, the next 10, 20 or 30 years, you will join forces, and help to forge closer bonds of friendship and cooperation in a manner that will make a major contribution not only to academia but also to permanent world peace. And, it is our hope that our symposia be continued as long as the fund permits, for several more years at least.

In conclusion, we sincerely trust that all the participants, both Japanese and foreign, will return home with pleasant memories of the symposium and of our enjoyable time together.

Thank you.

Preface

This volume contains papers presented at the Ninth Taniguchi Symposium on the Theory of Condensed Matter, which was held 14–18 November, 1986, at Susono, Japan. The topic of the symposium was "Quantum Monte Carlo Methods in Equilibrium and Nonequilibrium Systems".

The field of quantum Monte Carlo methods in equilibrium and nonequilibrium systems, namely quantum statistical Monte Carlo methods, has been studied only for about ten years. Quite recently many physicists began to work in this field, partly because larger-scale and higher-speed computers are now available, and partly because many important problems cannot yet be solved analytically. The basic idea of quantum statistical Monte Carlo methods is to transform the relevant d-dimensional quantum system into the corresponding (d+1)-dimensional classical system, using the generalized Trotter formula.

This volume consists of three parts. In the first paper of Part I, quantum statistical Monte Carlo methods are reviewed generally with great emphasis on methodology and some mathematical aspects such as the convergence and correction of the so-called ST transformation. The following four papers treat some general aspects of quantum systems, which are related to the quantum Monte Carlo approach. Part II contains ten papers which treat quantum spin systems using quantum statistical Monte Carlo methods, and two papers which treat quantum spin glasses. The five papers in Part III treat fermion systems, including nuclear systems.

The present Taniguchi Symposium was supported by the Taniguchi Foundation. Mr. Toyosaburo Taniguchi, who is the former president of Toyobo Co., Ltd., has supported various academic activities in the natural and human sciences for many years. His speech delivered at the reception in Kyoto after the symposium appears as the Foreword to this volume to illustrate the philosophy behind this support.

On behalf of all the participants we would like to express our hearty thanks to Mr. Taniguchi and the members of the Taniguchi Foundation.

Tokyo, 1 March 1987

Masuo Suzuki

First row (left to right)
N. Ito, M. Kolb, J.E. Hirsh, M. Suzuki, R. Kubo, D.J. Scalapino, J.W. Negele, J. Gubernatis, T. Onogi, K. Saitoh

Second row (left to right)
H. De Raedt, M.A. Novotny, H. Betsuyaku, M. Imada, T. Oguchi, S. Homma, I. Morgenstern

Third row (left to right)
S. Miyashita, H. Takano, H. Ishii, O. Nagai, M. Marcu

Fourth row (left to right)
M. Takasu, Y. Okabe, M. Katori

Contents

Part I

General Aspects of Quantum Monte Carlo Methods

General Review of Quantum Statistical Monte Carlo Methods

M. Suzuki

Department of Physics, Faculty of Science, University of Tokyo, Hongo, Bunkyo-ku, Tokyo 113, Japan

1. INTRODUCTION

The main purpose of the present paper is to explain the methodology of quantum statistical Monte Carlo Simulations to those who are coming into this rather new field of quantum Monte Carlo at finite temperatures.

Until quite recently, physics has been classified into the two categories, namely theoretical physics and experimental physics. Now there exists another category, so-called computational physics mainly based on computer simulations with use of Monte Carlo methods [1-5]. The classical Monte Carlo method was introduced by Metropolis et al. [6]. Numerical simulations of quantum systems at zero temperature were started in rather early days after quantum mechanics was established [5]. There have been proposed many methods such as the Green-function Monte-Carlo method by Kalos [7], variational Monte Carlo methods [5], and diffusion equation Monte Carlo methods [5], in order to study quantum-mechanical properties at zero temperature.

However, quantum Monte Carlo at finite temperatures had been believed to be quite difficult except a special symmetric system of the isotropic Heisenberg model [8], until the present author [9-12] proposed a general approach of "quantum statistical Monte Carlo" in which d-dimensional quantum systems are mapped onto (d+1)-dimensional classical systems using the following Trotter-like formula of exponential operators [9-12]:

$$e^{A_1+A_2+\cdots+A_p} = \lim_{n\to\infty}(e^{A_1/n}e^{A_2/n}\cdots e^{A_p/n})^n. \qquad (1.1)$$

The above general approach was first applied to quantum spin systems by Suzuki, Miyashita and Kuroda [11,13]. Barma and Shastry [14] applied this general idea to find equivalent classical models of one-dimensional fermi lattices. Hirsch et al. [15] performed explicitly Monte Carlo simulations in fermi systems. On the other hand, De Raedt and Lagendijk [16,17] studied extensively quantum spin, fermi and bose systems, and polaron problems, using the above general transformation method (1.1).

Many numerical investigations of the validity and convergence of the above general approach based on (1.1) have been reported by De Raedt-De Raedt [18], Wiesler [19], Cullen-Landau [20]. There have been reported many other interesting applications of the above idea to S = 1/2 spin systems by Satija-Wysin-Bishop [21], Marcu-Müller-Schmatzer [22], Sakaguchi-Kubo-Takada [23], to higher-spin systems by Takano [24] and Marcu-Wiesler [25], and to fermi gas by Takahashi and Imada [26]. Kolb [27] performed Monte Carlo renormalization group calculations in the two-dimensional quantum transverse Ising model by transforming it to the corresponding three-dimensional Ising model, as was suggested by the present author [28]. Betsuyaku [29] performed calculations of the linear quantum spin systems numerically by using the quantum transfer-matrix method [29,30]. Tsuzuki [31] applied this transfer-matrix method to the case of higher-spin chains using the cluster decomposition method [12,30]. Quite recently Nagai et al. [32] applied the above general method to the transverse Ising model.

In Section 2, basic ideas of the quantum statistical Monte Carlo will be explained in detail. Namely we discuss generalized Trotter-like approximations, decomposition formulae of exponential operators and convergence theorems [33], and moment- and cumulant-expansions of sliced exponential operators. The equivalence theorem by the present author [10,12] will be also reviewed in Section 2. In Section 3, the thermofield quantum Monte Carlo method [34-39] will be presented. In Section 4, the thermofield transfer-matrix methods will be discussed. The decomposition of the Trotter-like formula (1.1) is not unique and consequently it has the merit that cluster decompositions are also applicable in (1.1), as will be discussed in Section 6. In Section 7, the quantum Monte Carlo renormalization approach will be discussed. In Section 8, quantum spin glasses will be discussed briefly using some equivalent classical random spin systems. In Section 9, Fisher's finite-size scaling law is discussed in connection with Monte Carlo simulations. A finite-Trotter-number scaling ansatz is proposed in Section 10. This is originally proposed in the present paper. In Section 11, correlation identities by the present author [41] will be presented in a completely general form with possible applications to testing of Monte Carlo calculations. They take the following general form

$$\langle fg \rangle = \langle f \langle g \rangle_{\mathcal{H}_g} \rangle \tag{1.2}$$

for arbitrary classical functions f and g under the conditions that f and g do not contain common variables that $\langle g \rangle$ is the canonical average of g over the partial Hamiltonian $\mathcal{H}_g$ which is connected to the variables contained in g. This is a complete generalization of Callen's identity [42]. These correlation identities are very useful to test the accurary of Monte Carlo calculations, as was discussed by Mouritsen [3]. In Section 12, symmetry properties of Kubo's canonical correlations with respect to Trotter's number n will be studied in detail. In Section 13, quantum dynamics via Monte Carlo will be discussed. In Section 14, summary and future problems will be given.

2. GENERAL THEORY OF QUANTUM MONTE CARLO

2.1 Generalized Trotter-like Approximations

As was briefly mentioned in Section 1, the quantum statistical Monte Carlo method [10-13] is based on Trotter's formula [43] or the generalized Trotter formula (1.1). It is convenient to introduce here the following extremely generalized Trotter-like formula [44,45]

$$e^{-\beta\mathcal{H}} = (e^{-\beta\mathcal{H}/n})^n = \lim_{n\to\infty}(f_m[-\tfrac{1}{n}\beta\mathcal{H}])^n, \tag{2.1}$$

where the approximant $f_m[-(1/n)\beta\mathcal{H}]$ satisfies the following condition

$$e^{-\beta\mathcal{H}/n} = f_m[-\frac{1}{n}\beta\mathcal{H}] + O(\frac{1}{n^m}) \qquad (2.2)$$

for large n, with some positive integer m.

2.2 Decomposition Formulae of Exponential Operators and Convergence Theorems

Now we have the following theorem concerning the above generalized Trotter-like decomposition.

Theorem 1 [12]: With the condition (2.2), we have

$$e^{-\beta\mathcal{H}} = (f_m[-\frac{1}{n}\beta\mathcal{H}])^n + O(\frac{1}{n^{m-1}}), \qquad (2.3)$$

or more explicitly

$$\| e^{-\beta\mathcal{H}} - (f_m[-\frac{1}{n}\beta\mathcal{H}])^n \| \leq nM^{n-1} \| e^{-\beta\mathcal{H}/n} - f_m[-\frac{1}{n}\beta\mathcal{H}] \|, \qquad (2.4)$$

where

$$M \equiv \max\{\|\exp(-\frac{1}{n}\beta\mathcal{H})\|, \|f_m[-\frac{1}{n}\beta\mathcal{H}]\|\}. \qquad (2.5)$$

These are very useful in performing quantum Monte Carlo simulations of many-body quantum systems in equilibrium, as shown in previous papers [10-32].

The partition function of the relevant system described by the Hamiltonian is given by

$$Z = \mathrm{Tr}\, e^{-\beta\mathcal{H}} = \mathrm{Tr}(f_m[-\frac{1}{n}\beta\mathcal{H}])^n + O(\frac{1}{n^{m-1}}). \qquad (2.6)$$

A more non-uniformly generalized form of Eq. (2.3) is given by

$$e^{-\beta\mathcal{H}} = e^{-(\beta-\beta_n)\mathcal{H}} e^{-(\beta_n-\beta_{n-1})\mathcal{H}} \cdots e^{-(\beta_1-\beta_0)\mathcal{H}} e^{-\beta_0\mathcal{H}}$$

$$= f_m[-(\Delta\tau)\mathcal{H}] f_m[-(\Delta\tau_n)\mathcal{H}] \cdots f_m[-(\Delta\tau_0)\mathcal{H}] + O(\overline{\Delta\tau}^{m-1}), \qquad (2.7)$$

where

$$\overline{\Delta\tau} = \max\{\Delta\tau, \Delta\tau_n, ---, \Delta\tau_0\} \text{ and } \Delta\tau=\beta-\beta_n, \Delta\tau_n=\beta_n-\beta_{n-1}, ---, \Delta\tau_0=\beta_0. \qquad (2.8)$$

There are many different ways to express the approximate operator $f_m[-\beta\mathcal{H}/n]$ explicitly, and to evaluate $\mathrm{Tr}(f_m[-\beta\mathcal{H}/n])^n$ numerically, as in [10-32]. Many people have been trying to find new approximate expressions of $f_m[-\beta\mathcal{H}/n]$ and efficient evaluation methods of them in the above general scheme of generalized Trotter-like approximations (2.3).

For example, we have [9,10,12,33]

$$f_2[A_1+A_2+---+A_p] = e^{A_1}e^{A_2}---e^{A_p}. \qquad (2.9)$$

These Trotter-like approximations have been used frequently [10-33].

It will be instructive to summarize here some useful decomposition formulae.

<u>Theorem 2 [33]: For any set of operators $\{A_j\}$ in a Banach algebra (i.e., normed space),</u>

$$\left\| \exp\left(\sum_{j=1}^{p} A_j\right) - \left(\prod_{j=1}^{p} e^{(1/n)A_j}\right)^n \right\| \leq \frac{1}{2n}\left(\sum_{j>k} \| [A_j, A_k] \|\right) \exp\left(\sum_{j=1}^{p} \| A_j \|\right) \tag{2.10}$$

<u>with an arbitrary positive integer p.</u>

In particular, for p=2, we obtain Trotter's formula

$$e^{A+B} = \lim_{n\to\infty}(e^{A/n}\, e^{B/n})^n. \tag{2.11}$$

<u>Theorem 3 [33]: For any operators A and B in a Banach algebra,</u>

$$\left\| e^{A+B} - (e^{A/2n}\, e^{B/n}\, e^{A/2n})^n \right\| \leq \frac{1}{n^2}\Delta_2(A,B), \tag{2.12}$$

<u>where</u>

$$\Delta_2(A,B) = \frac{1}{12}\{\| [[A,B],B] \| + \frac{1}{2}\| [[A,B],A] \|\} \times \exp(\|A\|+\|B\|). \tag{2.13}$$

Now we discuss higher order decompositions of the general exponential operator $\exp(A_1+A_2+\cdots+A_p)$. For this purpose, we may find $f_m(\{\frac{1}{n}A_j\})$ which satisfies the condition

$$\exp\left(\frac{1}{n}\sum_{j=1}^{p} A_j\right) = f_m\left(\left\{\frac{1}{n}A_j\right\}\right) + O\left(\frac{1}{n^m}\right), \tag{2.14}$$

according to the generalized Trotter-like formula (2.1) with (2.2). It is convenient to note that the (m+1)th approximant defined by

$$f_{m+1}\left(\frac{A}{n},\frac{B}{n}\right) = e^{A/n}\, e^{B/n}\, e^{C_2/n^2} \cdots e^{n^{-m}C_m} \tag{2.15}$$

satisfies the condition (2.14), if the coefficients $\{C_j\}$ are given by [9]

$$C_2 = \frac{1}{2}[B,A],\ C_3 = \frac{1}{3}[C_2, A+2B], \cdots. \tag{2.16}$$

Thus, we can obtain formally any higher order decompositions of the exponential operator (2.14).

Up to now, we have discussed the convergence of decomposed <u>operators</u>. We are, however, practically interested in the trace of such decomposed operators. Then, we may have more convenient theorems which assure more rapid convergence. The following correction formula is now well known [15,30,33,44-46]:

$$\mathrm{Tr}e^{A+B} = \mathrm{Tr}(e^{A/n}e^{B/n})^n + O(n^{-2}). \tag{2.17}$$

This is generalized as follows [44]:

<u>Theorem 4</u>: <u>If $f_m(\{A_j\})$ satisfies the condition</u>

$$f_m(\{-A_j\})^{-1} = f_m(\{A_j\})^t \tag{2.18}$$

<u>then the approximant</u>

$$Z_m(n) \equiv \mathrm{Tr}[f_m(\{\tfrac{1}{n}A_j\})]^n \tag{2.19}$$

<u>is an even function of n, namely,</u>

$$Z_m(-n) = Z_m(n). \tag{2.20}$$

<u>Theorem 5 (corollary of Theorem 4)[44]</u>: <u>With the conditions (2.14) and (2.19), we have</u>

$$Z_{exact} = Z_{2m}(n) + O(\frac{1}{n^{2m}}). \tag{2.21}$$

In particular, if we put $f_m(\{A_j\})$ as (2.9) for symmetric $\{A_j\}$ (namely, $A^t{}_j = A_j$), then we have the following theorem:
<u>Theorem 6 [44]</u>: <u>If $\{A_j\}$ are symmetric operators</u> (i.e., $A^t{}_j = A_j$), <u>then we have</u>

$$Z_2(n) \equiv \mathrm{Tr}(e^{A_1/n}e^{A_2/n}\cdots e^{A_p/n})^n = Z_2(-n). \tag{2.22}$$

<u>Consequently we have</u>

$$Z_{exact} = Z_2(n) + O(\frac{1}{n^2}). \tag{2.23}$$

This was first obtained by Fye [45] for a general integer p in a different condition.

Theorem 6 suggests [44] the following new extrapolation method:

$$Z_{exact} \simeq Z(n) + \frac{a}{n^2+b}\ . \tag{2.24}$$

<u>Theorem 7 [43]</u>: <u>If $Q^t=Q$ with the condition (2.18), the average of any quantum operator Q defined by</u>

$$\langle Q\rangle_m(n) \equiv \mathrm{Tr}Q[f_{2m}(\{\tfrac{1}{n}A_j\})]^n/Z_{2m}(n) \tag{2.25}$$

<u>is an even function of n, namely</u>

$$\langle Q\rangle_m(-n) = \langle Q\rangle_m(n). \tag{2.26}$$

In particular, we have

$$\langle Q\rangle_{exact} = \langle Q\rangle_1(n) + O(\frac{1}{n^2}) \cdot \qquad (2.27)$$

As was proposed in the previous paper [44], it will be useful to make use of the following extrapolation formula:

$$Z_{2m}(n)(\text{or } \langle Q\rangle_m(n)) \simeq \frac{1}{n^{2m-2}} \frac{a}{n^2+b} + c. \qquad (2.28)$$

The coefficients a, b and c are determined by the least square method from Monte Carlo data. The parameter c thus obtained gives an extrapolation to the exact value of Z(or Q).

The above $1/n^2$ correction law explains well the numerical results by Betsuyaku [29] and by Marcu et al. [22,25].

2.3 Moment- and Cumulant-Expansions of Sliced Exponential Operators

The sliced exponential $\exp[-(\Delta\tau)\mathcal{H}]$ in (2.7) may be approximated by

$$e^{-(\Delta\tau)\mathcal{H}} = f_m[-(\Delta\tau)\mathcal{H}] + O((\Delta\tau)^m), \qquad (2.29)$$

as was discussed in (2.2). A typical example of $f_m[-(\Delta\tau)\mathcal{H}]$ is given by (2. 7), namely a product of partial Boltzmann factors.

There is another type of approximation, namely

$$f_m[-(\Delta\tau)\mathcal{H}] = \sum_{k=0}^{m} \frac{1}{k!}[-(\Delta\tau)\mathcal{H}]^k, \qquad (2.30)$$

which was introduced first by the present author [9] and was applied explicitly to quantum Monte Carlo by Hirsch et al. [15]. It was proven in [9] that

$$\left\| e^{-\beta\mathcal{H}} - (f_m[-\beta\mathcal{H}/n])^n \right\| \leq \frac{(\beta\|\mathcal{H}\|)^{m+1}}{n^m(m+1)!} e^{\beta\|\mathcal{H}\|}. \qquad (2.31)$$

Thus, the convergence of this approximation is very rapid, as is seen from the factor n^{-m} in (2.31).

In ordinary situations, the Hamiltonian is extensive and consequently the expansion (2.30) contains many terms of different orders in the system size N. Therefore, it will be more effective in practical applications to make a resummation of (2.30) in the following form:

$$f_m[-(\Delta\tau)\mathcal{H}] = (\exp f_{p,q}^{(m)}[-(\Delta\tau)\mathcal{H}]-(1-\delta_{p,q})). \qquad (2.32)$$

Here, the matrix elements $\{f_{p,q}^{(m)}[-(\Delta\tau)\mathcal{H}]\}$ are determined so that the expansion of (2.32) may agree with (2.30) up to the m-th order in $\Delta\tau$.

This is the so-called cumulant expansion [47]. In general $f_{p,q}^{(m)}[X]$ is expanded as

$$f_{p,q}^{(m)}[X] = \sum_{k=0}^{m-1} b_k;$$

$$b_k = -\sum_{k=k_1+2k_2+\cdots} \frac{(-1)^{k_1+k_2+\cdots}}{(k_1+k_2+\cdots)} \times \frac{(k_1+k_2+\cdots)!}{k_1!k_2!\cdots} a^{k_1}a^{k_2}\cdots, \tag{2.33}$$

where $a_k=\langle p|X^k|q\rangle/k!\equiv\langle X^k\rangle/k!$. For example, we have

$$f_{p,q}^{(1)}[X] = \langle p|X|q\rangle \equiv \langle X\rangle \equiv b_0,$$

$$f_{p,q}^{(2)}[X] = b_0 + b_1 \;;\; b_1 = \frac{1}{2}(\langle X^2\rangle - \langle X\rangle^2),$$

$$b_2 = \frac{1}{3!} \; (\langle X^3\rangle-3\langle X^2\rangle\langle X\rangle+2\langle X\rangle^3). \tag{2.34}$$

This scheme will be more effective than Eq. (2.32). Takano's approximation [24] based on the coherent representation may be interpreted as the first example of the above general resummed approximations.

For large system size N, however, it will be more convenient to make use of the formula (1.1) in any representation. For example, the following alternative formulation of Takano's coherent representation method [24] will be suggested here, namely we apply the formula (1.1) to the exponential operator $\exp \mathcal{H}(\Omega)$ in the coherent representation $|\Omega\rangle$ and we approximate it as

$$e^{\mathcal{H}(\Omega)} = \lim_{n\to\infty}(e^{\frac{1}{n}\mathcal{H}_1(\Omega)}\cdots e^{\frac{1}{n}\mathcal{H}_p(\Omega)})^n, \tag{2.35}$$

for the decomposition of the original Hamiltonian: $\mathcal{H}= \mathcal{H}_1 +\cdots+ \mathcal{H}_p$.

2.4 Equivalence Theorem

In the present subsection, we explain a general basic idea of quantum Monte Carlo at finite temperatures, namely "quantum statistical Monte Carlo method". It is based on the following equivalence theorem.
Equivalence Theorem: The partition function of the relevant d-dimensional quantum system is transformed into that of the corresponding (d+1)-dimensional classical system by using the decomposition formula (2.1) of exponential operators:

$$Z = \mathrm{Tr} \exp(-\beta\mathcal{H}) = \mathrm{Tr}^{(d+1)}\exp\mathcal{H}_{\mathrm{eff}}^{(d+1)}. \tag{2.36}$$

Here $\mathcal{H}_{\mathrm{eff}}^{(d+1)}$ denotes an effective (d+1)-dimensional Hamiltonian defined by

$$Z = \mathrm{Tr} \exp(-\beta\mathcal{H}) = \mathrm{Tr}\{\exp(-\frac{1}{n}\beta\mathcal{H})\}^n$$

$$= \lim_{n\to\infty} \mathrm{Tr}\{f_m[-\frac{1}{n}\beta\mathcal{H}]\}^n$$

$$= \lim_{n\to\infty} \sum_{|\alpha_1\rangle}\cdots\sum_{|\alpha_n\rangle} \langle\alpha_1|\rho_{m,n}|\alpha_2\rangle\langle\alpha_2|\rho_{m,n}|\alpha_3\rangle$$

$$\cdots \langle\alpha_n|\rho_{m,n}|\alpha_1\rangle \equiv \mathrm{Tr}^{(d+1)}\exp(\mathcal{H}_{\mathrm{eff}}^{(d+1)}), \tag{2.37}$$

with $\rho_{m,n} = f_m[-\frac{1}{n}\beta\mathcal{H}]$ defined by (2.2), where the states $\{|\alpha_j\rangle\}$ denote arbitrary orthogonal complete (or over-complete) sets.

New many-body interactions appear in equivalent (d+1)-dimensional effective lattices. The additional (d+1)th dimension plays a role of quantum effect, namely the non-commutativity of the relevant operators. Thus, the additional dimension is the "quantal dimension" (or Trotter's direction). Quantum dynamics in equilibrium can be expressed by correlations in this quantal dimension, as will be discussed in more detail in Section 13. Therefore, the introduction of the quantal dimension in the present formulation is essential to study the quantum effect of the relevant system.

3. Thermofield Quantum Monte Carlo Method

The thermofield Monte Carlo method [34-39] is a unification of the zero-temperature projector Monte Carlo [48] and the quantum statistical Monte Carlo methods[10-13].

The statistical average $\langle Q\rangle$ of a physical quantity Q at temperature T is expressed by the expectation value of the form

$$\langle Q\rangle \equiv \mathrm{Tr}Qe^{-\beta\mathcal{H}}/Z(\beta) = \langle 0(\beta)|Q|0(\beta)\rangle; \quad Z(\beta) = \mathrm{Tr}\, e^{-\beta\mathcal{H}}, \tag{3.1}$$

where $\mathcal{H}$ is the Hamiltonian of the relevant quantum system. The thermal quantum state $|0(\beta)\rangle$ is defined by [49,50]

$$|0(\beta)\rangle \equiv Z(\beta)^{-1/2}\sum_n e^{-\beta E_n/2}|n,\tilde{n}\rangle = Z(\beta)^{-1/2}e^{-(1/2)\beta\mathcal{H}}|I\rangle;$$

$$|I\rangle \equiv \sum_n |n,\tilde{n}\rangle \tag{3.2}$$

with an eigenvalue E_n and $|\tilde{n}\rangle$ is the corresponding eigenstate in the fictitious dynamical system $\tilde{\mathcal{H}}$ identical to the original system together with the conjugation rules [49,50]

$$\widetilde{AB} = \tilde{A}\tilde{B}, \quad (c_1A + c_2B)^{\sim} = c_1^*\tilde{A} + c_2^*\tilde{B},$$

$$(c_1|m\rangle + c_2|n\rangle)^{\sim} = c_1^*|\tilde{m}\rangle + c_2^*|\tilde{n}\rangle. \tag{3.3}$$

Here c_1 and c_2 are c-numbers and c_j^* denotes the complex conjugate of c_j. Owing to these conjugation rules, the state $|I\rangle$ is independent of representations, namely [34-40],

$$|I\rangle \equiv \sum_n |n,\tilde{n}\rangle = \sum_\alpha |\alpha,\tilde{\alpha}\rangle\,. \tag{3.4}$$

It is convenient from our point of view to take the number representation [35] for fermi and bose systems and the spin representation [35-40] for spin systems, because the state $|I\rangle$ can be easily given explicitly in these representations.

Our starting point is the following path-sum formulation [34-40] of the thermal quantum state

$$|0(\beta)\rangle = \lim_{n\to\infty} (Z_n(\beta)/Z(0))^{-1/2}(e^{-\beta\mathcal{H}_1/2n}\cdots e^{-\beta\mathcal{H}_p/2n})^n|0(0)\rangle, \quad (3.5)$$

where

$$|0(0)\rangle = Z(0)^{-1/2}|I\rangle, \quad \mathcal{H} = \mathcal{H}_1 + \mathcal{H}_2 + \cdots + \mathcal{H}_p, \quad (3.6)$$

and

$$Z_n(\beta) = \mathrm{Tr}\,(e^{-\beta\mathcal{H}_1/2n}\cdots e^{-\beta\mathcal{H}_p/2n})^{2n}\,. \quad (3.7)$$

Here we have used the generalized Trotter formula (1.1).

Here, we decompose the Hamiltonian $\mathcal{H}$ into $\mathcal{H}_1 + \mathcal{H}_2 + \cdots + \mathcal{H}_p$ so that the matrix elements $\langle\alpha|\exp(-\beta\mathcal{H}_j)|\alpha'\rangle$ may be obtained explicitly with use of effective local interactions.

From a practical point of view, it is more convenient to make use of the following "symmetric" expression [38] of the thermal quantum state:

$$|0(\beta)\rangle = Z(\beta)^{-1/2}\exp(-\tfrac{1}{2}\beta\mathcal{H}^s)|I\rangle; \quad \mathcal{H}^s = \tfrac{1}{2}(\mathcal{H} + \tilde{\mathcal{H}}), \quad (3.8)$$

when $[\mathcal{H}, \tilde{\mathcal{H}}] = 0$, as is usually the case. Correspondingly we have the following "symmetric" path-sum formulation

$$|0(\beta)\rangle = \lim_{n\to\infty}|\Psi_n\rangle, \quad (3.9)$$

where $|\Psi_n\rangle$ is given by

$$|\Psi_n\rangle = Z_n(\beta)^{-1/2}(e^{-\beta\mathcal{H}_1^s/2n}\cdots e^{-\beta\mathcal{H}_p^s/2n})^n|I\rangle. \quad (3.10)$$

Our basic idea of the thermo field Monte Carlo [34] is to calculate the thermal state $|\Psi_n\rangle$ as a diffusion process by extending the Kuti-Blanckenbecler-Sugar (KBS) Monte Carlo method at zero temperature [48,51] to that of finite temperatures, with important sampling [51].

First we select the initial thermal state $|I\rangle_i$ randomly, as a subspace of $|I\rangle$. Then, we apply the partial "canonical thermal operator" $\exp(-\beta\mathcal{H}_j^s/(2n))$ successively to the thermal state already obtained, namely

$$|0(\beta_j)\rangle \propto \exp[-(\beta_j-\beta_{j-1})\mathcal{H}]|0(\beta_{j-1})\rangle. \quad (3.11)$$

Thus, the thermal states can be calculated successively from each previous thermal state at a higher temperature [38]. This procedure makes a diffusion process in the double Hilbert space $(\mathcal{H}, \tilde{\mathcal{H}})$. That is, the thermal state can be obtained [38] as a "sample ensemble" of the form

$$|0(\beta)\rangle \simeq \sum_{p,q\in S} W_{p,q}(\beta)|p,\tilde{q}\rangle\,, \quad (3.12)$$

where the weight $W_{p,q}(\beta)$ is given by

$$W_{p,q}(\beta) = \mathcal{N}(\beta)\sum_{r\in S_0}\langle q,p|(e^{-\frac{\beta}{2n}\mathcal{H}_1^s}\cdots e^{-\frac{\beta}{2n}\mathcal{H}_p^s})^n|r,\tilde{r}\rangle \quad (3.13)$$

with some appropriate normalization $\mathcal{N}(\beta)$. Here, S_0 denotes the initial sample set in which all the states $|r,\tilde{r}>$ have an equal probability, and S denotes a final ensemble produced by the Monte Carlo method. Consequently, the thermal average of a physical quantity Q is given by

$$<Q> = <0(\beta)|Q|0(\beta)> \simeq \sum_{p,s\in S} <p|Q|s> \sum_{q\in S} W^{*}_{p,q}(\beta)W_{s,q}(\beta)\mathcal{N}^2(\beta). \quad (3.14)$$

An explicit application of this method has been reported by Suzuki, Miyashita and Takasu [39].

4. Thermofield Transfer-Matrix Method

Owing to the path-sum formulation (3.5) of the thermal quantum state $|0(\beta)>$, we can calculate it successively by the transfer-matrix method as follows:

$$|0(\beta)> = \mathcal{N} \lim_{n\to\infty} \mathcal{T}^{n}|0(0)>, \quad (4.1)$$

where $\mathcal{N}$ denotes the normalization and

$$\mathcal{T} = \mathcal{T}_1\mathcal{T}_2 \cdots \mathcal{T}_p; \quad \mathcal{T}_j = \exp(-\Delta\tau\ \mathcal{H}_j) \quad (4.2)$$

for the decomposition $\mathcal{H}=\mathcal{H}_1+\mathcal{H}_2+\cdots+\mathcal{H}_p$. Namely we obtain the following iterative formulation of the thermal quantum state:

$$|0(\beta_1)> = \mathcal{N}_1 \exp(-\tfrac{1}{2}\beta_1\mathcal{H})|0(0)>,$$

$$|0(\beta_2)> = \mathcal{N}_2 \exp[-\tfrac{1}{2}(\beta_2-\beta_1)\mathcal{H}]|0(\beta_1)>,$$

$$\vdots$$

$$|0(\beta_n)> = \mathcal{N}_n \exp[-\tfrac{1}{2}(\beta_n-\beta_{n-1})\mathcal{H}]|0(\beta_{n-1})>. \quad (4.3)$$

Thus, all the thermal quantum states can, in principle, be generated successively from each previous thermal quantum state at a higher temperataure [40]. For an explicit application of this method, see Ref. 40. It is remarkable that all the temperature range can be studied with the same accuracy in this method.

5. Folding Scheme and Combination of Quantum Monte Carlo and Transfer-Matrix Methods

In the present section, we explain the idea of "folding scheme" [34] inspired by the thermo field transfer-matrix method [34,40]. The folding scheme is based on the following expressions:

$$Z = \mathrm{Tr}\ e^{-\beta\mathcal{H}} = \Sigma|<p|\exp(-\tfrac{1}{2}\beta\mathcal{H})|q>|^2 \quad (5.1)$$

and

$$<Q> \equiv \mathrm{Tr}Qe^{-\beta\mathcal{H}}/Z = \mathrm{Tr}e^{-\beta\mathcal{H}/2}Qe^{-\beta\mathcal{H}/2}/Z$$

$$= \sum_{p,q,r}<p|Q|r><r|e^{-\beta\mathcal{H}/2}|q><q|e^{-\beta\mathcal{H}/2}|p>/Z, \quad (5.2)$$

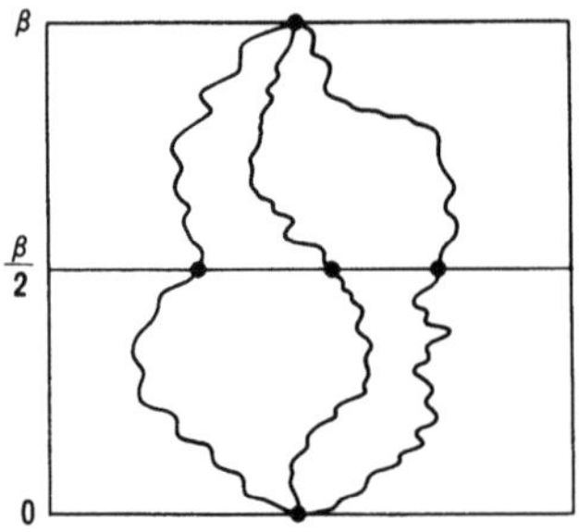

Fig. 1
Folding scheme of quantum statistical Monte Carlo.

as was discussed in [34]. The reason why this is called the "folding scheme" is that the temperature range $0<\beta'<\beta$ is folded into the half region $0<\beta'<\beta/2$ in this scheme as shown in Fig. 1.

The "canonical operator" $\exp(-\frac{1}{2}\beta\mathcal{H})$ in Eqs. (5.1) and (5.2) is approximated again as

$$\exp(-\frac{1}{2}\beta\mathcal{H}) = (f_m[-\frac{1}{2n}\beta\mathcal{H}])^n + O(\frac{1}{n^{m-1}}). \qquad (5.3)$$

For example, the expectation value of the relevant physical quantity Q is approximated by

$$\langle Q\rangle = Z^{-1}\Sigma_{r,s}\langle r|Q|r\rangle|\langle r|(f_m[-\frac{1}{2n}\beta\mathcal{H}])^n|s\rangle|^2 \qquad (5.4)$$

in the representation $\{|r\rangle\}$ to diagonalize Q. It should be remarked that the weight factor is always non-negative in the above scheme and that this property is very convenient in practical applications of this scheme.

There are also some different methods to evaluate the above expressions, namely Monte Carlo and transfer- matrix methods, and their modifications, i.e., the combination [52] of the former two fundamental methods.

6. Cluster Decompositions and their Convergence

As was briefly mentioned in Section 1, the decomposition of the exponential operator (1.1) is not unique, and consequently there is a possibility to find better decompositions. In fact, a cluster decomposition of cluster size $m(>2)$ shown, for example, in Fig. 2 is better than the pair decomposition originally used by Suzuki, Miyashita and Kuroda [11]. Larger Trotter's number n may compensate the smallness of the cluster size m. Clearly, the infinite limit $m\to\infty$ yields the exact solution of the relevant exponential operator. Such cluster decompositions will be particularly useful for frustrated quantum spin systems [12, 53-57], because the quantum frustration effect may be treated partially even inside each cluster.

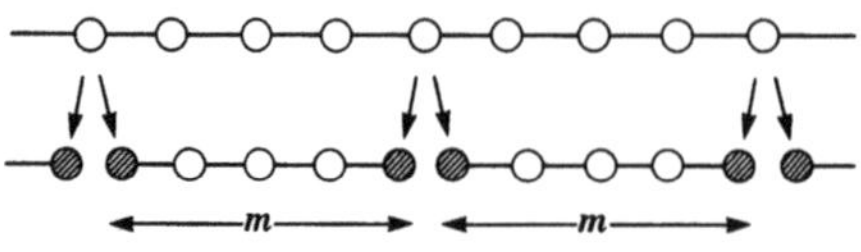

Fig. 2
A cluster decomposition with size m.

The approximate partition function $Z_{m,n}$ for the cluster size m and for Trotter's number n is shown [36] to satisfy

$$|Z-Z_{m,n}| \leq \frac{c(T)}{(m+a)(n^2+b)} \tag{6.1}$$

with some appropriate constatnts a and b under the condition that $\{\mathcal{H}(\mathbf{r})\}$ are symmetric, i.e., $\mathcal{H}^t(\mathbf{r}) = \mathcal{H}(\mathbf{r})$. The coefficient c(T) is a function of temperature independent of m and n. For more details of the m-dependence of $Z_{m,n}$, see the paper by Fye [45].

The above result suggests the following extrapolation formula

$$Z_{m,n} \simeq Z_{exact} + \frac{c(T)}{(m+a)(n^2+b)}, \tag{6.2}$$

as was already discussed in Ref. 12.

Some explicit applications of the above cluster decomposition have been reported already [31,57].

The above idea of cluster decomposition may be closely related to the decoupled cell method [58]. A more mathematical argument concerning the relation between the present cluster decomposition Monte Carlo and the decoupled cell methods will be an open question.

7. Quantum Monte Carlo Renormalization Group approach

As was suggested first by the present author [28], it is possible to apply Wilson's RG approach [59] or its modified Monte Carlo RG method in classical systems [60-65] to quantum systems with the use of the equivalence theorem (2.37). Our starting point is the following decomposition [28]

$$\exp(-\beta\mathcal{H}) = \lim_{n\to\infty}(\exp(-\frac{1}{n}\hat{\mathcal{H}}_1)\exp(-\frac{1}{n}\hat{\mathcal{H}}_2))^{Kn}, \tag{7.1}$$

where $K=\beta J$ and $\mathcal{H} = J(\hat{\mathcal{H}}_1 + \hat{\mathcal{H}}_2)$. As was discussed by the present author [28] and by Kolb [27], the accuracy of the above decomposition (7.1) is independent of temperature T (or $\beta = 1/k_BT$). This remark is substantial in applying the RG method to quantum systems, because the RG transformation makes the temperature T move to another one T'. The temperature-dependence comes [28] only through the thickness L (or Kn) in the (d+1)-th quantal direction of the equivalent lattice.

There are some tricky points to be solved near future concerning this quantum Monte Carlo RG method. For more details, see the papers by Kolb [27].

8. Quantum Spin Glasses and Equivalent Random Spin Systems

Since Edwards and Anderson [66] proposed a theory of spin glasses, many papers have appeared. It is, however, still a challenging problem. In contrast to the renormalization group prediction that the lower critical dimensinality of spin glasses is $d_c=4$, it has been shown using Monte Carlo simulations [67,68] that the three-dimensional (±J) Ising spin glass shows a phase transition with the negative divergence of the nonlinear susceptibility phenomenologically predicted by the present author [69].

Quite recently several authors [70-72] worked on quantum spin glasses. It will be instructive to remark briefly that quantum spin glass models are also transformed into the corresponding Ising random systems using the present generalized Trotter-like transformation (1.1) as follows.

For simplicity, we consider here the quantum transverse Ising spin glass model described by the Hamiltonian

$$\mathcal{H} = \mathcal{H}_1 + \mathcal{H}_2 = -\sum_{ij} J_{ij}\sigma_i^z\sigma_j^z - \mu_B H \sum_{j=1}^{M} \sigma_j^x \qquad (8.1)$$

with the Pauli matrices σ_j^x and σ_j^z. Here $\{J_{ij}\}$ are random interactions with some appropriate distribution function $P(J_{ij})$. Following Edwards and Anderson [66], we consider the following n-replicated partition function

$$Z^{(n)} = \langle \mathrm{Tr}\{\exp(-\beta\mathcal{H})\}^n \rangle_{av}, \qquad (8.2)$$

where $\langle\cdots\rangle_{av}$ denotes the average over the random distribution of J_{ij}, namely

$$\langle Q(\{J_{ij}\})\rangle_{av} = \int\cdots\int Q(\{J_{ij}\}) \prod_{\langle ij\rangle} P(J_{ij})dJ_{ij}. \qquad (8.3)$$

Now the replicated partition function is transformed into the following form

$$Z^{(n)} = \mathrm{Tr}\langle \exp(-\beta\sum_{k=1}^{n}\mathcal{H}^{(k)})\rangle_{av}, \qquad (8.4)$$

where $\mathcal{H}^{(k)}$ is the k-th replicated Hamiltonian and it is separated into the following two parts

$$\mathcal{H}^{(k)} = \mathcal{H}_1^{(k)} + \mathcal{H}_2^{(k)} \qquad (8.5)$$

corresponding to (8.1). Then, we have

$$Z^{(n)} = \lim_{m\to\infty} \mathrm{Tr}\langle \prod_{k=1}^{n}\{\exp(-\frac{\beta}{m}\mathcal{H}_1^{(k)})\exp(-\frac{\beta}{m}\mathcal{H}_2^{(k)})\}^m\rangle_{av}$$

$$= \lim_{m\to\infty} c_m \sum_{\{S_{j,\ell}^{(k)}=\pm1\}} \langle \exp\sum_{k=1}^{n}\mathcal{H}_{eff}(\{S_{j,\ell}^{(k)}\})\rangle_{av}, \qquad (8.6)$$

where $c_m=(\frac{1}{2}\sinh(2\beta\,\mu_B H/m))^{mM/2}$ and

$$\mathcal{H}_{eff}(\{S_{j,\ell}^{(k)}\}) = \sum_{\ell=1}^{m}\left(K_m\sum_{i=1}^{M} S_{i,\ell}^{(k)}S_{i,\ell+1}^{(k)} + \sum_{\langle ij\rangle} K_{ij}^{(m)} S_{i,\ell}^{(k)}S_{j,\ell}^{(k)}\right) \qquad (8.7)$$

with $K_{ij}^{(m)}=\beta J_{ij}/m$ and $K_m=\frac{1}{2}\log\coth(\beta\mu_B H/m)$. If we take the average over $\{J_{ij}\}$ in (8.6), then the following (d+2)-dimensional Hamiltonian appears:

$$\langle \exp \sum_{k=1}^{n} \mathcal{H}_{\mathrm{eff}}(\{S_{j,\ell}^{(k)}\}) \rangle_{\mathrm{av}} = \exp(K \sum_{\ell=1}^{m} \sum_{i=1}^{N} S_{i,\ell}^{(k)} S_{i,\ell+1}^{(k)})$$

$$\times \int \cdots \int \exp(\sum_{\langle ij \rangle} K_{ij}^{(m)} S_{i,\ell}^{(k)} S_{j,\ell}^{(k)}) \prod_{\langle ij \rangle} P(\{J_{ij}\}) dJ_{ij} \equiv \exp \mathcal{H}_{\mathrm{eff}}^{(d+2)}(\{S_{j,\ell}\}), \tag{8.8}$$

as shown in Fig. 3.

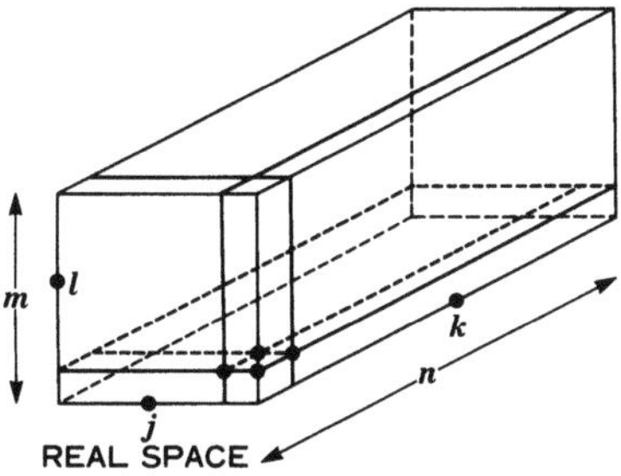

Fig. 3 Equivalent replica lattice: m denotes Trotter's number and n, the replica parameter.

For an explicit application of this formulation, see the paper by Ishii et al. [70].

9. Fisher's Finite-Size Scaling law and Monte Carlo Methods

Fisher[73] proposed the following finite-size scaling ansatz: the free energy F of a finite system with size L takes the asymptotic form

$$F(T,L) \simeq L^{-(2-\alpha)/\nu} F(\varepsilon L^{1/\nu}) \tag{9.1}$$

near the critical point T_c, where $\varepsilon=(T-T_c)/T_c$ and α and ν denote the critical exponents of the specific heat and the correlation length in the bulk limit, respectively. This ansatz was derived by the present author [74] using the RG method. This has been very useful in estimating critical exponents from Monte Carlo data [75].

This ansatz is also useful even in analyzing quantum Monte Carlo data. There appears, however, a new aspect of complications, i.e., there are two kinds of finite parameters in our quantum Monte Carlo approach, namely, the system size L and Trotter's number n. This will be discussed in the succeeding section.

10. Finite-Trotter-Number Scaling

As was briefly discussed in the preceding section, our equivalent classical system has two finiteness parameters such as the finite size L and finite Trotter's number n. It was shown by the present author [30] that the $n\to\infty$ and $L\to\infty$ limits are interchangeable. This was confirmed numerically by Marcu et al. [76]. Therefore, we may make an extrapolation for the limit $L \to \infty$ first and then we discuss the asymptotic behavior for $n \to \infty$.

Analogously to Fisher's finite-size scaling law, it seems to be natural to make an ansatz of finite-Trotter-number scaling as follows, though the reasoning of this ansatz may be quite different from that of Fisher's finite-size scaling law. The content of the present section is original.

As seen from the above general scheme, the two limits $\beta\to\infty$ and $n\to\infty$ are quite interesting and particularly important in one-dimensional quantum systems in which the critical point is given by $\beta=\infty$ (i.e., $T=0$). In this case, relevant physical quantities such as the susceptibility $\chi_0(T)$ and correlation length $\xi(T)$ may diverge as

$$\chi_0(T) \sim T^{-\gamma} \text{ and } \xi(T) \sim T^{-\nu}, \tag{10.1}$$

near T=0, as discussed by many authors [77-79].

If we apply the above general scheme to this problem, all the physical quantities are obtained as functions of the temperature T (or inverse temperature β) and Trotter's number n:

$$\chi_0 = \chi_0(T,n) \text{ and } \xi = \xi(T,n). \tag{10.2}$$

We are now interested in the asymptotic behaviors of these quantities for $T\to 0$ and $n\to\infty$. As seen from the above general scheme, or more explicitly from the following identity [80]

$$e^{-\beta\mathcal{H}} = \{(\exp(-\frac{\beta}{n}\mathcal{H}))^{n/\beta}\}^{\beta}, \tag{10.3}$$

larger β requires larger n. Under some appropriate conditions [44], the average $\langle Q\rangle$ is an even function of n and consequently is expanded as

$$\langle Q\rangle_n = \langle Q\rangle + \sum_{k=1}^{\infty} a_k(T)/n^{2k}. \tag{10.4}$$

The coefficients may diverge near T=0, when $\langle Q\rangle$ diverges as

$$\langle Q\rangle \sim \frac{1}{T^{\psi}}. \tag{10.5}$$

Here, we introduce the following scaling ansatz, namely $a_k(T) \sim T^{-(2k\Delta+\psi)}$. Then, Eq.(10.4) has the following scaling form

$$\langle Q\rangle_n \simeq \frac{1}{T^{\psi}} f^{(sc)}\left(\frac{1}{nT^{\Delta}}\right) \tag{10.6}$$

with some scaling exponent Δ. The above scaling law may be called "finite-Trotter-number scaling". This ansatz will be extremely convenient in analyzing numerical data obtained in the above general scheme, because the critical exponent ψ and scaling exponent Δ can be evaluated numerically from the scaling plot $T^{\psi}\langle Q\rangle_{eq}$ —— nT^{Δ} to find the universal scaling function $f^{(sc)}(x)$. Numerical data by Takada and Kubo [79] seem to be well fitted by this finite-Trotter-number scaling plot. More detailed analyses will be published elsewhere.

The above finite-Trotter-number scaling may be extended to critical phenomena at finite temperatures ($T_c\neq 0$), but quite different scaling structures might be required. This study will be reported in the near future, for example, in connection with the quantum KT transition in the two-dimensional quantum XY model [81].

11. Correlation Identity and Testing of Monte Carlo Calculations

There are many correlation identities in classical systems such as the Ising model. They are expressed in the following general abstract form

$$\langle fg\rangle = \langle f\langle g\rangle_{\mathcal{H}_g}\rangle, \tag{11.1}$$

as was briefly mentioned in Section 1, where f and g do not contain common variables and $\langle g\rangle_{\mathcal{H}_g}$ is the partial canonical average defined by

$$\langle g\rangle_{\mathcal{H}_g} = \mathrm{Tr}ge^{-\beta\mathcal{H}_g}/\mathrm{Tr}e^{-\beta\mathcal{H}_g}. \tag{11.2}$$

Here $\mathcal{H}_g$ is the partial Hamiltonian which contains all the interactions relevant to the variables contained in g, namely $\mathcal{H}' = \mathcal{H} - \mathcal{H}_g$ does not contain the variables appearing in g. Here $\mathcal{H}$ denotes the total Hamiltonian of the relevant "classical" Hamiltonian whose partial Hamiltonians commute with each other.

The proof of (11.1) is given quite analogously to that in Ref. 41. It may be intructive to give it here, because it is quite simple and suggestive in many respects:

$$\begin{aligned}\langle fg\rangle Z &= \mathrm{Tr}fge^{-\beta\mathcal{H}} = \mathrm{Tr}'e^{-\beta\mathcal{H}'}f\mathrm{Tr}_g ge^{-\beta\mathcal{H}_g}\\ &= \mathrm{Tr}'[e^{-\beta\mathcal{H}'}f\mathrm{Tr}_g e^{-\beta\mathcal{H}_g}(\mathrm{Tr}_g ge^{-\beta\mathcal{H}_g}/\mathrm{Tr}_g e^{-\beta\mathcal{H}_g})]\\ &= \mathrm{Tr}'\mathrm{Tr}_g e^{-\beta\mathcal{H}'}e^{-\beta\mathcal{H}_g}(f\langle g\rangle_{\mathcal{H}_g})\end{aligned} \tag{11.3}$$

with $Z=\mathrm{Tr}e^{-\beta\mathcal{H}}$ where we have used that $\mathcal{H}_g$ and $\mathcal{H}'$ commute with each other as well as other variables. This yields the general correlation identity (11.1).

For example, in the Ising model

$$\mathcal{H} = -\sum_{i,k} J_{ik}S_iS_k - \mu_B H\sum_j S_j\ , \tag{11.4}$$

where $S_j=\pm 1$, we have the following identity

$$\langle S_jS_k\rangle = \langle S_j\tanh\{\beta(\sum_i J_{ik}S_i+\mu_B H)\}\rangle. \tag{11.5}$$

This has been used very frequently [3,83]. This kind of correlation identities [41,42,83] are very useful in testing the validity and accuracy of Monte Carlo calculations, as was emphasized by Mouritsen [3].

12. Symmetry Properties of Kubo's Canonical Correlation

The above general scheme of quantum Monte Carlo is also useful in studying Kubo's canonical correlation [82]

$$\langle A;B\rangle = \frac{1}{\beta}\int_0^\beta \langle AB(\tau)\rangle d\tau, \tag{12.1}$$

where

$$B(\tau) = e^{\tau\mathcal{H}}Be^{-\tau\mathcal{H}}. \tag{12.2}$$

It will be instructive to note the following theorem.

Theorem 8: The symmetrized correlation function

$$\langle A,B;\tau\rangle_{sym} \equiv \frac{1}{2}(\langle A(\tau)B\rangle + \langle B(\tau)A\rangle) \tag{12.3}$$

is an even function of n, when it is approximated by the generalized Trotter-like formula as

$$\langle A(\tau)B\rangle = \mathrm{Tr}[(P_nAP_n^{-1})BP_n^{\beta/\tau}]/Z_n, \tag{12.4}$$

where

$$P_n = (f_m[-\frac{\tau}{n}\mathcal{H}])^n \text{ and } Z_n = \mathrm{Tr}P_n^{\beta/\tau} \tag{12.5}$$

and we have assumed the following conditions that $A^t=A$, $B^t=B$ (symmetric) and

$$P_{-n} = P_n^t, \tag{12.6}$$

as is satisfied in ordinary situations [44].

The proof is given as follows. First note that

$$Z_{-n}\langle A(\tau)B\rangle^{(-n)} \equiv \mathrm{Tr}[(P_{-n}AP_{-n}^{-1})BP_{-n}^{\beta/\tau}] = \mathrm{Tr}[(P_n^t A(P_n^t)^{-1}B(P_n^t)^{\beta/\tau}]$$

$$= \mathrm{Tr}(P_n^{-1}AP_n)^tB^t(P_n^{\beta/\tau})^t = \mathrm{Tr}[P_n^{\beta/\tau}B(P_n^{-1}AP_n)]^t = \mathrm{Tr}[P_n^{\beta/\tau}B(P_n^{-1}AP_n)]$$

$$= \mathrm{Tr}(P_nBP_n^{-1})AP_n^{\beta/\tau} = Z_n\langle B(\tau)A\rangle^{(n)}, \tag{12.7}$$

where $Z_{-n}=\mathrm{Tr}P_{-n}^{\beta/\tau}=\mathrm{Tr}(P_n^t)^{\beta/\tau}=Z_n$, as shown in [44]. Thus, we arrive at

$$\langle A,B;\tau\rangle_{sym}^{(-n)} \equiv \frac{1}{2}(\langle A(\tau)B\rangle^{(-n)} + \langle B(\tau)A\rangle^{(-n)})$$

$$= \frac{1}{2}(\langle B(\tau)A\rangle^{(n)} + \langle A(\tau)B\rangle^{(n)}) = \langle A,B;\tau\rangle_{sym}^{(n)}. \tag{12.8}$$

In particular, a symmetric choice of P_n yields the following theorem.

Theorem 9: If $P_n^t=P_n=P_{-n}$, then we have

$$\langle A;B\rangle^{(-n)} = \langle A;B\rangle^{(n)} = \text{even function of } n. \tag{12.9}$$

The proof is evident from

$$\langle A(\tau)B\rangle^{(-n)} = \langle A(\tau)B\rangle^{(n)}, \tag{12.10}$$

which is clear from (12.7) with the condition $P_n{}^t = P_n$. The above evenness property (12.9) was discussed [79] for a special choice $f_m[-\beta\mathcal{H}/n]$.

These properties will be of great use in quantum Monte Carlo and transfer-matrix methods.

13. Quantum Dynamics via Monte Carlo

According to Kubo's linear response theory [82], we need the time correlation function of the form

$$\langle A(t)A(0)\rangle = Z(\beta)^{-1}\mathrm{Tr}(e^{-\beta\mathcal{H}}e^{it\mathcal{H}/\hbar}Ae^{-it\mathcal{H}/\hbar}A) \tag{13.1}$$

in order to calculate transport coefficients [82].

In Section 12, we have discussed the imaginary time correlation function

$$C(z) = \langle e^{z\mathcal{H}}Ae^{-z\mathcal{H}}A\rangle \tag{13.2}$$

for real z, and it is possible to calculate it with the use of our quantum Monte Carlo method. It is, however, rather difficult to make a numerical analytic continuation of C(z) to the real time axis. This requires a very long computing time, because a large value of Trotter's number n has to be used to calculate explicitly the canonical operator $\exp(z\mathcal{H})$ on the basis of the equivalence theorem (2.36).

Recently Hirsch et al. [15] and Wolynes [84] tried to calculate the time correlation (13.1) using the quantum Monte Carlo method based on the equivalence theorem (2.36), by separating the real and imaginary parts of the temporal evolution operator $\exp(it\mathcal{H}/\hbar)$.

It will be an open question to find an efficient algorithm to calculate the dynamical correlation function (13.1).

14. Summary and Future Problems

The methodology of quantum statistical Monte Carlo has been reviewed with great emphasis on mathematical aspects of it. The generalized Trotter-like formula (2.1) is the basis of the quantum statistical Monte Carlo method. There are many choices of the approximants $f_m[-\beta\mathcal{H}/n]$ in (2.1). This is one of the great merits of our general approach. A typical example of such approximants is given by the generalized Trotter formula (1.1). The convergence of decomposition formulae has been discussed. The traces of decomposed exponentials have the evenness property with respect to Trotter's number n. This gives convenient extrapolation schemes of quantum Monte Carlo, as shown in (2.21), (2.24), (2.27), (2.28) and (2.29). Moment- and cumulant-expansions of sliced exponential operators have been newly proposed in §2.3. In particular, the cumulant-type expansion formula (2.32) with (2.33) and (2.34) will be of great use in practical applications. The equivalence theorem (2.36) is the foundation of our formulations. Thermofield quantum Monte Carlo and transfer-matrix methods have been discussed. Cluster decompositions and their convergence have been reviewed. The quantum Monte Carlo renormalization group approach based on the equivalence theorem (2.36) have also been briefly discussed. Quantum spin glass remains still a challenging problem. The solution of the quantum Sherrigton-Kirkpatrick model based on the equivalence theorem (2.36) will be reported in the near future.

The finite-Trotter-number scaling ansatz has been newly proposed in the present paper. The confirmation of it will be reported in the near future. General correlation identities (1.2) or (11.1) have been derived here with possible applications to equivalent classical systems.

Symmetry properties of Kubo's canonical correlations with respect to Trotter's number n has also been studied in detail. The above three results are original in the present paper. Quantum dynamics via Monte Carlo has also been briefly discussed.

There remains a still difficult problem, so-called negative sign problem in two- or higher-dimensional fermi systems or in frustrated quantum spin systems. A partial solution of this problem has been given by the thermofield Monte Carlo method or the folding scheme (or its modification), as discussed in the present paper. An alternative approach has also been proposed by Elser [85]. A general solution of this problem will, however, be an open question which is one of the most challenging problems in this field.

Quantum Monte Carlo simulations in fractal lattices will also be interesting in order to understand the role of the dimensionality of the relevant quantum system compared to the corresponding classical system.

Acknowledgements

The author would like to thank Prof. R. Kubo and Prof. T. Oguchi for useful discussions and encouragement and also thank Dr. H. Betsuyaku, Dr. Miyashita, Dr. T. Onogi and Miss M. Takasu for their collaboration on the present work.

The Taniguchi foundation is greatly acknowledged for its sponsoring the present International Symposium of Quantum Monte Carlo in Equilibrium and Non-Equilibrium Systems.

References

1. D. M. Ceperley, adn M. H. Kalos: in Monte Carlo Methods in Statistical Physics, ed. by K. Binder, Topics in Current Physics, Vol. 7 (Springer, New York 1979)
2. K. Binder, and D. Stauffer: in Applications of the Monte Carlo Method in Statistical Physics, ed. by K. Binder, Topics in Current Physics, Vol. 36 (Springer, New York 1984)
3. Ole G. Mouritsen, Computer Studies of Phase Transitions and Critical Phenomena, (Springer, New York 1984)
4. K. Binder: in Monte Carlo Methods in Statistical Physics, ed. by K. Binder Topics in Current Physics, Vol. 7 (Springer, New York 1979)
5. Monte Carlo Methods in Quantum Problems, ed. by M. H. Kalos, D. Reidel Pub. Co., Dordrecht, 1984
6. N. Metropolis, A. W. Rosenbluth, M. N. Rosenbluth, A., H. Teller and E. Teller, J. Chem. Phys. 21, 1087 (1953). See also L. D. Fosdick, J. Math. Phsy. 3, 1251 (1962).
7. M. H. Kalos, Phys. Rev. A2, 250 (1970)
8. D. C. Handscomb, Proc. Cambridge Philos. Soc. 58: 594 (1962); 60, 115 (1964). See also S. Chakravarty and D. B. Stein, Phys. Rev. Lett. 49, 582 (1982), J. W. Lyklema, Phys. Rev. Lett. 49, 88 (1982); Phys. Rev. B27: 3108 (1983). D. H. Lee, J. D. Joannopoulos, and J. W. Negele, Phys. Rev. B30, 1599 (1984)

9. M. Suzuki, Commun. Math. Phys. 51, 183 (1976); ibid. 57, 193 (1977)
10. M. Suzuki, Prog. Theor. Phys., 56, 1454 (1976)
11. M. Suzuki, S. Miyashita and A. Kuroda, Prog. Theor. Phys. 58, 1377 (1977)
12. M. Suzuki, J. Stat. Phys. Vol. 43, Nos. 5/6, 883 (1986)
13. M. Suzuki, Proceedings of International Conference on Frontiers of Theoretical Physics held at New Dehli, January 6-12, 1977, ed. by F. C. Auluck, L. S. Kothari and V. S. Nanda
14. M. Barma and B. S. Shastry, Phys. Lett. 61A, 15 (1977); Phys. Rev. B18, 3351 (1978)
15. J. E. Hirsch, D. J. Scalapino, R. L. Sugar and R. Blankenbecler, Phys. Rev. Lett. 47, 1628 (1981); Phys. Rev. B26, 5033 (1982); J. E. Hirsch, D. J. Scalapino, Phys. Rev. B27, 7169 (1983)
16. H. De Raedt and A. Lagendijk, Phys. Rev. Lett. 46, 77 (1981); J. Stat. Phys. 27, 731 (1982); Phys. Rev. Lett. 49: 1552 (1982); Phys. Rev. 24B. 463 (1981); H. De Raedt, B. De Raedt and A. Lagendijk, Z. Phys. B57, 209 (1984)
17. H. De Raedt and A. Lagendijk, Phys. Rep. 127, 233 (1985)
18. H. De Raedt and B. De Raedt, Phys. Rev. B10, 5325 (1984)
19. A. Wiesler, Phys. Lett. A89, 359 (1982)
20. J. J. Cullen and D. P. Landau, Phys. Rev. B27, 297 (1983)
21. I. Satija, G. Wysin and A. R. Bishop, Phys. Rev. B31, 3205 (1985)
22. M. Marcu, J. Müller and F. K. Schmatzer, J. Phys. A: Math. Gen. 18, 3189 (1985)
23. T. Sakaguchi, K. Kubo and S. Takada, J. Phys. Soc. Jpn. 54, 861 (1985)
24. H. Takano, Prog. Theor. Phys. 73, 332 (1985), and in this volume
25. M. Marcu and A. Wiesler, J. Phys. A: Math. Gen. 18, 2479 (1985)
26. M. Takahashi and M. Imada, J. Phys. Soc. Jpn. 53, 963 and 3765 (1984)
27. M. Kolb, Phys. Rev. Lett. 51, 1696 (1983), and in this volume
28. M. Suzuki, Prog. Theor. Phys. 58, 755 (1977)
29. H. Betsuyaku, Phys. Rev. Lett. 53, 629 (1984), Prog. Theor. Phys. 73, 319 (1985)
30. M. Suzuki, Phys. Rev. B31, 2957 (1985)
31. T. Tsuzuki, Prog. Theor. Phys. 72, 956 (1984)
32. O. Nagai et al., J. Mag. Mat. 54-57, 687 (1985), and in this volume
33. M. Suzuki, J. Math. Phys. 26, 601 (1985)
34. M. Suzuki, Phys. Lett. 111A, 440 (1985)
35. M. Suzuki, J. Phys. Soc. Jpn. 54, 4483 (1985)
36. M. Suzuki, "Progress in Quantum Field Theory" ed. by H. Ezawa and S. Kamefuchi (North-Holland Pub.) 1986
37. M. Suzuki, Proceedings of the International Symposium on "Quantum Field Theory" held at Positano, Salerno, Italy, June 5-7, 1985, ed. by F. Mancini
38. M. Suzuki, J. Stat. Phys. 42, 1047 (1986)
39. M. Suzuki, S. Miyashita and M. Takasu, Phys. Rev. B35, 3569 (1987)
40. M. Suzuki, and H. Betsuyaku, Phys. Rev. B34, 1829 (1986)
41. M. Suzuki, Phys. Lett., 19A, 267 (1965)
42. H. B. Callen, Phys. Lett. 4A, 161 (1963)
43. H. F. Trotter, Proc. Am. Math. Soc. 10, 545 (1959). S. T. Butler and M. H. Friedman, Phys. Rev. 98, 287 (1955)
44. M. Suzuki, Phys. Lett. 113A, 299 (1985)
45. R. Fye, Phys. Rev. B33, 6271 (1986)

46. H. De Raedt and B. De Raedt, Phys. Rev. 28A, 3575 (1983)
47. R. Kubo, J. Phys. Soc. Jpn. 17, 1100(1962)
48. R. Blankenbecler and R. L. Sugar, Phys. Rev. D27, 1304 (1983)
49. Y. Takahashi and H. Umezawa, Collect. Phenom. 2, 55 (1975)
50. H. Umezawa, H. Matsumoto and M. Tachiki, Thermo Field Dynamics and Condensed States (North-Holland, Amsterdam, New York, London, 1982)
51. J. Kuti, Phys. Rev. Lett. 49, 183 (1982)
52. M. Imada, and M. Takahashi, preprint and M. Imada, in this volume
53. P. W. Anderson, Mat. Res. Bull. 8. 153 (1973). P. Fazekas and P. W. Anderson, Phil. Mag. 30, 423 (1974)
54. K. Hirakawa, H. Kadowaki and K. Ubukoshi, J. Phys. Soc. Jpn. 54, 3526 (1985); I. Yamada, K. Ubukoshi and K. Hirakawa, J. Phys. Soc. Jpn. 54, 3571 (1985)
55. L. G. Marland and D. D. Betts, Phys. Rev. Lett. 43, 1618 (1979)
56. T. Oguchi, H. Nishimori and Y. Taguchi, J. Phys. Soc. Jpn. 55, 323 (1986)
57. M. Takasu, S. Miyashita and M. Suzuki, Prog. Theor. Phys. 75, 1254 (1986)
58. S. Homma, H. Matsuda and N. Ogita, Prog. Theor. Phys. 72, 1245 (1984), and in this volume
59. K. G. Wilson, Phys. Rev. B4, 3174, 3184 (1971)
60. S. -k, Ma, Phys. Rev. Lett. 37, 461 (1976)
61. R. H. Swendsen, Phys. Rev. Lett. 42, 859 (1979)
62. G. S. Pawley, R. H. Swendsen, D. J. Wallace and K. G. Wilson, Phys. Rev. B29, 4030 (1984)
63. M. Kikuchi and Y. Okabe, Prog. Theor. Phys. 75, 192 (1986)
64. K. Binder, Z. Phys. B43, 119 (1981)
65. M. N. Barber and W. Selke, J. Phys. A15, L617 (1982)
66. S. F. Edwards, and P. W. Anderson, J. Phys. F: Metal Phys. 5, 965 (1975)
67. A. T. Ogielski, and I. Morgenstern, Phys. Rev. Lett. 54, 928 (1985); J. Appl. Phys. 57, 3382 (1985)
68. A. T. Ogielski, Phys. Rev. 32B, 7384 (1985)
69. M. Suzuki, Prog. Theor. Phys. 58, 1151 (1977) See also M. Katori and M. Suzuki, Prog. Theor. Phys. 74, 1175 (1985)
70. H. Ishii and T. Yamamoto, J. Phys. C: 18, 6225 (1985)
71. K. D. Usadel, Solid State Commun. Vol. 58, 629 (1986)
72. W. Wiethege and D. Sherrington, J. Phys. C: 19, 6983 (1986)
73. M. E. Fisher, J. Vac. Sci. and Tech. 10, 665 (1973). M. E. Fisher, and M. N. Barber, Phys. Rev. Lett. 28, 1516 (1972)
74. M. Suzuki, Prog. Theor. Phys. 58, 1142 (1977)
75. M. N. Barber, in Phase Transition and Critical Phenomena, Vol. 8 ed. by C. Domb and J. L. Lebowitz, Academic Press, London, 1983
76. M. Marcu, J. Müller, and F. K. Schmatzer, Phys. Lett. 116A, 447 (1986)
77. D. Schlottmann, Phys. Rev. Lett. 54, 2131 (1985). M. Takahashi and M. Yamada, J. Phys. Soc. Jpn. 54, 2808 (1985)
78. M. Inoue and M. Suzuki, Prog. Theor. Phys. 75, 1066 (1986) and references cited therein
79. S. Takada and K. Kubo, J. Phys. Soc. Jpn. 55, 1671 (1986)
80. M. Suzuki, Prog. Theor. Phys. 58, 755 (1977) See Eq. (4.8)
81. T. Onogi, S. Miyashita and M. Suzuki, Prog. Theor. Phys. 73, 833 (1985), and J. Stat. Phys. 45, Nos. 1/2, 183 (1986)
82. R. Kubo, J. Phys. Soc. Jpn. 12, 570 (1957)
83. M. Suzuki, in preparation. (Review article of the correlation identity)
84. P. G. Wolynes, in Ref. 5, and references cited therein
85. V. Elser, Phys. Rev. A34, 2293 (1986)

Interpolation Between Dimensions

M.A. Novotny

IBM Bergen Scientific Center, Allégaten 36, N-5000 Bergen, Norway

A method is presented which allows one to interpolate between integer dimensions in a translationally invariant fashion using finite size transfer matrices. For the ferromagnetic Ising model an exact solution is presented for interpolation between isolated spins and $d=1$, and finite strip calculations are presented for interpolation between $d=1$ and $d=2$. A possible application of this interpolation scheme is the numerical study of quantum spin systems.

I. INTRODUCTION

The idea of continuing the spatial dimension d to non-integer dimensions for critical phenomena has led to greater understanding of the physics of experimental systems, which of necessity have integer dimensions. The ideas of continuing in dimension have been particularly useful in implementing renormalization group calculations, both in $4-\epsilon$ dimensions [1] and in $1+\epsilon$ dimensions [2]. It has also been possible to analytically continue series expansions in terms of the dimension d [3,4]. Although such ideas have been useful theoretically, there has not as yet been any possibility to check the renormalization group or series expansion results for non-integer dimensions. The only attempt to perform such a comparison through numerical methods has been the study of critical phenomena on fractal lattices. However, since fractals do not possess translational symmetry other parameters can determine the universality class [5].

Here I present a method of introducing into the transfer matrix formalism a parameter which can continuously interpolate between two dimensions. The method of interpolation will preserve translational invariance. The critical behavior of such an interpolation between $d=1$ and $d=2$ will be studied explicitly for the $q=2$ Potts model (the Ising model). I will also argue why the study of such an interpolation may be useful for the study of quantum spin systems.

In the next section I will present the general formalism used for interpolating between integer dimensions. Section III contains numerical transfer matrix studies of the $q=2$ Potts model interpolating between dimensions 1 and 2. Section IV contains discussions and and conclusions. In particular Section IV contains arguments describing the connection between the dimensional interpolation and the numerical study of quantum spin systems.

II. FORMULATION FOR DIMENSIONAL INTERPOLATION

The notation used in this paper was introduced previously by the author [6]. In particular, it is important to remember that any array of matrices surrounded by braces, { }, refers to the matrix product introduced in ref. 6. Any array of elements which are surrounded by parenthesis, ( ), denotes a matrix. The matrix **I** is the identity under normal matrix multiplication (which is denoted by the matrices being adjacent to each other), while the matrix **J** denotes the identity under element by element (Hadamard) matrix multiplication (which will be denoted by $\odot$). Thus all elements of **J** are equal to unity, while **I** has all diagonal elements equal to unity and all non-diagonal elements equal to zero. Any blank spaces in an array denoting the { } matrix product should be understood to stand for a matrix **J** of the appropriate size. The reader is referred to Ref. 6 for a description of the mathematical properties of the matrix product, including its relation to regular matrix multiplication, Hadamard matrix multiplication, and the direct (Kronecker) product for matrices. I will use the language of magnetic spin systems throughout the paper, but the methods are more general.

a. Construction of Dimensional Interpolation

Let **a** be a two-particle transfer matrix between magnetic spins with the same number of states. Then **a** is a symmetric matrix with its elements equal to the Boltzmann weight of the energy associated with the two spins being in the appropriate states. For example, for the $q=2$ ferromagnetic Potts model (the Ising model) studied in this paper

$$\mathbf{a} = \begin{pmatrix} e^K & 1 \\ 1 & e^K \end{pmatrix}, \tag{2.1}$$

where $K = J/k_B T$, J is the strength of the spin-spin coupling, T is the temperature, and k_B is Boltzmann's constant. The partition function for a chain of M spins with periodic boundary conditions is then given in the standard way [7,8] by

$$Z = \mathrm{Tr}(\mathbf{a}^M). \tag{2.2}$$

The partition function for N non-interacting M-spin chains can be written using the direct product of matrices as

$$Z_0 = \mathrm{Tr}(\mathbf{A}^M) \tag{2.3}$$

with the matrix **A** the direct product of N **a** matrices, i.e.

$$\mathbf{A} = \left\{ \begin{matrix} \mathbf{a} & & & \\ & \mathbf{a} & & \\ & & \ddots & \\ & & & \mathbf{a} \end{matrix} \right\}. \tag{2.4}$$

If the N chains of M spins are arranged such that the spins occupy the sites of a square lattice with periodic boundary conditions, then the partition function can be written in the standard way [8,9] as

$$Z_1 = \mathrm{Tr}\big((\mathbf{D}_1\,\mathbf{A})^M\big) \tag{2.5}$$

with $\mathbf{A}$ given by Eq. 2.4 and $\mathbf{D}$ a diagonal matrix. The matrix $\mathbf{A}$ contains the interactions going from one column to another, while the matrix $\mathbf{D}_1$ contains the interactions within a column. In fact, all of the dimensional information can be regarded as being in the structure of the diagonal matrix $\mathbf{D}$. For example, for a d dimensional hypercubic lattice, the structure of $\mathbf{D}$ is such that it corresponds to interactions between a $d-1$ dimensional hypercubic lattice. Clearly for any $\mathbf{D}$ the problem possesses translational invariance in the direction having M spins.

The structure of the diagonal matrix is $\mathbf{D}_0 = \mathbf{I}$ for $d=1$, while for $d=2$ it has the form

$$\mathbf{D}_1 = \mathbf{I} \odot (\mathbf{PA}), \tag{2.6}$$

where $\mathbf{A}$ is given by Eq. 2.4 and $\mathbf{P}$ is a permutation matrix given by

$$\mathbf{P} = \begin{Bmatrix} \mathbf{J} & \mathbf{I} & \mathbf{J} & \mathbf{J} & \cdots & \mathbf{J} \\ \mathbf{J} & \mathbf{J} & \mathbf{I} & \mathbf{J} & \cdots & \mathbf{J} \\ \mathbf{J} & \mathbf{J} & \mathbf{J} & \mathbf{I} & & \mathbf{J} \\ \vdots & \vdots & \vdots & \ddots & \ddots & \vdots \\ \mathbf{J} & \mathbf{J} & \mathbf{J} & \cdots & \mathbf{J} & \mathbf{I} \\ \mathbf{I} & \mathbf{J} & \mathbf{J} & \cdots & \mathbf{J} & \mathbf{J} \end{Bmatrix}. \tag{2.7}$$

For the Ising model $\mathbf{P}$ is the $2^N \times 2^N$ matrix introduced in early transfer matrix studies [10]. The matrix $\mathbf{P}$ is sometimes called the generalized flip matrix since it has the property that it flips in a cyclic permutation the spins in a vector constructed using the direct product of vectors, $\big[1 \leftarrow 2 \leftarrow 3 \cdots \leftarrow N \leftarrow 1\big]$. The elements of the diagonal matrix for the $d=2$ Ising model given by Eq. 2.6 are identical to the diagonal matrix given in the formulation of the Ising model in Ref. 8. For a d dimensional hypercubic lattice with screw boundary conditions in the $d-1$ dimensions that have N spins, the diagonal matrix would be given by

$$\mathbf{D}_{d-1} = \mathbf{I} \odot \Big(\mathbf{PA}\Big) \odot \Big(\mathbf{P}^{N^{\frac{1}{d-1}}}\mathbf{A}\Big) \odot \Big(\mathbf{P}^{N^{\frac{2}{d-1}}}\mathbf{A}\Big) \odot \cdots \odot \Big(\mathbf{P}^{N^{\frac{d-2}{d-1}}}\mathbf{A}\Big) \tag{2.8}$$

when $N^{\frac{1}{d-1}}$ is an integer. Note that translational invariance in the direction(s) with N spins can be interpreted as the statement that $\mathbf{D}$ commutes with $\mathbf{P}$.

For $d=2$ the diagonal matrix is given by Eq. 2.6, while for $d=1$ the diagonal matrix can be considered to be any function times the identity matrix $\mathbf{I}$. Thus, one can formally write

$$\mathbf{D}_\eta \equiv \mathbf{I} \odot (\mathbf{P}^\eta \mathbf{A}) \tag{2.9}$$

and

$$Z_\eta = \mathrm{Tr}\big((\mathbf{D}_\eta \mathbf{A})^M\big), \tag{2.10}$$

which gives the $d=2$ Ising model for $\eta = 1$ and the $d=1$ Ising model for $\eta = 0$ (since $\mathbf{P}^0 = \mathbf{I}$). The parameter η can be used to interpolate between $d=1$ and $d=2$. (Note that for any non-zero integer value of η one again gets a $d=2$ Ising model — or depending on the values of η and N a number of decoupled $d=2$ Ising models.) The matrix $\mathbf{P}^\eta$ can be obtained by using the spectral decomposition

$$\mathbf{P}^{\eta} = \sum_{i} \lambda_i^{\eta} \vec{\mathbf{v}}_i \vec{\mathbf{v}}_i^{\dagger} \tag{2.11}$$

with λ_i the eigenvalues of $\mathbf{P}$ and $\vec{\mathbf{v}}_i$ the associated right eigenvector of $\mathbf{P}$. The left eigenvectors of $\mathbf{P}$ are the Hermitian conjugates of the right eigenvectors. Since $\mathbf{P}$ is a permutation matrix, its eigenvalues are just given by roots of unity [11]. In the next section, the justification for the definition in Eq. 2.10 will be given for the special case of the $q=2$ Potts model.

b. The Ising Model

In the rest of this paper, I will concentrate on the specific case of the ferromagnetic $q=2$ Potts model (the Ising model) for $0 \leq \eta \leq 1$. The two-particle transfer matrix is given by Eq. 2.1 and the partition function is given by Eq. 2.10. Note that the eigenvalues and the elements of the eigenvectors of $\mathbf{P}$ are complex numbers — yet to make contact with normal statistical mechanics all the elements of $\mathbf{D}_\eta$ should be real and positive. By constructing explicitly the elements of $\mathbf{D}_\eta$ for the Ising model, I will show that all the elements of $\mathbf{D}_\eta$ are indeed real and positive whenever N is an odd integer and the phase ϕ_i of the eigenvalues

$$\lambda_i = e^{\imath\phi_i} \tag{2.12}$$

of $\mathbf{P}$ is chosen such that $-\pi < \phi_i \leq \pi$.

For a lattice with N Ising spins in the η direction the phases ϕ_i are the N roots of unity. If N is a prime number, then each of the N roots of unity appear $(2^N-2)/N$ times, except for the eigenvalue $\lambda_i = 1$ which has a degeneracy two greater than the other eigenvalues. The statement that 2^N-2 is divisible by N is a statement of Fermat's theorem on primes, and the degeneracy of the eigenvalues of $\mathbf{P}$ can also be concisely written for non-prime values of N [12]. Let $\langle i|$ be the projection operator onto state i. Then the diagonal elements of $\mathbf{D}_\eta$ are given by

$$\langle i|\mathbf{D}_\eta|i\rangle = \sum_{j=1}^{2^N} e^{\imath\eta\phi_j} \langle i|\vec{\mathbf{v}}_j|1\rangle\langle 1|\vec{\mathbf{v}}_j^{\dagger}\mathbf{A}|i\rangle. \tag{2.13}$$

However, since $\mathbf{A}$ commutes with $\mathbf{P}$ one can choose the eigenvectors such that

$$\vec{\mathbf{v}}_j^{\dagger}\mathbf{A} = \hat{\kappa}_j \vec{\mathbf{v}}_j^{\dagger}, \tag{2.14}$$

where $\hat{\kappa}_j$ is an eigenvalue of $\mathbf{A}$. Thus one obtains

$$\langle i|\mathbf{D}_\eta|i\rangle = \sum_{j=1}^{2^N} e^{\imath\eta\phi_j}\, \hat{\kappa}_j |\langle i|\vec{\mathbf{v}}_j|1\rangle|^2. \tag{2.15}$$

Hence, the fact that the eigenvectors of $\mathbf{P}$ have elements from the field of complex numbers does not affect the positivity of the elements of $\mathbf{D}_\eta$.

The eigenvectors of $\mathbf{P}$ can be constructed from the eigenvectors of $\mathbf{a}$. Let $\mathbf{a}\vec{\mathbf{u}}_i = \kappa_i \vec{\mathbf{u}}_i$ with $\vec{\mathbf{u}}_1 = \frac{1}{\sqrt{2}}\begin{pmatrix}1\\1\end{pmatrix}$ and $\vec{\mathbf{u}}_2 = \frac{1}{\sqrt{2}}\begin{pmatrix}1\\-1\end{pmatrix}$ so $\kappa_1 = e^K + 1$ and $\kappa_2 = e^K - 1$. Then an eigenvector associated with the eigenvalue $\hat{\kappa}_j = \kappa_1^m \kappa_2^{N-m}$ with $(0 \le m \le N)$ of $\mathbf{A}$ and $e^{\imath\phi_j}$ of $\mathbf{P}$ can easily be constructed. Start with any eigenvector $\vec{\mathbf{U}}_j$ of $\mathbf{A}$ with $\vec{\mathbf{U}}_j$ formed from the direct product of m vectors $\vec{\mathbf{u}}_1$ and $N - m$ vectors $\vec{\mathbf{u}}_2$. Then an (unrenormalized) eigenvector of $\mathbf{P}$ associated with eigenvalue $\omega_\ell = e^{\imath\phi_\ell}$ is given by

$$\vec{\mathbf{v}}_{j\ell} = \sum_k \omega_\ell^k \, \mathbf{P}^k \vec{\mathbf{U}}_j. \tag{2.16}$$

The sum in Eq. 2.16 is over all distinct permutations of the vectors formed by the construction $\mathbf{P}^k\vec{\mathbf{U}}_j$. Using this construction with different $\vec{\mathbf{U}}_j$ gives a full set of eigenvectors. Since the construction of Eq. 2.16 is to be done for all distinct values of ℓ, it is clear that for N odd any given starting vector $\mathbf{U}_j$ generates a pair of eigenvectors of $\mathbf{P}$ which are complex conjugates of each other — the exception being only the case where $\omega = 1$, which is not paired. Hence grouping together these pairs of eigenvalues in the sum in Eq. 2.15 makes every term in the sum positive, so all diagonal elements of $\mathbf{D}_\eta$ are positive. A slight generalization of this proof works for the ferromagnetic q-state Potts model with integer values of q.

c. Interpolation Between $d = 0$ and $d = 1$

One question which must be asked is what type of 'couplings' are generated by the procedure discussed in Sect. II.b. In this section an exact solution is presented for the Ising model when the interpolation is between $d = 0$ and $d = 1$. The solution shows that any long range couplings which are generated are not sufficient to produce a mean field type of critical behavior. In Section II.d, the form of the generated Hamiltonians will be presented.

The decoupling of the η direction can be accomplished by letting the partition function be equal to the sum of all the elements of $\mathbf{D}_\eta$ given by Eq. 2.13. Let $\mathbf{V}$ be a matrix with all the columns equal to the eigenvectors of $\mathbf{P}$ and $\mathbf{A}$ described in the previous section. Then the uncoupled partition function is given by

$$Z_u = \mathrm{Tr}(\mathbf{D}_\eta) = \mathrm{Tr}\big((\mathbf{V}\rho^\eta\mathbf{V}^\dagger\mathbf{A}) \odot \mathbf{I}\big), \tag{2.17}$$

where $\mathbf{P} = \mathbf{V}\rho\mathbf{V}^\dagger$ with ρ a diagonal matrix containing the eigenvalues of $\mathbf{P}$. Since the eigenvectors of $\mathbf{P}$ are also the eigenvectors of $\mathbf{A}$ one has $\mathbf{V}^\dagger\mathbf{A} = \kappa\mathbf{V}^\dagger$ with κ a diagonal matrix containing the eigenvalues of $\mathbf{A}$. Using the cyclical properties of the trace gives

$$Z_u = \mathrm{Tr}(\mathbf{V}\rho^\eta\kappa\mathbf{V}^\dagger) = \mathrm{Tr}(\rho^\eta\kappa), \tag{2.18}$$

where the redundant operations of the trace and $\odot\mathbf{I}$ have been removed. For N prime (and $N \ge 3$), a simple formula emerges from Eq. 2.18. For N prime with a given starting vector $\vec{\mathbf{U}}_j$ one can construct N different eigenvectors associated with the N eigenvalues of $\mathbf{P}$ using Eq. 2.16 — except for the two vectors which are direct products of either all $\vec{u}_1$ or all $\vec{u}_2$ vectors. Then it is easy to see that the partition function is given by

$$Z_u = \left[1 - f(\eta)\right]\left(\kappa_1^N + \kappa_2^N\right) + f(\eta)\left(\kappa_1 + \kappa_2\right)^N, \tag{2.19}$$

where

$$f(\eta) = \frac{1}{N}\left[1 + \sum_{\ell=1}^{\frac{N-1}{2}} \cos\left(\frac{2\pi\ell\eta}{N}\right)\right] \tag{2.20}$$

and the κ_i are the eigenvalues of the two particle transfer matrix **a**.

Taking the thermodynamic limit for fixed $\eta \neq 1$ gives

$$\lim_{N\to\infty} \frac{1}{N}\ln(Z_u) = K + \ln(2) + \lim_{N\to\infty} \frac{1}{N}\ln(f(\eta)), \tag{2.21}$$

where in the thermodynamic limit

$$f(\eta) = \frac{\sin(\pi\eta)}{\pi\eta}. \tag{2.22}$$

Of course, taking the thermodynamic limit for fixed $\eta = 1$ gives the normal solution to the $d = 1$ Ising model. Note from Eq. 2.21 that there is no finite temperature phase transition in the uncoupled model. Hence the model does not possess a mean field type of critical behavior due to any infinite range couplings introduced by the interpolation procedure.

d. Construction of a Hamiltonian

One can ask about the effective Hamiltonian in the η direction which is generated by the interpolation scheme. That is, one can write a Hamiltonian which contains all possible terms allowed by the translational symmetry, up-down symmetry, and reflection symmetry. This Hamiltonian has the form

$$\mathcal{H} = \mathcal{H}_0 + \mathcal{H}_{nn} + \mathcal{H}_{nnn} + \cdots + \mathcal{H}_{all}, \tag{2.23}$$

where the subscripts denote the constant term,nearest neighbor term, next-nearest neighbor term, $\cdots$, all spins interacting term. The number of terms of course depends on N. For small N a fixed solution is possible, but for large N a numerical solution to the linear equations is required. In general, the strength of the individual terms depends on η and T ($J/k_B = 1$). The nn term for small lattices is shown in Fig. 1(a) for fixed η as a function of T, and in Fig. 1(b) for fixed T as a function of η. Note that although long-range terms are present for all $0 < \eta < 1$, they are not large enough to produce a finite temperature phase transition, as was

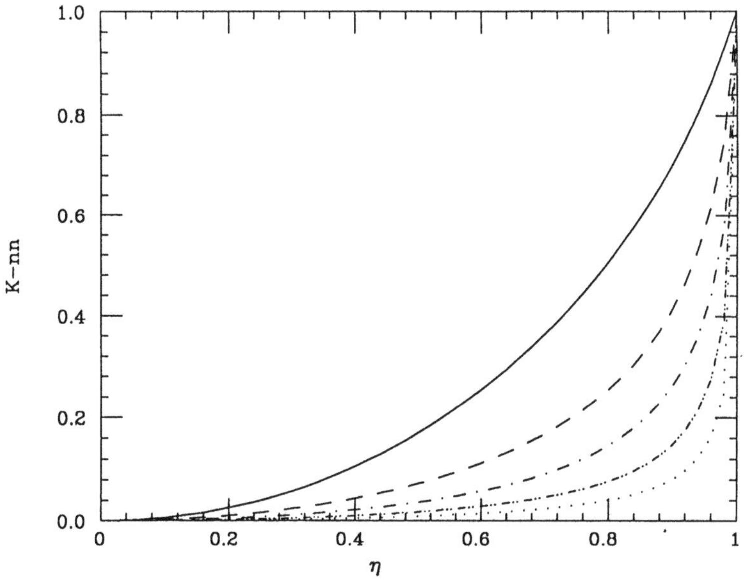

Fig. 1 The nearest neighbor coupling term of the Hamiltonian is shown as a function of η for $T = 1$. The lines are $N = 3$ solid, $N = 5$ dashed, $N = 7$ dot dash, $N = 9$ dot dot dash, and $N = 11$ dotted.

shown in the previous section. It should also be noted that numerically all terms turned out to be ferromagnetic — although no proof of this statement has yet been discovered.

III. DATA AND ANALYSIS

The partition function Z_η defined in Eq. 2.10 has been studied on finite strips for the zero field ferromagnetic $q=2$ Potts model. The free energy per spin

$$\mathcal{F} = \frac{1}{MN\beta} \ln(Z_\eta), \tag{3.1}$$

internal energy per spin $U = -\frac{1}{MN}\frac{\partial \ln(Z_\eta)}{\partial \beta}$, and specific heat $C = \frac{\partial U}{\partial T}$ were calculated for different strip widths N. Here $\beta = \frac{1}{k_B T}$. The $M \to \infty$ limit was taken, so Z_η corresponds to the largest eigenvalue of the transfer matrix. For all values of $0 < \eta < 1$ studied, a finite temperature first-order transition was seen. Figure 2 shows the internal energy and specific heat as a function of temperature for $\eta = \frac{1}{2}$. Note the clear evidence for a first order transition. This can be seen from the fact that the height of the specific heat maximum scales with lattice size exponentially fast — as seen in Fig. 3. A fit to the finite size scaling form for first-order transitions [13]

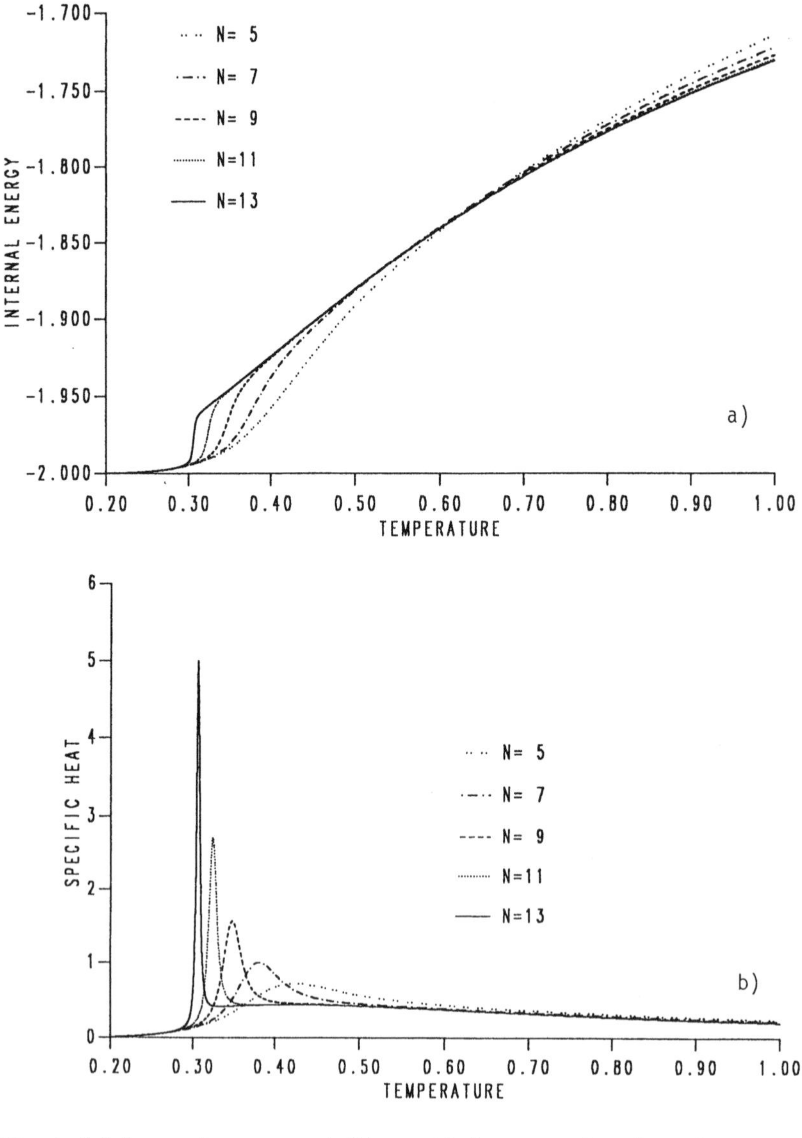

Fig. 2 (a) Internal energy and (b) specific heat as a function of temperature for $\eta = \frac{1}{2}$. The evidence for a first order transition is clearly seen.

$$C_{max} = A\exp(cN) + b \tag{3.2}$$

was performed. For $\eta = 0.8$, the data scale very well with the form of Eq. 3.2, especially for $N > 7$ (the largest system studied had $N = 17$). For values of η close to either 0 or 1 the data seem to be in a cross-over regime for the $N < 17$. By using the scaling form of Eq. 3.2, it is possible to obtain the interfacial tension

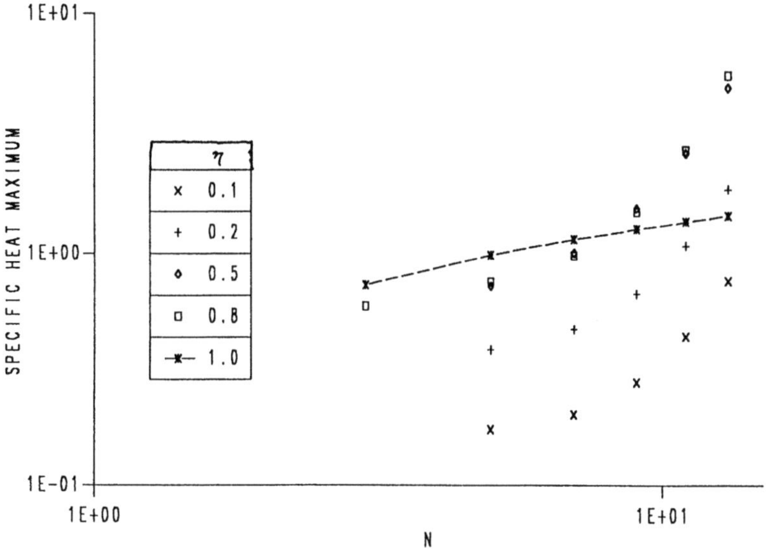

Fig. 3 Specific heat maximum as a function of lattice size for different values of η. The fact that the specific heat maximum scales exponentially fast with lattice size indicates a first order transition. Note the logarithmic dependence with lattice size for $\eta = 1$ $(d=2)$.

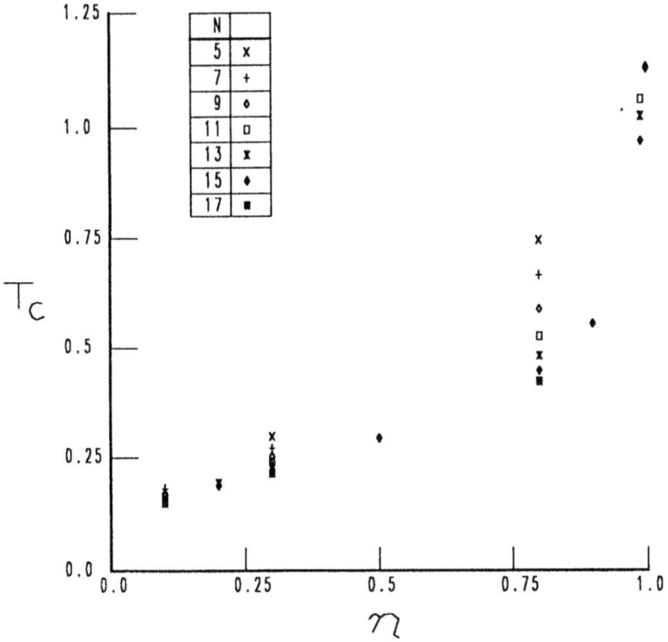

Fig. 4 The temperature at which the specific heat maximum occurs as a function of η is shown for various lattice sizes N.

[13] (or domain wall free energy) and compare with the 1+ϵ renormalization group [2]. This study is currently underway for small values of η. Work to determine the magnetic susceptibility as a function of N is also in progress — this would eliminate the non-divergent part from the scaling form, and hence allow a better determination of the interfacial tension. It is clear from the preliminary data that the critical temperature, T_c, decreases with η. As shown in Fig. 4, $T_c \rightarrow 0$ as $\eta \rightarrow 0$.

IV. DISCUSSION AND CONCLUSIONS

The interpolation between dimensions using finite size transfer matrix calculations is still in its infancy. Consequently this work raises many more questions that it answers.

When this method is applied to the ferromagnetic Ising model with an interpolation procedure between 1 and 2 dimensions, the resulting system exhibits many of the properties one would desire of a system in non-integer dimensions. In particular, for any non-integer value of η the model exhibits a finite temperature phase transition, with the critical temperature decreasing with η when the system is between 1 and 2 dimensions. Conversely, when the system is between 0 and 1 dimension no finite temperature phase transition exists. Does the interpolated model have other properties of non-integer dimensional systems, such as convexity violations [14]? Although no such violations were seen, some of the properties of the interpolated system are different from what is expected for integer dimensional statistical systems. For example, Fig. 5 shows the free energy per spin as a function of lattice size for finite strips. Note that at high temperatures the free energy approaches its infinite lattice value much slower than would be expected in integer

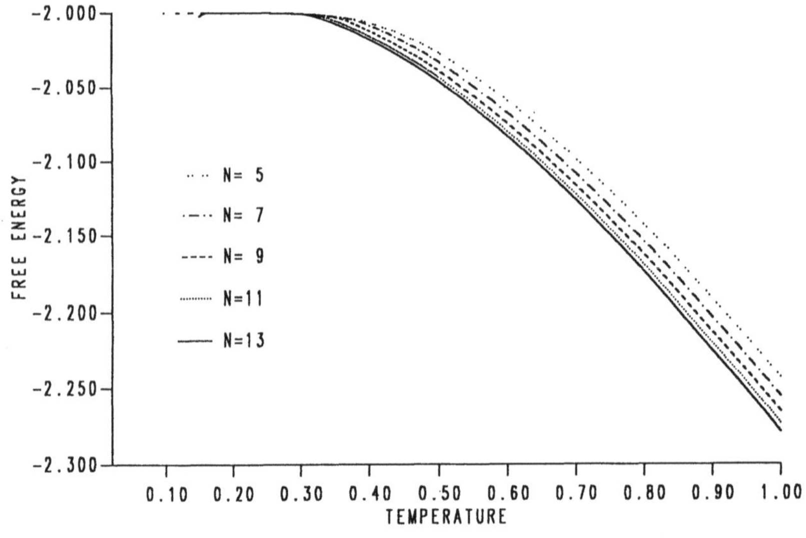

Fig. 5 The free energy per spin is shown as a function of temperature for $\eta = \frac{1}{2}$.

dimensional systems (where an exponentially fast convergence would be expected). Further work is also necessary, and is in progress, to calculate from finite size scaling at the first-order transition the magnetic susceptibility and interfacial tension as a function of η.

Where is the connection with quantum spin systems? As discussed in other contributions to the proceedings of this symposium, in order to study a quantum spin system using Monte Carlo techniques one must replace the quantum system by a 'classical equivalent'. One way of doing this is to utilize the Trotter formula [15]

$$\exp\Big(\mathbf{A}_1 + \mathbf{A}_2 + \cdots + \mathbf{A}_p\Big) \approx \Big(\exp(\frac{\mathbf{A}_1}{n})\ \exp(\frac{\mathbf{A}_2}{n})\ \cdots \exp(\frac{\mathbf{A}_p}{n})\Big)^n. \quad (4.1)$$

This has been exploited in many Monte Carlo and transfer matrix studies of quantum systems [16], with the 'classical equivalent' being in a dimension one greater than the dimension of the quantum system. The question which naturally arises is whether there exists for some quantum systems a 'classical equivalent' which can be interpreted as being in a non-integer number of dimensions. The answer to this question is still very much open.

REFERENCES

[1] K. G. Wilson and M. E. Fisher, Phys. Rev. Lett. **28** 240 (1972).

[2] D. J. Wallace and R. K. P. Zia, Phys. Rev. Lett. **43** 808 (1979); D. Forster and A. Gabriunas, Phys. Rev. A **23** 2627 (1981).

[3] M. E. Fisher and D. S. Gaunt, Phys. Rev. A **133** 224 (1964). R. Abe, Prog. Theor. Phys. (Kyoto) **47** 62 (1972).

[4] G. A. Baker, Jr. and L. P. Benofy, J. Stat. Phys. **29** 699 (1982).

[5] Y. Gefen, A. Aharony, Y. Shapir, and B. B. Mandelbrot, J. Phys. A **17** 435 (1984).

[6] M. A. Novotny, J. Math. Phys. **20** 1146, 1979.

[7] H. E. Stanley, *Introduction to Phase Transitions and Critical Phenomena* (Oxford University Press, New York, 1971).

[8] K. Huang, *Statistical Mechanics* (John Wiley & Sons, Inc., New York, 1963).

[9] L. Onsager, Phys. Rev. **65** 117 (1944).

[10] H. A. Kramers and G. H. Wannier, Phys. Rev. **60** 252 (1941).

[11] M. Marcus and H. Minc, *A Survey of Matrix Theory and Matrix Inequalities* (Allyn and Bacon, Inc., Boston, 1964).

[12] H. Gutfreund and W. A. Little, Am. J. Phys. **50** 219 (1982).

[13] V. Privman and M. E. Fisher, J. Stat. Phys. **33** 385 (1983).

[14] R. B. Griffiths and P. D. Gujrati, J. Stat. Phys. **30** 563 (1983).

[15] H. F. Trotter, Proc. Am. Math. Soc. **10** 545 (1959); M. Suzuki, Commun. Math. Phys. **51** 183 (1976), **57** 193 (1977).

[16] M. Suzuki, J. Stat. Phys. **43** 883 (1986), and references therein.

High-Speed Algorithm for Quantum Monte Carlo Simulation

Y. Okabe and M. Kikuchi

Department of Physics, Tohoku University, Sendai 980, Japan

This paper is concerned with a high-speed algorithm for Monte Carlo simulation of the quantum spin system using the Suzuki-Trotter formula. We introduce a binary-number spin representation using logical commands. The calculation is fully vectorized by decomposing into appropriate sublattices. The one-dimensional antiferromagnetic Heisenberg model (S=1/2) is studied for the Trotter number $m \leq 32$ to check the efficiency of our algorithm. We deal with both the pair-spin decomposition and the cluster-spin decomposition approximations.

1. Introduction

With the development of a computer, marked progress has been made in the speedup of Monte Carlo simulation of classical systems recently /1-6/. The use of vector computers and the development of the efficient algorithm have made the speedup possible. The speeds of Monte Carlo simulation of the three-dimensional Ising model with periodic boundary conditions are often compared. Applying the newly developed algorithm by BHANOT, DUKE and SALVADOR /5/, we have realized the speed of 251×10^6 spin-flip attempts per second with a vector computer NEC SX-2 /6/.

The Monte Carlo simulation is also an attractive method for the study of various quantum systems. Since the proposal of SUZUKI /7/ in 1976, the generalized Trotter formula (the Suzuki-Trotter formula) has been successfully used to realize the quantum Monte Carlo simulation /8,9/. In this formula a d-dimensional quantum system is mapped on to a (d+1)-dimensional classical system. The partition function is numerically evaluated not only by the Monte Carlo method /10-15/ but also by the transfer-matrix method /16-20/. The quantum system is exactly reproduced when the Trotter number m tends to infinity. The considerably large m should be used at low temperatures where quantum effects are important. However, it is difficult to employ large Trotter numbers in the numerical calculations, due to the memory size of a computer in the case of the transfer-matrix method, and due to the computation time in the case of the Monte Carlo method.

It is desirable that a technique of speedup for classical systems is applied to quantum statistical Monte Carlo simulations. To answer such a demand, we implemented a high-speed algorithm for Monte Carlo simulation of the quantum spin system /15/. In order to check the efficiency of our algorithm, we studied the one-dimensional (1-d) XY model, and obtained the excellent agreement with the exact solution of KATSURA /21/.

In this paper we briefly explain our algorithm of the vectorized Monte Carlo simulation of a quantum spin system. As an example, we study the 1-d antiferromagnetic (AF) Heisenberg model (S=1/2). We treat both the pair-spin decomposition and the cluster-spin decomposition approximations.

2. High-Speed Quantum Monte Carlo Simulation

Let us consider the 1-d quantum spin system described by the generalized Hamiltonian

$$\mathcal{H} = -2 \sum_{i=1}^{N} (J_x s_i^x s_{i+1}^x + J_y s_i^y s_{i+1}^y + J_z s_i^z s_{i+1}^z) . \quad (1)$$

According to the Suzuki-Trotter formula /7/ the partition function is approximated as

$$Z = \mathrm{Tr}\, \exp(-\beta\mathcal{H}) \simeq \mathrm{Tr}\, [\prod_j \exp(-\beta\mathcal{H}_j/m)]^m , \quad (2)$$

where $\mathcal{H}_j$ are the local sub-Hamiltonians such that $\mathcal{H}=\sum\mathcal{H}_j$, and m is called the "Trotter number". It is to be noted that how to decompose the Hamiltonian into the local sub-Hamiltonians is arbitrary. We first treat the pair-decomposition approximation for simplicity. Then, the present system is mapped onto the two-dimensional ($N\times 2m$) Ising model with four-spin interactions by the use of checkerboard decomposition /22/. The graphical representation of the checkerboard pair-spin decomposition is given in Fig. 1, where the shaded squares denote the four-spin interaction. Only eight four-spin configurations are allowed for S=1/2, and they are graphically shown in Fig. 2. For the special symmetrical case $J_x=J_y$ the matrix elements of two four-spin configurations become zero. Let us regard a four-spin configuration as a single information and call it a "plaquette" spin, p. There are $N\times m$ plaquette spins as a total in our system. We assign 0-7 to each state of the plaquette spin. The binary-number representation of the assignment is also given in Fig. 2. The spin flip is simply realized by the logical command XOR (exclusive OR) in our plaquette spin representation. The plaquette spin has three bits, and the command XOR is operated on each bit. The manipulation, such as p_{new}=XOR(p_{old},3), gives the spin-flipping process. Since the plaquette spin has the information on the four-spin configuration, we can readily calculate the energy, the specific heat, the magnetization, etc. /12,22/.

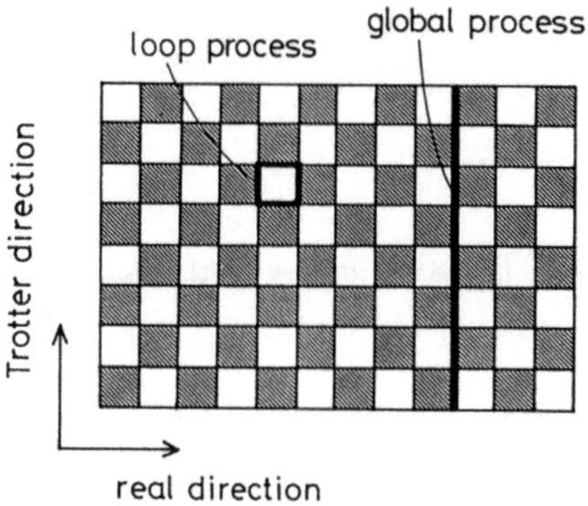

Fig. 1. Graphical representation of the checkerboard pair-spin decomposition. The shaded squares denote the four-spin interaction. Periodic boundary condition is imposed along the Trotter direction. The loop and global spin-flip processes are also illustrated

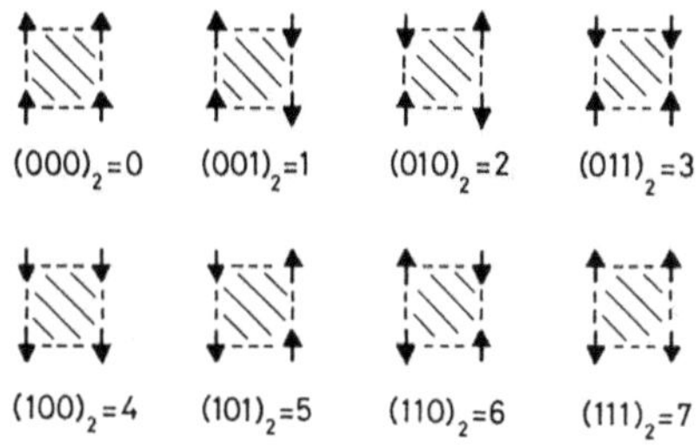

Fig. 2. "Plaquette" spin representation of the four-spin configuration

Next consider the vectorization of Monte Carlo simulations. Generally, the calculation is fully vectorized /1-3,5,6/ if the following condition is satisfied: One divides the lattice into interpenetrating sublattices so that one can update spins on each sublattice independently. Such a decomposition is always possible if the interaction is of short range. In our Monte Carlo simulation of the 1-d quantum spin system, we should treat multi-spin-flip processes /23/. We employ both loop and global processes to update the spin configuration /24/. Both processes are graphically shown in Fig. 1. All the spins on the bold line are flipped according to the transition probability determined by the change of local energy at the spin flip. We can easily vectorize the calculation for both processes by decomposing into appropriate sublattices.

3. One-Dimensional Antiferromagnetic Heisenberg Model

We have carried out the Monte Carlo simulation of the 1-d AF Heisenberg model, $J_x=J_y=J_z=-J$ ($J>0$), by the vector computer NEC SX-1 of the computer center of Tohoku University. The number of the spins and the Trotter number are $N=64$, and $m=8$, 16 and 32. We define a single Monte Carlo step by updating all the mapped classical spins ($N\times 2m$). The number of Monte Carlo steps is 2×10^5, 4×10^5 and 8×10^5 for $m=8$, 16 and 32, respectively. For extremely low temperatures we have employed 8×10^5, 16×10^5 and 32×10^5 Monte Carlo steps for each m. The ratio of the loop process to the global process is chosen as 4:1. We measure the energy, the specific heat, and the susceptibility. We monitor these quantities for each five Monte Carlo steps. In the case of $N=64$ and $m=16$, it takes about 54 seconds for the calculation of 4×10^5 Monte Carlo steps. It means that we update 15.2×10^6 classical spins per second. We have achieved the comparable speed with the classical case.

We show the temperature dependence of energy per spin in Fig. 3. We measure the energy and temperature in units of J. The statistical errors are smaller than the mark. We have estimated the statistical errors by the short-time calculations. The solid curve is obtained by extrapolating short chain diagonalization data of BONNER and FISHER /25/. We see that the calculated energy is close to that of BONNER and FISHER /25/ for large m even at extremely low temperatures. The exact value of the ground-state energy is known as $-2\ln 2 + 1/2 \simeq -0.8863$. SUZUKI /26/ proved that the convergent rate of the average of the quantum operator is proportional to $1/m^2$. We give the $1/m^2$ dependence of the energy for fixed T in Fig. 4. We make the quadratic estimate,

$$E(m) = E(\infty) + a/m^2 + b/m^4 \quad ,$$

using the data for $m=32$, 16 and 8. In Table I, the estimates of the $m\to\infty$ value (for $0.15\leq T\leq 0.4$) are tabulated.

Table I. Estimates of the $m\to\infty$ values of the energy, the specific heat and the susceptibility

T	E	C	χ
0.15	-0.8826	0.0589	0.2353
0.2	-0.8796	0.0690	0.2355
0.3	-0.8703	0.1157	0.2366
0.4	-0.8569	0.1573	0.2415

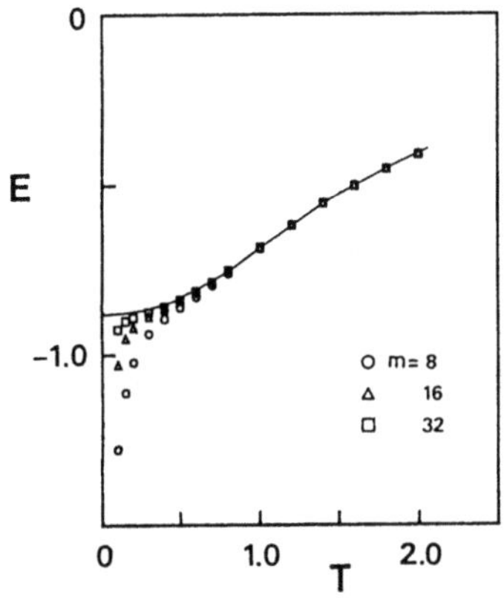

Fig. 3. Temperature dependence of energy for the 1-d AF Heisenberg model for m=8, 16 and 32 in units of J. The solid curve is obtained by extrapolating short chain diagonalization data of BONNER AND FISHER /25/

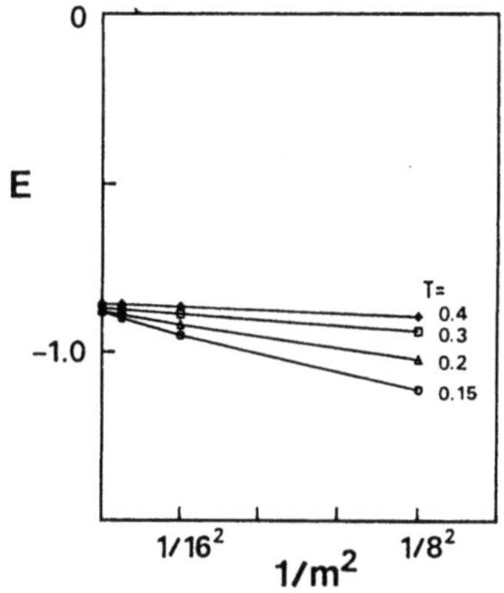

Fig. 4. Energy versus $1/m^2$ for the 1-d AF Heisenberg model at fixed temperatures

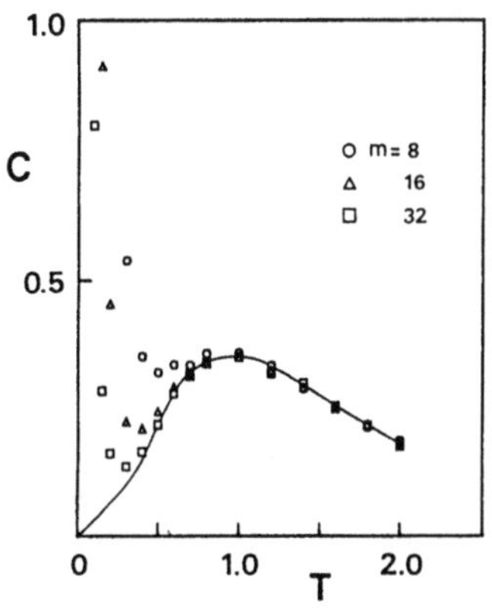

Fig. 5. Temperature dependence of specific heat for the 1-d AF Heisenberg model for m=8, 16 and 32. The solid curve is the calculation of BONNER and FISHER /25/

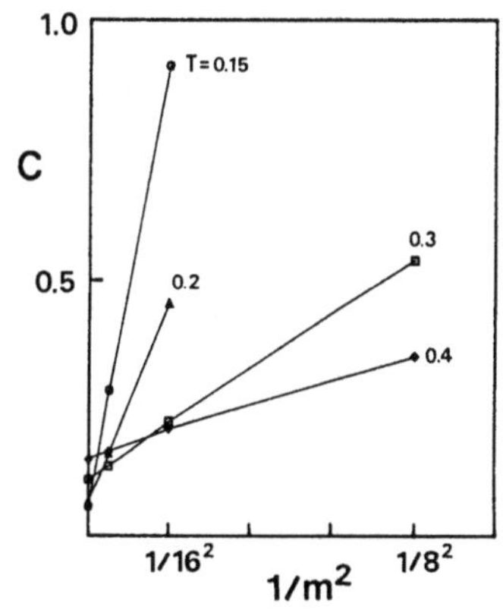

Fig. 6. Specific heat versus $1/m^2$ for the 1-d AF Heisenberg model at fixed temperatures

The temperature dependence of specific heat per spin is shown in Fig. 5. The error bars are given in the figure when the errors are larger than the mark. We see from the figure that the calculated specific heat is very close to that of BONNER and FISHER /25/ even at low temperatures. We show the $1/m^2$ dependence of the specific heat in Fig. 6. We give the extrapolated values of specific heat in Table I.

We show the temperature dependence of the susceptibility in Fig. 7. At low temperatures we find the deviations from the result of BONNER and

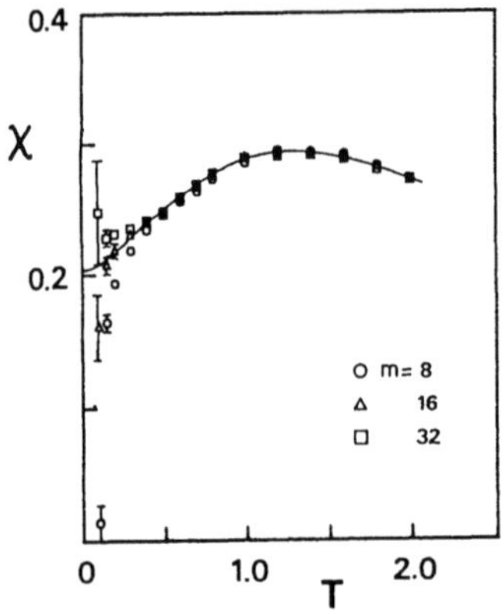

Fig. 7. Temperature dependence of susceptibility for the 1-d AF Heisenberg model for m=8, 16 and 32. The solid curve is the calculation of BONNER and FISHER /25/

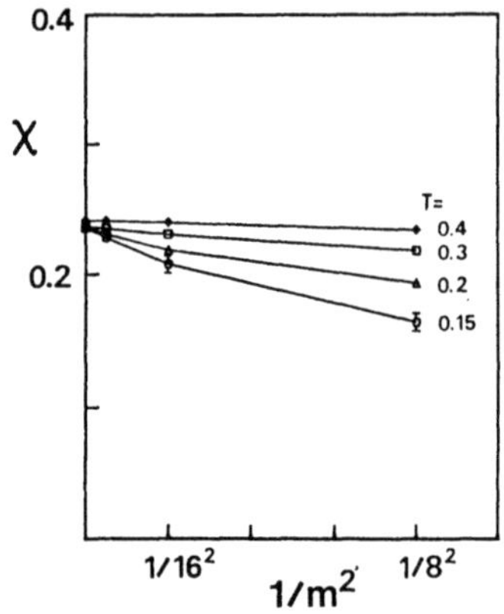

Fig. 8. Susceptibility versus $1/m^2$ for the 1-d AF Heisenberg model at fixed temperatures

FISHER /25/. The zero-temperature value of BONNER and FISHER /25/ corresponds to GRIFFITHS' conjecture /27/, $2/\pi^2 \simeq 0.2026$. Using the cluster-spin transfer matrix method, TSUZUKI /28/ also observed the deviations from the result of BONNER and FISHER /25/. One possibility to reconcile our low-temperature behavior and GRIFFITHS' conjecture /27/ is that there exists a $\sqrt{T}$ singularity near T=0. Such a $\sqrt{T}$ expansion was found by TAKAHASHI and YAMADA /29/ for the ferromagnetic Heisenberg model. However, more detailed studies are necessary to elucidate this point. The $1/m^2$ dependence is graphically shown in Fig. 8, and the extrapolated values of susceptibility are tabulated in Table I.

4. Cluster-Spin Decomposition

As was stated before, how to decompose the Hamiltonian $\mathcal{H}$ into local sub-Hamiltonians $\mathcal{H}_i$ is arbitrary. The approximation with larger-cluster decomposition is better than the smaller one, if the Trotter number m is fixed. TSUZUKI /17/ and BETSUYAKU /19/ showed that the calculation of the transfer-matrix method is much improved by the cluster-spin decomposition. In this section we apply the cluster-spin decomposition to the Monte Carlo simulation. We consider the four-spin cluster, and the decomposition scheme is illustrated in Fig. 9. The motivation for treating this four-spin cluster is to study the next-nearest neighbor coupling system. This model is interesting because it gives an example for a competing system and because one encounters a negative-sign problem for the quantum simulation. KIMURA, HARADA and TONEGAWA /30/ studied this model by the transfer-matrix method. We use the Monte Carlo simulation and the next-nearest coupling is chosen as zero in this study. The present system is mapped onto the two-plate Ising model with eight-spin interactions, shown in Fig. 10. We can easily extend the binary-number representation to this case. The spin flip is also realized by the logical command XOR. We follow three types of updating processes, the loop, local and global processes. They are shown in Fig. 10. It goes without saying that all the updating processes can be fully vectorized.

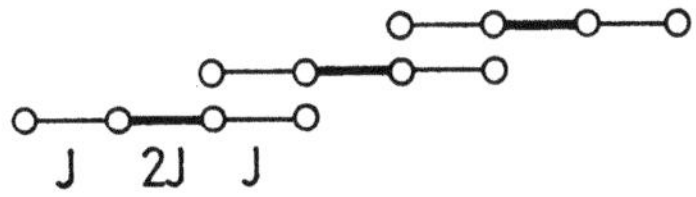

Fig. 9. Schematic illustration of the four-spin cluster

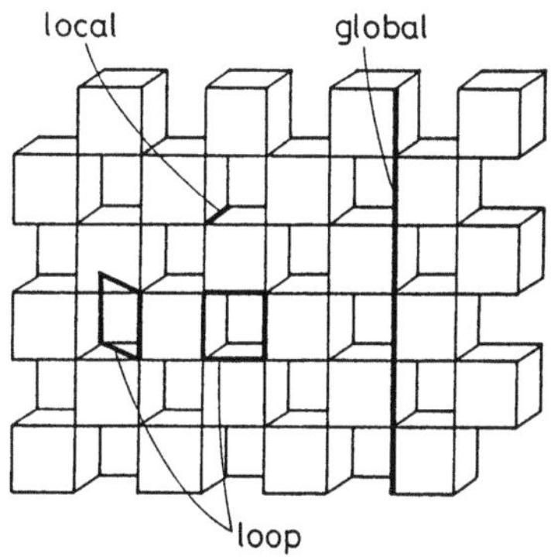

Fig. 10. Graphical representation of the four-spin cluster decomposition. the loop, local and global spin-flip processes are also shown

Next turn to the result of the calculation. The number of spins and the Trotter numbers are N=64 and m=4, 8 and 16. The number of Monte Carlo steps is 2.4×10^5, 4.8×10^5 and 9.6×10^5 for m=4, 8 and 16, respectively. The ratio of the loop, local and global processes is chosen as 4:1:1. We have achived the 5.4×10^6 classical-spin updates per second. We only show the result of energy in Fig. 11. We also show the $1/m^2$ dependence in Fig. 12. It is clearly seen that the calculation is much improved for the cluster-spin decomposition compared with the pair-spin decomposition.

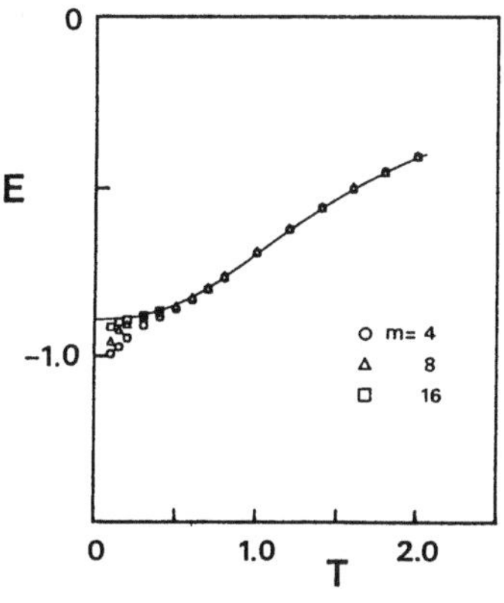

Fig. 11. Temperature dependence of energy for the 1-d AF Heisenberg model for m=8, 16 and 32 obtained by four-spin cluster decomposition Monte Carlo method. The solid curve is the calculation of BONNER and FISHER /25/

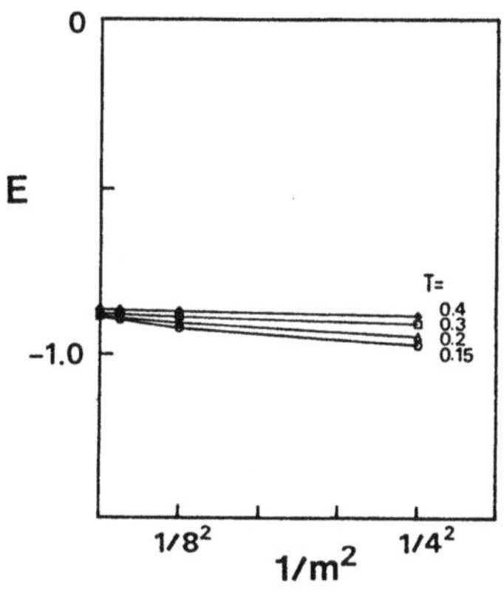

Fig. 12. Energy versus $1/m^2$ for the 1-d AF Heisenberg model obtained by four-spin cluster decomposition Monte Carlo method at fixed temperatures

5. Discussions

We have explained a high-speed algorithm for the Monte Carlo simulation of the quantum spin system /15/. The Monte Carlo calculation is generally vectorized by decomposing the lattice into appropriate sublattices in each updating process. Our plaquette spin representation enables us to make a simple and fast calculation. We have simulated the 1-d AF Heisenberg model using the high-speed algorithm. We have treated both the pair-spin decomposition and the cluster-spin decomposition. We have shown that the m-th approximant of the energy, the specific heat and the susceptibility converges with the rate of $1/m^2$ as was proved by SUZUKI /26/. It is desirable that the high-speed technique will be applied to the quantum statistical Monte Carlo simulations of various fields, especially to reveal the low-temperature properties where quantum effects are important.

Acknowledgments

We wish to thank I. Harada, T. Tonegawa and T. Tsuzuki for informing us of their works prior to publication and for valuable discussions. Thanks are also due to H. Betsuyaku and H. Nishimori for fruitful discussions.

References

1. S. Wansleben, J.G. Zabolitsky and C. Kalle: J. Stat. Phys. 37, 271 (1984); C. Kalle: J. Phys. A17, L801 (1984)
2. M. Kikuchi and Y. Okabe: Prog. Theor. Phys. 74, 458 (1985); J. Phys. Soc. Jpn. 55, 1359 (1986)
3. S. Wansleben: to be published
4. G.O. Williams and M.H. Kalos: J. Stat. Phys. 37, 283 (1984)
5. G. Bhanot, D. Duke and R. Salvador: Phys. Rev. B33, 7841 (1986)
6. M. Kikuchi and Y. Okabe: to be published
7. M. Suzuki: Prog. Theor. Phys. 56, 1454 (1976)
8. H. De Raedt and A. Lagendijk: Phys. Rep. 127, 233 (1985)
9. M. Suzuki: J. Stat. Phys. 43, 883 (1986)
10. M. Suzuki, S. Miyashita and A. Kuroda: Prog. Theor. Phys. 58, 1377 (1977)
11. J.E. Hirsch , R.L. Sugar, D.J. Scalapino and R. Blankenbecler: Phys. Rev. B26, 5033 (1982)
12. J.L. Cullen and D.P. Landau: Phys. Rev. B27, 297 (1983)
13. T. Sakaguchi, K. Kubo and S. Takada: J. Phys. Soc. Jpn. 54, 861 (1985)
14. M. Marcu and A. Wiesler: J. Phys. A18, 2479 (1985); M. Marcu, J. Muller and F.K. Schmatzer: J. Phys. A18, 3189 (1985)
15. Y. Okabe and M. Kikuchi: Phys. Rev. B34, No. 11 (1986)
16. H. Betsuyaku: Phys. Rev. Lett. 53, 629 (1984); Prog. Theor. Phys. 73, 319 (1985)
17. T. Tsuzuki: Prog. Theor. Phys. 73, 1352 (1985); 75, 225 (1986)
18. T. Yokota and H. Betsuyaku: Prog. Theor. Phys. 75, 46 (1986)
19. H. Betsuyaku: Prog. Theor. Phys. 75, 774 (1986)
20. H. Betsuyaku and T. Yokota: Prog. Theor. Phys. 75, 808 (1986)
21. S. Katsura: Phys. Rev. 127, B1508 (1962)
22. M. Barma and B.S. Shastry: Phys. Rev. B18, 3351 (1978)
23. A. Wiesler: Phys. Lett. 89A, 359 (1982)
24. Instead of using the term, the 'local' process, as in Ref. 15, we here use the 'loop' process for the consistency with the case of the cluster-spin decomposition
25. J.C. Bonner and M.E. Fisher: Phys. Rev. 135, A640 (1964)
26. M. Suzuki: Phys. Rev. B31, 2957 (1985); J. Math. Phys. 26, 601 (1985); Phys. Lett. 113A, 289 (1985)
27. R.B. Griffiths: Phys. Rev. 133, A768 (1964)
28. T. Tsuzuki: private communication
29. M. Takahashi and M. Yamada: J. Phys. Soc. Jpn. 54, 2808 (1985); M. Yamada and M. Takahashi: J. Phys. Soc. Jpn. 55, 2024 (1986)
30. T. Kimura, I. Harada and T. Tonegawa: private communication

Monte Carlo Renormalization Group for Quantum Systems

M. Kolb

Laboratoire de Physique des Solides, Bat. 510, Université de Paris-Sud, F-91405 Orsay, France

The Monte Carlo Renormalization Group provides a general and flexible method to calculate critical properties. Quantum systems can be treated in the same way by means of classical representations based on the Trotter formula. This concept is used to calculate static and dynamic critical exponents for the Ising model in a transverse field in one and two dimensions and for the XY/Heisenberg model in one dimension. The advantage of different classical representations and the role of the symmetries are discussed.

1. Introduction

The Renormalization Group Transformation (RG) [1] has provided both conceptual insight towards understanding second order phase transitions and a powerful computational tool to determine critical exponents and scaling functions. Universality and scaling properties which were studied extensively in the 60's were given a mathematical foundation with the idea of a fixed point and the transformation properties of the RG near this fixed point. There is a very large body of work covering the many applications of the RG and excellent reviews exist on the subject [2-5].

A classical system near a second order transition point has fluctuations on all length scales up to the correlation length, the only relevant length in the problem. Exactly at the critical point, the correlation length diverges. The system is said to be scale invariant. For quantum systems at temperature T=0 the quantum fluctuations play a role analogous to the thermal fluctuations of classical systems at a finite temperature. This suggests that in quantum systems one expects to observe second order phase transitions induced by quantum fluctuations.

The close relation between quantum mechanics and classical mechanics dates back to Feynmans path integral formulation in the 40's, as reviewed, for example, in [6]. Comparing quantum and classical transitions one can formally identify the quantities corresponding to each other:

classical, T>0	*quantum, T=0*
d+1 spatial dimensions	d space and one-time dimensions
thermal fluctuations	quantum fluctuation
free energy	ground-state energy
correlation length	inverse energy gap
correlation function	ground-state correlation function
temperature	ordering field

The low-temperature behaviour of the quantum system additionally provides information on the dynamics of the transition. The first model investigated specifically for a quantum transition has been the Ising model in a transverse field.

The exact solution in one dimension clearly shows the close relation between classical and quantum critical points [7,8].

The Monte Carlo (MC) method has become an increasingly powerful way to investigate classical systems in statistical mechanics and elsewhere [9-11]. For quantum systems its implementation is complicated by the fact that the Boltzmann factor is not known explicitly due to the non-commutative character of the Hamiltonian. It was Suzuki [12-14] who proposed a general method based on an operator identity - the Trotter formula - to construct systematic approximants of quantum systems that are amenable to MC simulations. This representation also provides a way to study quantum phase transitions classically and to implement the Monte Carlo Renormalisation Group (MCRG) for quantum transitions.

In this article we review the MCRG as originally proposed (§2), sketch real space RG method specifically designed for quantum models (§3), introduce the Trotter formula (§4) and then define the MCRG for quantum systems and illustrate it for the Ising model in a transverse field in one and two dimensions and the XY/Heisenberg model in one dimension (§5). In §6 the extension of these ideas to other systems is discussed.

2. The Monte Carlo Renormalisation Group

Following the sucessful direct implementation of the Kadanoff cell picture as a RG in real space [15] Ma proposed to combine the RG and the MC-methods [16] . The idea will be explained for the kinetic Ising model with nearest neighbor interactions with Glauber dynamics in two dimensions: A system of spins is simulated by a standard Metropolis algorithm such that in equilibrium each configuration occurs with probability $P(s)=e^{K\Sigma_{nn} s_i s_j}$. Following the block rule [15], as defined in Fig. 1, every configuration corresponds to a renormalised configuration on a coarse grained lattice.

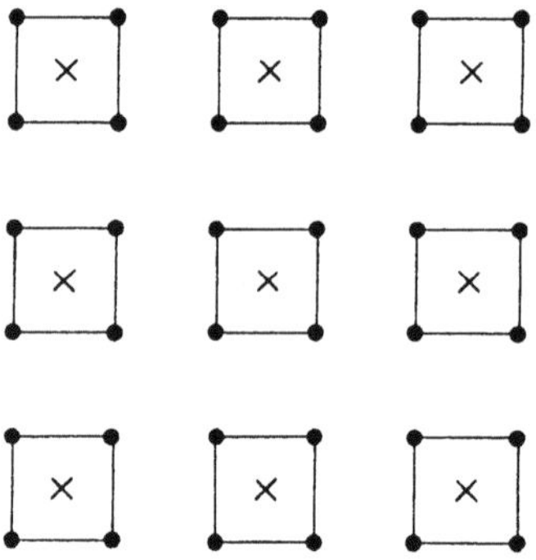

Figure 1
Block transformation on a square lattice. A block spin (X) is associated with a cell of four spins (•) on the original lattice.

The simulation thus simultaneously scans the phase space of the original lattice and - through the directly implemented RG - the phase space of the renormalised lattice, each with the appropriate statistical weight [16]. The major advantage of this approach in comparison with analytical methods in real space is that the RG is not truncated (the system size can be chosen sufficiently large that finite size corrections are negligible). The renormalised couplings are extracted as follows: suppose the transformed Hamiltonian has first and second neighbor interactions. Then the configuration on the left of Fig. 2 occurs with probability $P^+ = P_0 e^{4K'+4L'}$ and the one on the right of Fig. 2 with probability $P^- = P_0 e^{-4K'-4L'}$. The expression $\log(P^+ / P^-) = 8K'+8L'$ is therefore directly related to the renormalized nearest (K') and next nearest neighbor (L') couplings.

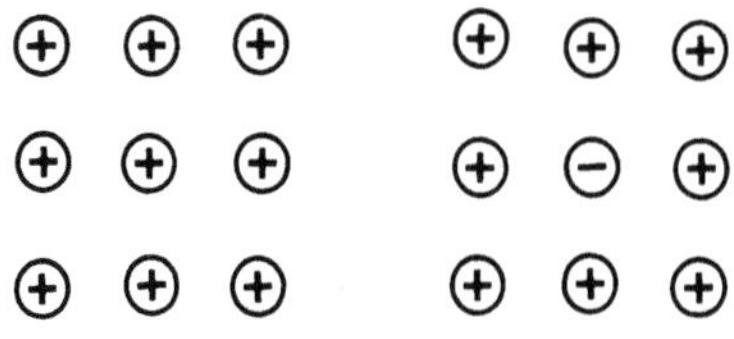

Figure 2
Local measurements used to extract couplings in the Hamiltonian. One compares probabilities of configurations with the central spin up resp. down with the surrounding spins held fixed.

Other relations of the same type are obtained by returning one or several of the neighboring spins. Two such relations are sufficient to determine K' and L'. Because of the proliferation of coupling constants more than two couplings are actually needed to characterize the renormalised states of the system. The procedure outlined in Fig. 2 thus contains a truncation to two parameters. The overdetermination serves as a method to estimate the error of the truncation [17]. It is easy to see that the procedure can be extended to more couplings by monitoring a larger environment around the central spin. The derivatives of the RG near a fixed point can also be calculated from Fig. 2; they simply correspond to the expectation values of the corresponding operators [16]. In order to estimate the statistical accuracy of the method one not only determines the renormalised couplings this way but also computes the original exactly known couplings with the method of Fig. 2.

This whole approach is very much in the spirit of the RG idea: the block transformation is local and the measuring procedure is local - utilising the basic hypothesis of the RG that local transformations are analytical and do not generate long-range interactions. As long as one is only interested in equilibrium properties the dynamics can be chosen at will, preferentially suitable for fast convergence. If one is interested in dynamical critical properties the MCRG is also suited to calculate them. Ma suggested to filter short-time fluctuations for the block spins and reinterpret the block dynamics in terms of a renormalised Glauber model from the measured flip rates in configurations such as shown in Fig. 2 [16]. Due to the large amount of freedom in defining the dynamics this still is a largely unresolved problem.

Two modifications have improved the numerical accuracy of the method in some cases of practical interest. The first change consists in working with the correlation functions instead of the interactions in the Hamiltonian. At a fixed point not only the couplings but also the expectation values of short-range operators are invariant. The transformation matrix can be obtained without knowing the couplings K',L', ... directly from the following identity [18-21]

$$d/dK_\beta \langle S_\alpha'\rangle = \Sigma_\gamma \left(\langle S_\alpha' S_\gamma'\rangle - \langle S_\alpha'\rangle\langle S_\gamma'\rangle\right) dK_\gamma'/dK_\beta = \langle S_\alpha' S_\beta\rangle - \langle S_\alpha'\rangle\langle S_\beta\rangle ,$$

where S_α are sums of local operators and K_α the corresponding couplings. The second modification consists in iterating the RG several times and in comparing lattices of different initial sizes to reduce the finite size effects. Multiple iterations formally have the effect that irrelevant operators die out and hence one gets closer to the fixed point. On the other hand, after several iterations the finite size effects become important. These effects can be reduced by comparing systems which differ in size and in the number of RG iterations.

3. The Renormalisation Group for Quantum Systems

Quantum mechanical phase transitions are expected to manifest themselves only at zero temperature, leading to a quantum-classical crossover at very low temperatures. This gives one physical motivation to develop RG's for quantum systems. More formally, there is an interest in quantum models as there is often a close relation between d-dimensional quantum systems and d+1 dimensional classical systems. For many problems in statistical mechanics this provides two alternative approaches to compute critical properties.

The RG method for quantum systems was introduced by the SLAC group [22,23] and by the Dutch group [24,25] independently, and is inspired by the classical real space RG - modified to account for the quantum problem with its dynamics. The following ingredients characterise these approaches. Let's consider a one dimensional spin one half quantum chain: divide the chain in groups of - say - three spins. Solve the Hamiltonian for this intra-cell problem exactly (or numerically exactly for larger cells). Then the series of energy levels is truncated to the lowest two in order to map the block system on the same spin 1/2 problem one started with. The new couplings between the block-spins are determined from the intercell operators between the retained low-lying states. The motivation for this procedure is that higher energy excitations contribute little to the critical properties. This and similar methods have been quite sucessful in calculating critical exponents (including dynamical properties) for a variety of models [26-31]. An important consideration in these calculations is that the symmetries are correctly dealt with: the RG should be defined such that the transformation has at least the symmetry of the model investigated [26,28].

The T=0 real space RG method provides qualitatively correct critical properties for many models (there are exceptions - notably for transitions of type essential singularity [30]). The precision of the results can be improved by choosing larger blocks and/or by retaining higher levels within a block. This necessitates numerical work. The T=0 methods have natural finite temperature extensions, as shown in [31]. They consist in treating the inter-block interactions perturbatively (high temperature expansion).

4. Classical Representation of Quantum Systems: The Trotter Formula

Classical formulations of quantum mechanics have been discussed and used since quantum mechanics was postulated. In the context of statistical mechanics there has been a renewed interest in this problem since very efficient methods were developed to perform numerical simulations and also because the computing power has increased considerably.

For classical systems one implements the Monte Carlo method easily by introducing a dynamics such that each state $|\alpha>$ of the system occurs with its equilibrium probability $P(\alpha)=e^{-\beta H(\alpha)}$ where H is the classical Hamiltonian. P can be calculated explicitly. For quantum systems $P(\alpha)=<\alpha|e^{-\beta H(\alpha)}|\alpha>$ is not known. Suzuki [12-14,32,33] proposed to evaluate $P(\alpha)$ approximately by means of the following operator identity by Trotter [34] for the non-commuting operators A and B

$$e^{A+B}=\lim_{m\to\infty}\left[e^{A/m}\,e^{B/m}\right]^m$$

The partition function of the quantum system is written as

$$Z=\sum_\alpha <\alpha|e^{-\beta H(\alpha)}|\alpha> \cong \sum_\alpha <\alpha|\left[e^{-\beta H_1/m}\,e^{-\beta H_2/m}\right]^m|\alpha>$$

$$=\sum_\alpha \prod_k <\alpha_k|\,e^{-\beta H_1/m}\,|\alpha_k'><\alpha_k'|e^{-\beta H_2/m}\,|\alpha_{k+1}> ,$$

where in the first step the Hamiltonian has been divided into two parts and the Trotter formula was used to get an approximate expression for Z which approaches Z uniformly as m goes to infinity. In the second step 2m-1 intermediate states were introduced and the partition function in d dimensions has been represented by a classical partition function in d+1 dimensions. The usefulness of this formulation depends on the breakup of $H=H_1+H_2$. The matrix elements

$$\langle\alpha | e^{-\beta H_i/m} |\alpha'\rangle = e^{H_{i,eff}(\alpha,\alpha')}$$

can be evaluated explicitly for an appropriate choice of H_i , i=1,2. The non-uniqueness of the decomposition is useful to find a most convenient representation for the actual simulations. A sucessful choice for H_1 and H_2 has been a sum of local operators which commute with each other. Different ways to decompose the Hamiltonian were discussed in great detail by Suzuki [33], who also gave exact bounds for the rate of convergence of Z_m. Note that the corrections using this approach are in terms of higher commutators (with increasing m). This is in contrast with the path integral formulation which is essentially a high temperature expansion. If the operators commute, the Trotter approximants are exact for any m. The exact solution of the m=1 approximant for the two-dimensional XY-model shows that even the lowest approximation contain nontrivial quantum effects [35].

Both spatial and temporal expectation values can be evaluated by MC simulations using the Trotter formula [13]. The correlation functions in the spatial directions correspond to approximations of spatial quantum mechanical correlations and the ones in the Trotter direction corresponds to dynamic correlations (temperature =imaginary time). Finite resp. zero temperature corresponds to a finite or an infinite size in the time direction (for a given fixed m). The boundary condition in the x-directions are at will, in the time direction they are always periodic.

The flexibility of this approach has been extensively used to do direct simulations on quantum systems[36-41]. In particular, the numerical convergence with m is quite rapid [42]. There are of course other approaches to quantum systems, some of which are partially based on similar ideas [43-51].

5. Quantum Monte Carlo Renormalisation Group Transformation (QMCRG)

The route via a classical representation has been used to devise a QMCRG [52]. This is sketched in Fig. 3.

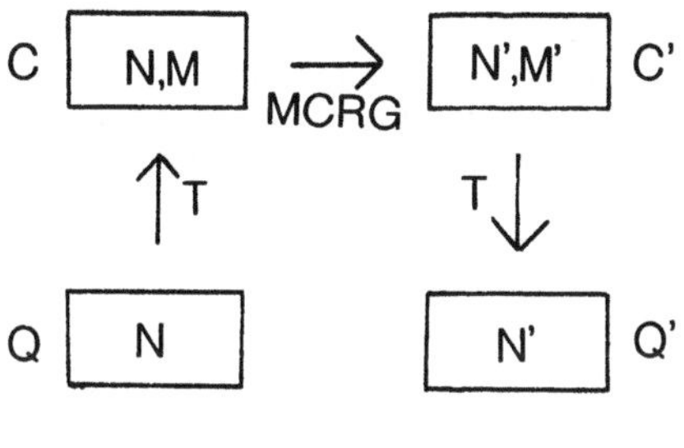

Figure 3
Monte Carlo Renormalisation Group for quantum systems. The connections between (Q)uantum (sizes N,N'=N/b) and (C)lassical (sizes N,M resp. N',M') systems is made via the Trotter formula (T). The classical RG is a MCRG.

The quantum partition function Z_Q is mapped on a classical partition function Z_m which is classically renormalised to $Z'_{m'}$. $Z'_{m'}$ is in turn identified with a renormalised quantum system $Z'_{Q'}$. The classical RG differs from standard MCRG in an important way: the role of space and time directions has to be distinguished and this distinction should be conserved during the RG. The classical system generated by the Trotter formula is highly anisotropic (the anisotropy increases with increasing m); the renormalization should not destroy this property.

This procedure will now be illustrated for the Ising model in a transverse field [7,8]. The quantum Hamiltonian is given by

$$H = \Sigma_{nn}\, s_i^x s_j^x + \Gamma\Sigma_i s_i^z = H_1+H_2 \ ,$$

where the first summation is over nearest neighbors and the second over all sites on the lattice. s^α are Pauli spin matrices and the transverse field Γ takes the role of the temperature. For $\Gamma=0$ the ground state is ordered and doubly degenerate, for $\Gamma=\infty$ it is a singlet. In one dimension the transition takes place at $\Gamma=1$. The Hamiltonian is split into the longitudinal (diagonal) and the transverse part. The resulting classical system is an anisotropic Ising model with different nearest neighbor couplings in the spatial directions and the time direction, namely

$$\beta_x=1/mT \quad ,$$

$$\beta_t=-1/2 \log \tanh(\Gamma/mT) \quad .$$

The MCRG of the classical system used is a standard block RG, but treating space and time separately. In this calculation isotropic blocks were used ($b=b_x=b_t$) as for this model space and time scale the same way. More generally, for systems with a dynamic exponent $z\neq1$, $b_x\neq b_t$ is more appropriate. The renormalized system is analysed locally as suggested in Fig. 2. From the two couplings β_x' and β_t' the new temperature T' and the new field Γ' are obtained from the same formulas but now with m replaced by $m'=m/b_t$. This means that the transformed system is truncated and interpreted again as an anisotropic Ising model.

The results are presented in Fig.4 and Table 1. Figure 4 shows the flow of the RG trajectories for temperature T versus field Γ for the two-dimensional transverse Ising model. There are two nontrivial fixed points, one for T=0 at $\Gamma=3.2$ - the quantum transition of the model - and the other for $\Gamma=0$ and T=2.26 - the classical Ising model in two dimensions. The flow between the two fixed points determines the phase boundary between paramagnetic (P) and ferromagnetic (F) phases. The temperature is a relevant variable at any field, in agreement with the fact that quantum effects are irrelevant at finite temperature.

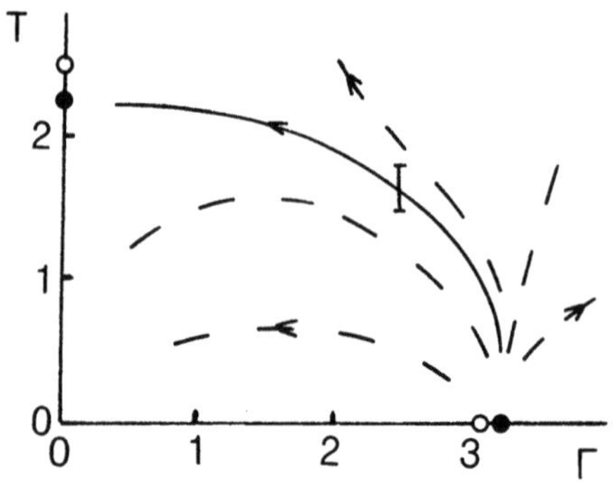

Figure 4
Flowlines of the RG for the two-dimensional transverse Ising model. The flow between the two fixed points separates paramagnetic (P) and ferromagnetic (F) phases.

Table 1
Critical indices as a function of the Trotter parameter m for a rescale factor $b=b_x=b_t=2$. Typical values from other calculations are $\Gamma_c=3.08$, $y_\Gamma=1.59$, $y_h=2.48$ and z=1. From [53].

m	b	Γ_c	y_Γ	y_h	z
4	2	3.26(20)	1.3(2)	2.7(3)	0.91(10)
6	2	3.20(20)	1.4(2)	2.8(3)	0.84(10)
8	2	3.04(30)	1.4(2)	2.6(5)	0.86(15)

The values of the critical coupling as a function of the Trotter parameter m are shown in table 1. The thermal (y_Γ) and the field (y_h) critical exponents are determined from the derivatives at the fixed point, the latter by adding a small longitudinal field. The dynamic exponent z follows from the temperature renormalization.

Note that the Trotter formula performs better at higher temperature. With other words to keep the approximation constant one has to increase m with decreasing temperature. One natural way to acomplish this is to keep β_x fixed and to lower the temperature by increasing the system size in the time direction. This is the procedure which was used here. Zero temperature can never been reached as it corresponds to $m=\infty$. In the simulations this was approximated by choosing the size in the spatial direction sufficiently small that the finite extent in the time direction was not felt.

The general procedure outlined can be applied to other models as well. Let me give some results for the XY/Heisenberg model and suggest alternative ways to implement the quantum MCRG. The way to break up the one-dimensional XY/Heisenberg Hamiltonian $H= \Sigma_i(s_i^x s_{i+1}^x + s_i^y s_{i+1}^y)+ \lambda \Sigma_i s_i^z s_{i+1}^z$ has been used by many workers after the first implementation by Suzuki. It is a decomposition into even and odd neighbor pairs on the chain, which leads to a classical representation with a plaquette interaction as shown in Fig.5.

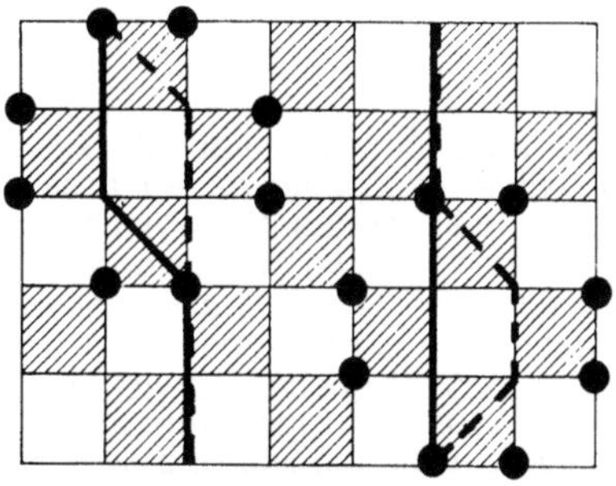

Figure 5
Plaquette interaction generated by the Trotter formula for the one-dimensional anisotropic Heisenberg model. The solid/dotted 'world-lines' (spin up states) show how one extracts the couplings from local measurements. The ratio of the probability of the solid vs. the dotted line with the spins (•) fixed determines the coupling constants, analogous to Fig. 2.

The important difference with the transverse Ising model are the conservation laws. The configurations on the classical system now are restricted: the number of up-spins (or fermions if interpreted this way) is conserved. Furthermore for $\lambda=1$ the system is fully rotationally invariant. Let us define a RG which conserves the number of fermions: 1) Remove two out of three world lines and 2) define block plaquettes as shown in Fig. 6.

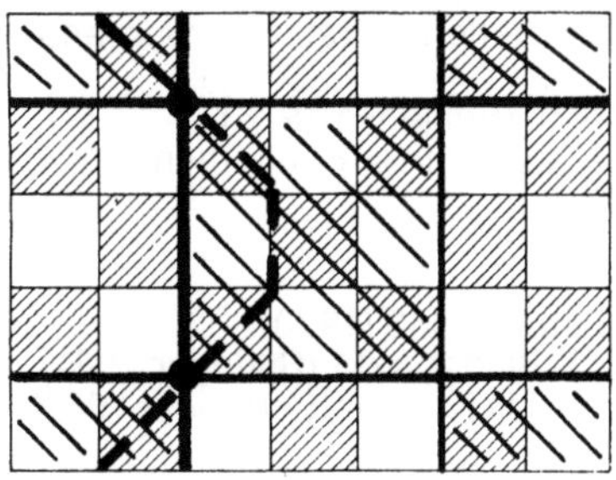

Figure 6
RG defined for the classical plaquette representation: 1) remove 2/3 of the world-lines and 2) define new world-lines on block-plaquettes (•) from the remaining lines. This way the plaquette character and particle conservation are respected.

The RG now maps a plaquette system onto itself. In order to extract the new coupling λ and the temperature T' one measures locally as shown in Fig. 5. The fixed point λ_c depends strongly on which configurations are used, λ_c =1.25(40) for m=4 and λ_c =1.18(60) for m=6. The corresponding exponent y_Γ is y_Γ =0.31 resp. y_Γ=0.28. The exact value λ=1 is not recovered as the RG does not satisfy the rotational symmetry.

There are other ways to implement the QMCRG. Two of them that are actually being studied shall be sketched briefly. Using Swendsen's idea [20] one works with the correlation functions instead of the coupling constants. This has also been used for the transverse Ising model in [53]. One performs a single mapping onto a classical representation and then renormalises theclassical system any number of times. The distinction betweeen space and time has to be conserved, in particular for models with conservation laws.

Another way to implement the Trotter formula and to perform QMCRG is inspired by the quantum RG discussed in §3. Divide the Hamiltonian in intra-cell and intercell parts. Change basis such that the intra-cell Hamiltonian is diagonal. Then use the even/odd site breakup - in the new basis - and simulate the resulting classical (plaquette) model. The RG now consists in retaining only the lowest two states and extracting the new couplings as indicated above. The nice feature of this method is that 1) it satisfies the symmetries of the model (i.e. for the isotropic Heisenberg model) and 2) it follows the intuition, that lowlying states are more important than higher ones. At the same time, because of the Trotter representation, it is not simply a high-temperature expansion in the inter-cell interactions.

6. Discussion

Monte Carlo calculations for quantum systems have become feasable since Suzuki proposed using the Trotter formula to define a mapping on a classical system. This opened the way to perform real space RG's for quantum systems by the MC method. So far the method was applied to one and two-dimensional spin models. Extensions to higher dimension and especially to fermions are feasible [36,37]. The major difficulty lies in properly treating the conservation laws and symmetry properties - which cause difficulties in the convergence and yield poor results if not properly dealt with. Another question is how to parametrize these models best? It may be helpful to pursue approaches along the line of a dynamic lattice RG proposed by Ma [54].

The calculations were performed on a Fujitsu VP200 at the C.I.R.C.E. in Orsay, France.

References

1 K.G.Wilson, Phys.Rev. B4, 3174,3184 (1971)
2 K.G.Wilson and J.Kogut, Phys.Repts. C12, 75 (1974)
3 P.Pfeuty and G.Toulouse, 'Introduction to the Renormalization Group and to Critical Phenomena' (Wiley 1975)
4 S.K.Ma, 'Modern Theory of Critical Phenomena' (Benjamin 1976)
5 C.Domb and M.S.Green, 'Phase Transitions and Critical Phenomena', Vol. 6 (Academic Press 1976)
6 J.Kogut, Rev. Mod. Phys. 51, 659 (1979)
7 P.Pfeuty, Ann. Phys. 57, 79 (1970)
8 R.J.Elliot, P.Pfeuty and C.Wood, Phys. Rev. Lett. 25, 443 (1970)
9 'Monte Carlo Methods in Statistical Physics', K.Binder (ed.) (Springer 1981)
10 'Monte Carlo Methods in Quantum Problems', M.H.Kalos (ed.) (Reidel 1984)
11 K.Binder in Vol. 5b of [5]
12 M.Suzuki, Prog. Theor. Phys. 46, 1337 (1971)
13 M.Suzuki, Prog. Theor. Phys. 56, 1454 (1976)
14 M.Suzuki, Prog. Theor. Phys. 58, 1377 (1977)

15 'Real Space Renormalisation', T.W.Burkhardt and J.M.J.van Leeuwen (eds.) Springer (1982)
16 S.K.Ma, Phys. Rev. Lett. 37, 461 (1976)
17 M.Kolb, thesis (University of California 1979)
18 R.H.Swendsen, Phys.Rev.Lett. 42, 859 (1979)
19 R.H.Swendsen, in [15]
20 R.H.Swendsen, Phys.Rev.Lett. 49, 1302 (1982)
21 R.H.Swendsen, Phys.Rev. B27, 391 (1983)
22 S.D.Drell, M.Weinstein and S.Yankielowicz, Phys. Rev. D14, 487 (1976)
23 S.D.Drell, M.Weinstein and S.Yankielowicz, Phys. Rev. D16, 1769 (1977)
24 H.P.van de Braak, W.J.Caspers, C. de Lange and M.W.M.Willanse, Physica 87A, 354 (1977)
25 W.J.Caspers, Phys.Repts 63, 223 (1980)
26 R.C.Brower, F.Kuttner, M.Nauenberg and K.Subbarao, Phys.Rev.Lett. 38, 1231 (1977)
27 J.N.Fields, H.W.J.Blöte and J.C.Bonner, J.Appl.Phys. 50, 1807 (1979)
28 J.M.Rabin, Phys.Rev. B21, 2027, B22, 2420 (1980)
29 P.Pfeuty, R.Jullien and K.A.Penson, in [15]
30 M.Kolb, R.Jullien and P.Pfeuty, J.Phys. A15, 3799 (1982)
31 A.L.Stella, C.Vanderzande and R.Dekeyser, Phys.Rev. B27, 1812 (1983)
32 M.Suzuki, Comm.Math.Phys. 51, 183 (1976)
33 M.Suzuki, J.Stat.Phys. 43, 883 (1986)
34 H.F.Trotter, Proc.Am.Math.Soc. 10, 545 (1959)
35 A.Langendijk and H.deRaedt, Phys.Rev.Lett. 49, 602 (1982)
36 M.Barma and B.S.Shastry, Phys.Lett. 61A, (1977)
37 M.Barma and B.S.Shastry, Phys.Rev. B18, 3351 (1978)
38 H. de Raedt and A.Langendijk, Phys.Rev.Lett. 46, 77 (1981)
39 H. de Raedt, A.Langendijk and J.Fivez, Z.Phys. B46, 261 (1982)
40 H. de Raedt and A.Langendijk, Phys.Repts. 127, 233 (1985)
41 Cullen and D.P.Landau, Phys.Rev. B27, 297 (1983)
42 A.Wiesler, Phys.Lett. 89A, 359 (1982)
43 D.J.Scalapino and R.L.Sugar, Phys.Rev.Lett. 46,519 (1981)
44 O.Martin and S.Ott, Nucl.Phys. B203, 297 (1982)
45 J.W.Lyklema, Phys.Rev.Lett. 49, 88 (1982)
46 J.E.Hirsch, Phys.Rev.Lett. 51, 1900 (1983)
47 H.Betsuyaku, Phys.Rev.Lett. 53,629 (1984)
48 J.E.Hirsch, Phys.Rev. B31, 4403 (1985)
49 H.Betsuyaku, Prog.Theor.Phys. 73, 319 (1985)
50 H.Takano, Prog. Theor. Phys. 73, 332 (1985)
51 M.H.Kalos, in [10]
52 M.Kolb, Phys.Rev.Lett. 51, 1696 (1983)
53 M.A.Novotny and D.P.Landau, J.Appl.Phys. 55, 2447 (1984)
54 S.K.Ma, Phys.Rev. B19, 4824 (1979)

Quantum Transfer-Matrix Method and Its Application to Quantum Spin Systems

H. Betsuyaku

Japan Atomic Energy Research Institute, Tokai, Ibaraki 319-11, Japan

1. Introduction

It is well known that the transfer-matrix method is very useful in the statistical mechanics of classical systems [1,2]. For example, the partition function of the three-dimensional (3D) Ising model is formulated in terms of the transfer matrix. If the size of the lattice is $N=l\times m\times n$, the partition function can be written as

$$Z = \mathrm{Tr}\,\mathcal{T}^n , \qquad (1.1)$$

where the transfer matrix $\mathcal{T}$ acts on the spin states of the 2D layer of $l\times m$ sites sliced from the 3D lattice. The transfer matrix can be pictured as an operator which governs the evolution of spin states from layer to layer. Since each layer has 2^{lm} spin states, the size of the matrix $\mathcal{T}$ is $2^{lm}\times 2^{lm}$. In the limit $n\to\infty$ the partition function is dominated by the maximum eigenvalue Λ of $\mathcal{T}$:

$$Z \sim \Lambda^n . \qquad (1.2)$$

Since the free energy of the system is given by

$$F/N = -k_BT \ln Z/N = -k_BT \ln \Lambda/lm , \qquad (1.3)$$

the thermodynamical quantities of the 3D Ising model of size $l\times m\times\infty$ can be calculated from the maximum eigenvalue of the transfer matrix. This is one of the great merits of the transfer-matrix method.

It has been recognized recently [3-13] that the transfer-matrix method also provides a powerful means for the numerical calculation of the thermodynamical quantities of quantum systems. The basis of the quantum transfer-matrix method comes from the following equivalence theorem [14,15]: a d-dimensional quantum spin system is mapped onto a $(d+1)$-dimensional Ising system with many-spin interactions; the partition function Z of a Hamiltonian $\mathcal{H}$ is expressed as

$$Z = \mathrm{Tr}^{(d)}\exp(-\beta\mathcal{H}) = \mathrm{Tr}^{(d+1)}\exp\mathcal{H}_{eff} , \qquad (1.4)$$

where $\mathcal{H}_{eff}$ denotes an effective $(d+1)$-dimensional Hamiltonian; the explicit expression of $\mathcal{H}_{eff}$ depends on a classical representation chosen. This equivalent partition function can be formulated in terms of the transfer matrix. The computational merit of this approach is that the thermodynamical quantities can be evaluated without diagonalizing the full Hamiltonian of the system; the knowledge required is about a local transfer matrix of very small size which transfers one state of a two-spin cluster to another state.

In the following the quantum transfer-matrix method is formulated and is applied to study the 1D quantum spin system described by the Hamiltonian:

$$\mathcal{H} = -2\sum_{i=1}^{N}(J_x S_i^x S_{i+1}^x + J_y S_i^y S_{i+1}^y + J_z S_i^z S_{i+1}^z) \equiv \sum_{i=1}^{N}\mathcal{V}_i \quad , \tag{1.5}$$

where $\mathbf{S}_i$ is the spin operator on site i of $S=\frac{1}{2}$ or 1, and N is the number of spins. The Hamiltonian (1.5) includes the following three important cases: (*1*) the ferromagnetic isotropic Heisenberg model $J_x=J_y=J_z=J$; (*2*) the anti-ferromagnetic isotropic Heisenberg model $J_x=J_y=J_z=-J$; and (*3*) the *XY* model $J_x=J_y=J$, $J_z=0$.

2. Quantum Transfer-Matrix Method

2.1 "Real-Space" Transfer Matrix

The partition function of the quantum system is expressed as

$$Z = \mathrm{Tr}\left[\exp(-\beta\mathcal{H})\right] = \mathrm{Tr}\left[\prod_{r=1}^{m}\exp(-\Delta\tau\mathcal{H})\right] \quad ; \quad \beta = 1/k_BT \quad , \tag{2.1}$$

where β is broken up into m subintervals of width $\Delta\tau=\beta/m$; m is an arbitrary positive integer which stands for the size along the Trotter direction. On insertion of m orthogonal complete sets, Z is written as

$$Z = \sum_{\alpha_1}\sum_{\alpha_2}\cdots\sum_{\alpha_m}\prod_{r=1}^{m}\langle\alpha_{r+1}|\exp(-\Delta\tau\mathcal{H})|\alpha_r\rangle \quad \text{with} \quad |\alpha_{m+1}\rangle\equiv|\alpha_1\rangle . \tag{2.2}$$

In the representation in which each S_i^z is diagonal, the states $|\alpha_r\rangle$ are simultaneous eigenstates of all S_i^z operators:

$$S_i^z|\alpha_r\rangle = S_{i,r}|\alpha_r\rangle \quad (S_{i,r} = -S, -S+1, \cdots, S) \quad , \tag{2.3}$$

where r is a label along the new Trotter direction, and each α_r runs over $(2S+1)^N$ states. The "real-space" transfer matrix is defined by

$$\langle\alpha'|\mathcal{T}|\alpha\rangle \equiv \langle\alpha'|\exp(-\Delta\tau\mathcal{H})|\alpha\rangle \quad , \tag{2.4}$$

which evolves spin states from $|\alpha\rangle$ to $|\alpha'\rangle$ along the Trotter direction. Its eigenvalues are denoted by $\{\mu_j\}$. Then the partition function is given in terms of $\mathcal{T}$ by

$$Z = \mathrm{Tr}\,\mathcal{T}^m = \sum_j \mu_j^m \quad . \tag{2.5}$$

Here it should be remembered that (2.5) holds exactly for any m, but that the eigenvalues $\{\mu_j\}$ depend on m in a singular way, and so all μ_j have to be retained in the summation as has been pointed out by SUZUKI [5] .

2.2 CBD Representation

The simplest classical representation of Z is the checkerboard decomposition (CBD) introduced by BARMA and SHASTRY [16] . It decomposes the Hamiltonian into two parts:

$$\mathcal{H} = \mathcal{H}_1 + \mathcal{H}_2 ; \quad \mathcal{H}_1 = \sum_{i\in A}\mathcal{V}_i \quad , \quad \mathcal{H}_2 = \sum_{i\in B}\mathcal{V}_i \quad , \tag{2.6}$$

where $A(B)$ is the set of odd(even) integers. The following relation [17]

$$\exp[-\Delta\tau(\mathcal{H}_1+\mathcal{H}_2)] = \exp(-\tfrac{1}{2}\Delta\tau\mathcal{H}_1)\exp(-\Delta\tau\mathcal{H}_2)\exp(-\tfrac{1}{2}\Delta\tau\mathcal{H}_1)+O((\Delta\tau)^3) \tag{2.7}$$

is used to decompose the transfer matrix $\mathcal{T}$:

$$\mathcal{T} = \mathcal{T}_1^{\frac{1}{2}}\mathcal{T}_2\mathcal{T}_1^{\frac{1}{2}} \ ; \qquad \mathcal{T}_1 = \prod_{i\in A} t_i \ , \ \mathcal{T}_2 = \prod_{i\in B} t_i \ . \tag{2.8}$$

Here

$$t_i \equiv \exp(-\Delta\tau\mathcal{V}_i) \tag{2.9}$$

is a local transfer matrix which evolves spin states of two-spin cluster (S_i, S_{i+1}) along the Trotter direction; its size is $(2S+1)\times(2S+1)$, and the explicit matrix elements are given in [4] for $S=\frac{1}{2}$ and in [8] for $S=1$. By substituting (2.8) into (2.2) and rearranging the order of $\mathcal{T}_1$ and $\mathcal{T}_2$, the mth approximant of the partition function is given by

$$Z_m^{\mathrm{CBD}} = \sum_{\alpha_1}\sum_{\alpha_2}\cdots\sum_{\alpha_m}\prod_{r=1}^{m}\langle\alpha_{2r+1}|\mathcal{T}_2|\alpha_{2r}\rangle\langle\alpha_{2r}|\mathcal{T}_1|\alpha_{2r-1}\rangle \quad \text{with } |\alpha_{2m+1}\rangle\equiv|\alpha_1\rangle . \tag{2.10}$$

The decomposition error is proportional to $1/m^2$ at a fixed temperature:

$$Z_m^{\mathrm{CBD}} = Z+O(\beta^3/m^2) \ . \tag{2.11}$$

This $1/m^2$ dependence enables us to estimate the limiting $m\to\infty$ behavior accurately.

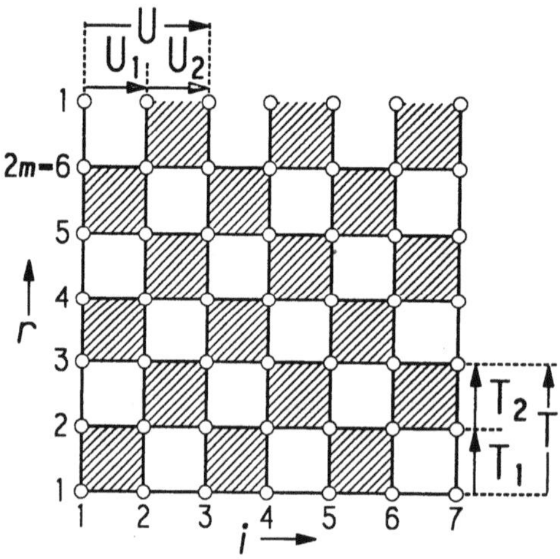

Fig. 1 Graphical representation of the checkerboard decomposition (CBD). An Ising spin $S_{i,r}$ is assigned to each site (i,r). The local transfer matrix $t_{i,r}$ transfers one state of two spins on the lower edges of a shaded square to another state on the upper edges. The "real-space" transfer matrix $\mathcal{T}=\mathcal{T}_2\mathcal{T}_1$ and the "virtual-space" transfer matrix $\mathcal{U}=\mathcal{U}_2\mathcal{U}_1$ are indicated in the figure.

The classical representation Z_m^{CBD} derived on the basis of two-spin cluster decomposition can be interpreted as the partition function of an Ising model with a checkerboard-like lattice structure; the 2D square lattice is defined in which rows are labeled by $r(1\leqq r\leqq 2m)$ and columns by $i(1\leqq i\leqq N)$ as shown in Fig.1. The hatched square denotes a local transfer matrix $t_{i,r}$, and the "real-space" transfer matrices, $\mathcal{T}_1$ and $\mathcal{T}_2$, are indicated in the figure. Periodic boundary conditions are required along the Trotter direction because of the trace operation. Free chain is assumed in the present study. It is seen that the decomposition (2.8) breaks the translational invariance.

2.3 *RSD Representation*

Another classical representation of Z is the real-space decomposition (RSD) introduced by SUZUKI [14]. The transfer matrix $\mathcal{T}$ is decomposed as

$$\mathcal{T} \equiv \exp(-\Delta\tau\mathcal{H}) = \prod_{i=1}^{N} \exp(-\Delta\tau\mathcal{V}_i) = \prod_{i=1}^{N} t_i \ . \tag{2.12}$$

By substituting (2.12) into (2.2), the partition function in RSD is given by

$$Z_m^{\mathrm{RSD}} = \sum_{\alpha_1}\sum_{\alpha_2}\cdots\sum_{\alpha_m}\prod_{r=1}^{m}\langle \alpha_{r+1} | \prod_{i=1}^{N} t_{i,r} | \alpha_r \rangle \ . \tag{2.13}$$

After introducing another m complete sets as intermediate states and relabeling r sequentially, Z_m^{RSD} can be interpreted as the partition function of the 2D Ising model with four-spin interaction as shown in Fig.2. All the lattice sites are equivalent in this decomposition.

It has been proved mathematically that Z_m^{CBD} transforms into Z_m^{RSD} in the case of free chain [18]. This transformation is easily seen by comparing Figs.1 and 2. If we relabel r indices in Z_m^{RSD} column by column as

$$r\text{-}1 \rightarrow r \text{ for } i=1 \quad \text{and} \quad r+i-2 \rightarrow r \text{ for } i \geqq 3 \ , \tag{2.14}$$

and use the periodic boundary conditions along the Trotter direction, we find that Z_m^{RSD} transforms into Z_m^{CBD}. Thus both decompositions give the same results for the thermodynamical properties.

However, the spin correlation functions, $\langle S_i^\alpha S_j^\alpha \rangle$ $(\alpha=x,y,z)$, are not invariant under the transformation. Figure 3 shows the interrelation between two kinds of correlation functions in RSD and CBD; $\langle S_i^\alpha S_{i+j}^\alpha \rangle^{\mathrm{RSD}}$ and $\langle S_i^\alpha S_{i+j}^\alpha \rangle^{\mathrm{CBD}}$ where two spins are apart by distance of j. It should be noticed that the correlation functions in RSD are translationally invariant, while those in CBD are not. This affects the rate of convergence largely, although the difference of the two correlations vanishes as $m \rightarrow \infty$.

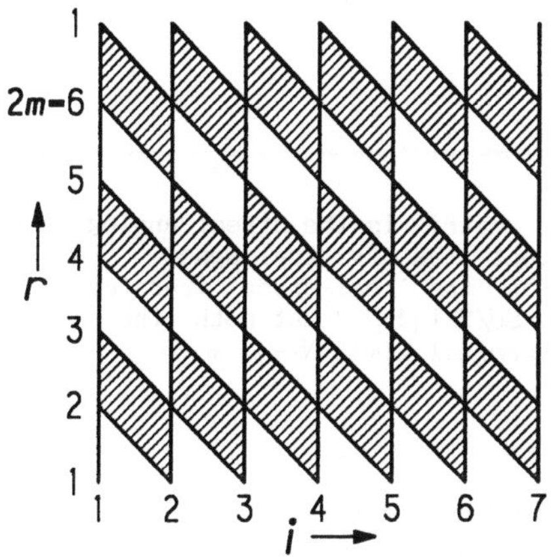

Fig. 2 Graphical representation of the real-space decomposition (RSD) in which the equivalent classical lattice is represented by a skewed checkerboard lattice. Each shaded portion indicates a local transfer matrix $t_{i,r}$.

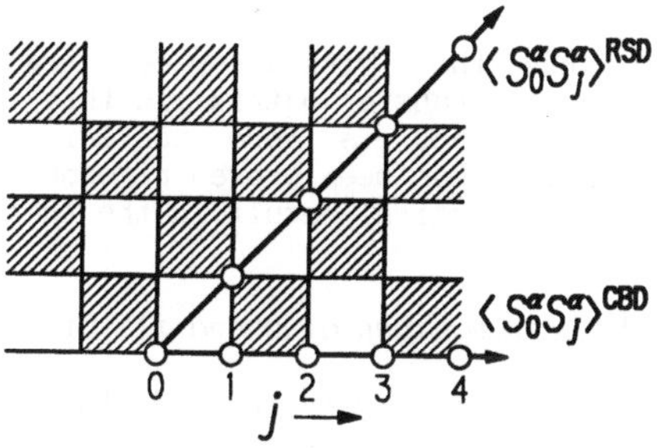

Fig. 3 Interrelation between the correlation functions in CBD and RSD, $\langle S_0^\alpha S_j^\alpha \rangle^{\mathrm{CBD}}$ and $\langle S_0^\alpha S_j^\alpha \rangle^{\mathrm{RSD}}$, respectively. In the 2D lattice for CBD, $\langle S_0^\alpha S_j^\alpha \rangle^{\mathrm{CBD}}$ are defined along the horizontal line while $\langle S_0^\alpha S_j^\alpha \rangle^{\mathrm{RSD}}$ are defined along the diagonal line.

2.4 "Virtual-Space" Transfer Matrix

As indicated in Fig.1, the local transfer matrix $\mathcal{L}_{i,r}$ transfers the state $(S_{i,r}, S_{i+1,r})$ to $(S_{i,r+1}, S_{i+1,r+1})$ along the Trotter direction. It is rearranged to give the "virtual-space" local transfer matrix $u_{i,r}$ which transfers the state $(S_{i,r}, S_{i,r+1})$ to $(S_{i+1,r}, S_{i+1,r+1})$ along the real axis. We define in terms of $u_{i,r}$ the "virtual-space" transfer matrices, $\mathfrak{U}_1$ and $\mathfrak{U}_2$, which evolve spin states from the ith column to the $(i+1)$th column as also indicated in Fig.1. They are given explicitly by the direct products of u's:

$$\mathfrak{U} = \mathfrak{U}_2\mathfrak{U}_1 \;; \quad \mathfrak{U}_1 = \prod_{r\in A'} u_{1,r} \;, \quad \mathfrak{U}_2 = \prod_{r\in B'} u_{2,r} \;, \tag{2.15}$$

where A' (B') is the set of odd(even) r.

The partition function can be rewritten in terms $\mathfrak{U}_1$ and $\mathfrak{U}_2$ as

$$Z_m^{\rm CBD} = \sum_{c_1}\sum_{c_2}\cdot\cdot\cdot\sum_{c_N}\prod_{k=1}^{N/2}\langle c_{2k+1}|\mathfrak{U}_2|c_{2k}\rangle\langle c_{2k}|\mathfrak{U}_1|c_{2k-1}\rangle \quad \text{with } |c_{N+1}\rangle \equiv |c_1\rangle , \tag{2.16}$$

where $|c_i\rangle$ represents the spin states on the ith column, and each c_i runs over $(2S+1)^{2m}$ states. $Z_m^{\rm CBD}$ is expressed in terms of the eigenvalues $\{\lambda_j\}$ of $\mathfrak{U}$ as

$$Z_m^{\rm CBD} = {\rm Tr}\; \mathfrak{U}^{N/2} = \sum_j \lambda_j^{N/2} \;. \tag{2.17}$$

In the thermodynamical limit $N\to\infty$ with m fixed, $Z_m^{\rm CBD}$ is dominated by the maximum eigenvalue λ of $\mathfrak{U}$:

$$Z_m^{\rm CBD} \sim \lambda^{N/2} \;. \tag{2.18}$$

Then the free energy and the internal energy are given by

$$F_m^{\rm CBD}/N = -k_BT \ln Z_m^{\rm CBD}/N = -k_BT \ln \lambda/2 \tag{2.19}$$

$$E_m^{\rm CBD}/N = -\frac{\partial}{\partial\beta}\ln Z_m^{\rm CBD}/N = -\frac{\partial}{\partial\beta}\ln \lambda \;. \tag{2.20}$$

Thus the "virtual-space" transfer matrix is the direct analog of the classical transfer matrix. In this respect the "virtual-space" transfer-matrix method is more convenient than the "real-space" transfer-matrix method. It has been proved rigorously by SUZUKI [5] that both the methods give an identical result in the thermodynamical limit $N\to\infty$, $m\to\infty$.

2.5 Maximum Eigenvalue and its Eigenstates

The maximum eigenvalue and its corresponding eigenstate can be obtained by the iterative operation of the matrix $\mathfrak{U}$ on an arbitrary state $|\chi\rangle$ which is not orthogonal to the maximum eigenstate. Because of the periodicity of $\mathfrak{U}$ along the real axis, it may be defined equally as $\mathfrak{U}=\mathfrak{U}_1\mathfrak{U}_2$ instead of $\mathfrak{U}=\mathfrak{U}_2\mathfrak{U}_1$ defined by (2.15). Consequently, it follows that there exist two eigenstates:

$$|\varphi_A\rangle \sim (\mathfrak{U}_2\mathfrak{U}_1)^N|\chi\rangle \;, \qquad |\varphi_B\rangle \sim (\mathfrak{U}_1\mathfrak{U}_2)^N|\chi\rangle \;, \tag{2.21}$$

which satisfy the following equations:

$$\mathfrak{U}_2\mathfrak{U}_1|\varphi_A\rangle = \lambda|\varphi_A\rangle \;, \qquad \mathfrak{U}_1\mathfrak{U}_2|\varphi_B\rangle = \lambda|\varphi_B\rangle \;. \tag{2.22}$$

Here $|\varphi_A\rangle$ is the eigenstate for the odd column $(i\in A)$ and $|\varphi_B\rangle$ is for the even column $(i\in B)$. They are transformed into each other by the relations:

$$\lambda^{\frac{1}{2}}|\varphi_B\rangle = \mathfrak{U}_1|\varphi_A\rangle \ , \qquad \lambda^{\frac{1}{2}}|\varphi_A\rangle = \mathfrak{U}_2|\varphi_B\rangle \ . \tag{2.23}$$

These $|\varphi_k\rangle$ are the linear combination composed of the set of $(2S+1)^{2m}$ direct product states:

$$\left\{|S_1,S_2,\cdots,S_{2m}\rangle;\quad S_r=-S,-S+1,\cdots,S\right\} , \tag{2.24}$$

which are simultaneous eigenstates of all spin variables S_r^z on one column along the Trotter direction. If $|\varphi_A\rangle$ is expressed symbolically as

$$|\varphi_A\rangle = \varphi(S_1,S_2,\cdots,S_{2m}) \ , \tag{2.25a}$$

then $|\varphi_B\rangle$ is given by the cyclic permutation of spin variables:

$$|\varphi_B\rangle = \varphi(S_2,S_3,\cdots,S_{2m},S_1) \ . \tag{2.25b}$$

Then it follows from (2.23) that

$$\lambda^{\frac{1}{2}}\varphi(S_2,S_3,\cdots,S_{2m},S_1) = \mathfrak{U}_1\varphi(S_1,S_2,\cdots,S_{2m}) \ . \tag{2.26}$$

Thus $\mathfrak{U}_1$ transfers the spin variable $S_{i,r}$ on the ith column to $S_{i+1,r+1}$ on the $(i+1)$th column. Likewise $\mathfrak{U}_2$ transfers the spin variables in the same manner.

The above transformation property on $\mathfrak{U}_1$ and $\mathfrak{U}_2$ implies that the translational symmetry holds for the group of spins belonging to each column in RSD as seen from Figs.2 and 3. However, the symmetry is broken in CBD.

2.6 *Internal Energy and Spin Correlation Functions*

The internal energy is related with the nearest-neighbor spin correlation functions by the relation:

$$E \equiv \langle\mathcal{H}\rangle = -2\sum_{i=1}^{N}\left(J_x\langle S_i^xS_{i+1}^x\rangle + J_y\langle S_i^yS_{i+1}^y\rangle + J_z\langle S_i^zS_{i+1}^z\rangle\right) \ . \tag{2.27}$$

The mth approximant to the energy is obtained by substituting the following two correlation functions into (2.27):

(*1*) The nearest-neighbor correlation function in RSD given in terms of the maximum eigenstates by

$$\langle S_i^\alpha S_{i+1}^\alpha\rangle^{\mathrm{RSD}} = \langle\varphi_B|S_{i+1,r+1}^\alpha\mathfrak{U}_1S_{i,r}^\alpha|\varphi_A\rangle/\langle\varphi_B|\mathfrak{U}_1|\varphi_A\rangle \ . \tag{2.28}$$

This correlation function takes proper account of the translational invariance, and so is expected to converge rapidly to the real quantum correlation function.

(*2*) The correlation function in CBD given by

$$\langle S_i^\alpha S_{i+1}^\alpha\rangle^{\mathrm{CBD}} = \langle\varphi_B|S_{i+1,r}^\alpha\mathfrak{U}_1S_{i,r}^\alpha|\varphi_A\rangle/\langle\varphi_B|\mathfrak{U}_1|\varphi_A\rangle \ . \tag{2.29}$$

This correlation function is not translationally invariant, which causes the slow convergence of E^{CBD} .

2.7 $\beta\to\infty$ Behavior

The mth approximant to the transfer matrix $\mathcal{T}_m$ is given by (2.8) for CBD, and by (2.12) for RSD. Its eigenvalues are denoted by $\{\mu_j^{(m)}\}$. Then the free energy in the $\beta\to\infty$ limit with m fixed is given by

$$F_m = -\frac{1}{\beta}\ln\{\mathrm{Tr}[(\mathcal{T}_m)^m]\} = -\frac{1}{\beta}\ln\{\sum_j(\mu_j^{(m)})^m\} \sim -\ln(\mu_0^{(m)})/\Delta\tau , \tag{2.30}$$

where $\mu_0^{(m)}$ is the maximum eigenvalue. Similarly, the energy is given by

$$\langle\mathcal{H}\rangle_m \equiv \mathrm{Tr}\{\mathcal{H}(\mathcal{T}_m)^m\}\Big/\mathrm{Tr}\{(\mathcal{T}_m)^m\} = \sum_j\langle j|\mathcal{H}|j\rangle(\mu_j^{(m)})^m\Big/\sum_j(\mu_j^{(m)})^m \sim \langle\dot{0}|\mathcal{H}|0\rangle , \tag{2.31}$$

where $|0\rangle$ is the maximum eigenstate corresponding to $\mu_0^{(m)}$.

In CBD, the maximum eigenvalue of $\mathcal{T}_m^{\mathrm{CBD}}$ is given in the $\beta\to\infty$ limit by [7,8]

$$\mu_0^{(m)} \sim \exp(-\Delta\tau E_0^{(2)}) , \tag{2.32}$$

where $E_0^{(2)}$ is the sum of the ground-state energy of a two-spin cluster, and $|0\rangle$ is the direct product of two-spin cluster states. Thus the energies, derived from (2.30) and (2.31), coincide and tend to $E_0^{(2)}$ irrespective of m. Therefore, E^{CBD} deviates markedly from the real ground-state energy at low temperatures.

In RSD, the maximum eigenstate of $\mathcal{T}_m^{\mathrm{RSD}}$ is a coherent state mixed up with all two-spin clusters as seen from (2.12), which explains why E^{RSD} derived from (2.31) stands for a good approximant of the real ground-state energy.

Recently, FYE [19] has analyzed generally the β dependence of the free energy and the internal energy for the Trotter-like decompositions.

3. Numerical Results Obtained by "Virtual-Space" Transfer-Matrix Method

3.1 Computational Procedures

The "virtual-space" transfer-matrix method has been applied to study several 1D quantum spin systems [3,4,6-8]. The basis of the "virtual-space" transfer matrices is chosen to be the $(2S+1)^{2m}$ direct product states defined by (2.23) for the Trotter size m. The maximum eigenvalue and eigenstate of $\mathcal{U}=\mathcal{U}_2\mathcal{U}_1$ are obtained by the iterative operation of $\mathcal{U}$ on an initial state $|\chi\rangle$:

$$|\psi\rangle \sim \mathcal{U}^N|\chi\rangle , \quad \mathcal{U}|\psi\rangle = \lambda|\psi\rangle , \tag{3.1}$$

where $|\chi\rangle$ is chosen to be

$$|\chi\rangle = (|\uparrow\uparrow\uparrow\cdots\rangle+|\downarrow\downarrow\downarrow\cdots\rangle) , \tag{3.2}$$

which is apparently not orthogonal to the maximum eigenstate.

The numerically exact transfer-matrix method [3,4] is used to perform the iterative matrix operation; the direct products of (2.15) are evaluated by multiplying u numerically term by term. The iterations are terminated when the relative error of λ becomes smaller than 10^{-10}. All the computations have been done by use of FACOM VP-100 array processor of our Institute. Only limitation of the present method is the storage requirement for the factor $(2S+1)^{2m}$.

The internal energies, E^{RSD} and E^{CBD}, are computed for several values of m as a function of T for $T \geqq 0.05$. The largest values of the Trotter size m calculated within the ability of our computer are up to 10 for $S=\frac{1}{2}$ and up to 6 for $S=1$. The specific heat is obtained by differentiating E numerically with respect to T.

3.2 Energy and Specific Heat Calculated in CBD

The internal energy is calculated by two different methods: (*1*) it is derived from the partition function by differentiating numerically with respect to β by use of (2.20); and (*2*) it is calculated in terms of the correlation function by use of (2.27) and (2.29). It is confirmed that both methods always provide the same numerical results.

Figure 4 shows the results for E^{CBD} of the 1D $S=\frac{1}{2}$ XY model. The thin lines are the calculations of m=1,2,4,8,10, which show a steady progression toward the exact result by Katsura [20] (thick line) with increasing m. Data points are the Monte Carlo results by Cullen and Landau [21]. The convergence is rapid and the actual estimate is quite good for temperatures T>0.30.

The ground-state energy in CBD is -1.0 irrespective of m, which is the ground-state energy of a two-spin cluster as has been discussed in *2.7*. The large difference of the ground-state energies between the classical system

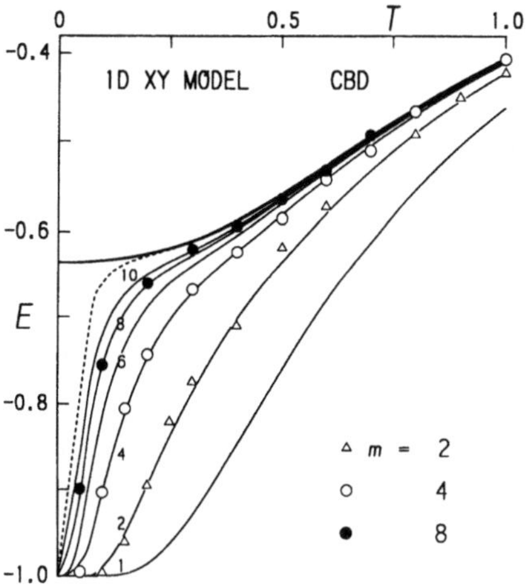

Fig. 4 Temperature dependence of E^{CBD} for the 1D $S=\frac{1}{2}$ XY model. The thin lines are the calculations of m=1,2,4,6,8,10; the number attached to each line. The thick line is the exact result by Katsura. Data points are the Monte Carlo results by Cullen and Landau. The dashed line is the $m\to\infty$ estimation.

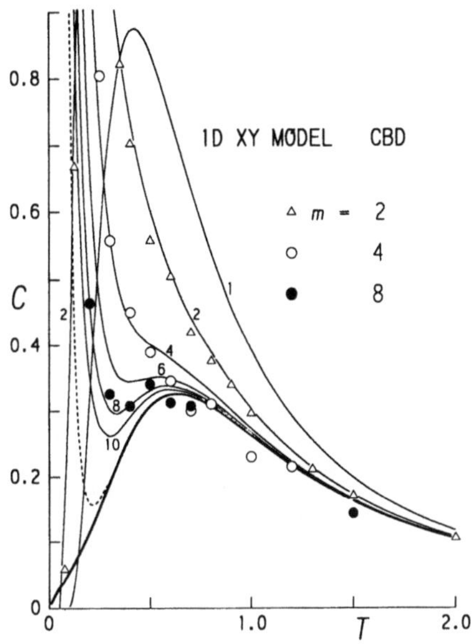

Fig. 5 Specific-heat results in CBD for the 1D $S=\frac{1}{2}$ XY model. The thin lines are the calculations of m=1,2,4,6,8,10. The thick line is the exact result by Katsura. Data points are the Monte Carlo results by Cullen and Landau. The dashed line is the $m\to\infty$ estimation.

in CBD and the real quantum system emphasizes the failure to reproduce the strong quantum effects at extremely low temperatures.

The specific-heat results are displayed in Fig.5. The progression toward the exact result (thick line) is quite systematic, and for m=10 we can see the peak in approximately the correct position. As expected from the energy results, the low-temperature results for the specific heat are poor.

The calculated results showed a remarkable agreement with the Monte Carlo results at every stage of m at all temperatures; the specific-heat results of the Monte Carlo method have rather poor statistics as m increases, while the results of the present method are numerically exact. This agreement might be self-evident, since both the methods based on the same classical representation of the free energy, F^{CBD} .

3.3 Energy and Specific Heat Calculated in RSD

Results for E^{RSD} of the 1D $S=\frac{1}{2}$ XY model are shown in Fig.6. The thin lines are the calculations of m=2,4,8. The m=8 calculation is very close to the exact result by Katsura (thick line) all over the temperature range studied. By comparing E^{RSD} with E^{CBD} at every stage of m, it is clearly seen that E^{RSD} converges much faster than E^{CBD} . The specific-heat results are shown in Fig.7. As expected from the energy results, they show a rapid convergence to the exact result by Katsura (thick line).

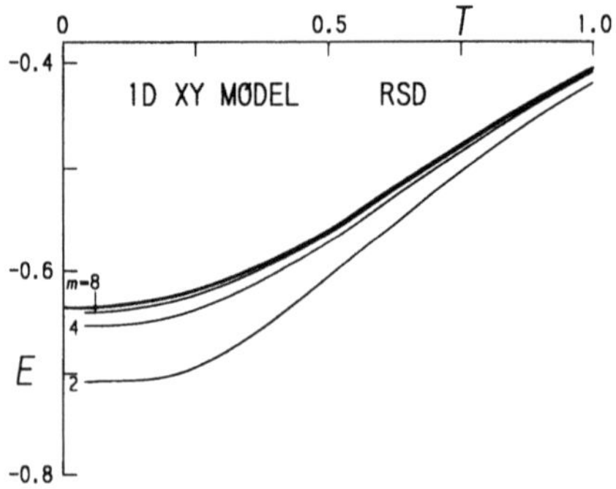

Fig. 6 Temperature dependence of E^{RSD} for the 1D $S=\frac{1}{2}$ XY model. The thin lines are the calculations of m=2,4,8. The thick line is the exact result by Katsura.

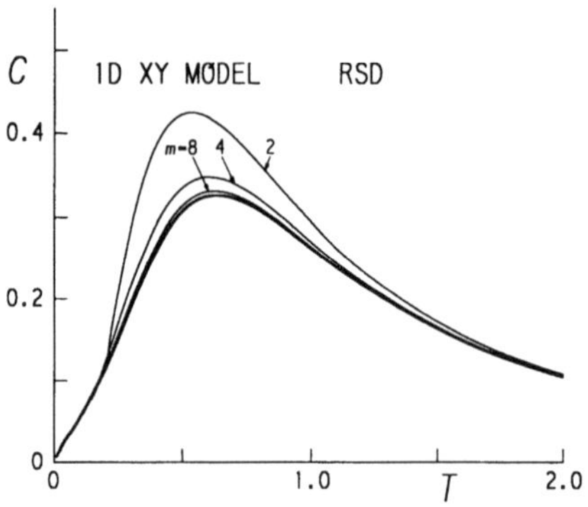

Fig. 7 Specific-heat results in RSD for the 1D $S=\frac{1}{2}$ XY model. The thick lines are the calculations of m=2,4,8. The thick line is the exact result by Katsura.

3.4 Extrapolation to Infinite Trotter Size

In Fig.8 the energies, E^{RSD} and E^{CBD}, are plotted versus $1/m^2$. It is evident that the same limiting energy is approached linearly with $1/m^2$ but the convergence rate of CBD is much slower at low temperatures than RSD; at T=0.2 E^{CBD} deviates markedly from the linear relation with $1/m^2$ suggesting the need to include higher order corrections.

The decomposition error of the partition function is given by (2.11), which suggests the following extrapolation formula:

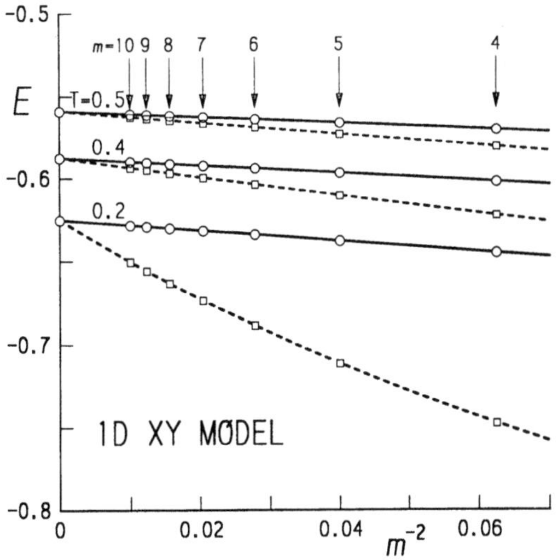

Fig. 8 Energies versus $1/m^2$ for the 1D $S=\frac{1}{2}$ XY model at several temperatures: E^{CBD} (squares) and E^{RSD} (circles).

$$E(m) = E(\infty)+\alpha/m^2 \ . \tag{3.3}$$

This relation holds quite accurately for RSD; the same relation holds for CBD but less accurately at low temperatures because of slow convergence. The limiting energies are tabulated in Table 1; these values are extrapolated from m=7,8,9,10 for RSD, and from m=8,9,10 for CBD at low temperatures. The limiting energy in CBD is included in Fig.4 (dashed line). The agreement with the exact value is very excellent for RSD, which indicates that the strong quantum effects at very low temperatures are well reproduced by E^{RSD}. However, the agreement is somewhat worse for CBD. This poor convergence of E^{CBD} arises from the fact that the CBD breaks the translational invariance. It has been proved recently [19,22] that the extrapolation formula (3.3) holds for the general Trotter-like decompositions.

The extrapolation formula (3.3) is seen to be complementary to the finite-chain extrapolation of Bonner and Fisher [23]. Their numerical method is to diagonalize the full Hamiltonian for finite chains, and so it corresponds to $m=\infty$. The properties of infinite chain can be estimated from data on finite chains by using the following extrapolation formula:

$$E(N) = E(\infty)+\beta/N^2 \ . \tag{3.4}$$

Table 1 Energies extrapolated from finite-m calculations. E_{exact} is the exact value by Katsura.

T	E_{exact}	E^{RSD}	E^{CBD}
0.0	-0.63662	———	———
0.1	-0.63398	-0.63417	-0.66130
0.15	-0.63060	-0.63069	-0.63767
0.2	-0.62570	-0.62574	-0.62816
0.25	-0.61906	-0.61907	-0.62010
0.3	-0.61049	-0.61049	-0.61100
0.4	-0.58768	-0.58768	-0.58784
0.5	-0.55897	-0.55897	-0.55903

Thus the numerical results extrapolated from finite Trotter size are as exact as those extrapolated from finite chains. This complementarity reflects the fact that the matrix elements (2.9) are symmetrical with respect to i and r.

4. Summary and Discussion

The quantum transfer-matrix method provides a powerful means for the numerical study of the thermodynamical properties of quantum systems. In this method the quantum transfer matrix can be formulated in two different ways depending on the direction of transfer: (*1*) "virtual-space" transfer matrix, and (*2*) "real-space" transfer matrix.

The "virtual-space" transfer matrix evolves spin states on the Trotter axis along the real axis. This matrix offers the direct analog of the classical transfer matrix, and the thermodynamical properties can be calculated in terms of the maximum eigenvalue and eigenstate of the transfer matrix. The behavior in the limit of infinite Trotter size can be estimated accurately from data on finite Trotter sizes in terms of a new type of extrapolation scheme. It is complementary to the finite-chain extrapolation scheme of Bonner and Fisher which estimates the behavior of infinite chain from data on finite chains.

The "real-space" transfer matrix evolves spin states on the real axis along the Trotter axis. This matrix reduces to a projection operator onto the ground state in the low-temperature limit, and the ground-state quantities can be evaluated accurately [6,9,11]. The "real-space" transfer matrix is also used as a thermofield transfer matrix when the temperature is finite [10,24]. This method provides a unification of the "virtual-space" transfer-matrix method and the projector method.

It is to be hoped that the present method is used to study various one- and two-demensional quantum systems.

Acknowledgements

The author would like to thank Professor M. Suzuki for stimulating discussions throughout this work, and Mr. T. Yokota for useful discussions. The author is grateful for the assistance of members of the computer center of JAERI.

References

1. H.A. Kramers and G.H. Wannier: Phys. Rev. **60**, 252 (1941)
2. G.F. Newell and E.W. Montroll: Rev. Mod. Phys. **25**, 353 (1953)
3. H. Betsuyaku: Phys. Rev. Lett. **53**, 629 (1984)
4. H. Betsuyaku: Prog. Theor. Phys. **73**, 319 (1985)
5. M. Suzuki: Phys. Rev. B **31**, 2957 (1985)
6. T. Yokota and H. Betsuyaku: Prog. Theor. Phys. **75**, 46 (1986)
7. H. Betsuyaku: Prog. Theor. Phys. **75**, 774 (1986)
8. H. Betsuyaku and T. Yokota: Prog. Theor. Phys. **75**, 808 (1986)
9. H. Betsuyaku and T. Yokota: Phys. Rev. B **33**, 6505 (1986)
10. M. Suzuki and H. Betsuyaku: Phys. Rev. B **34**, 1829 (1986)
11. H. Betsuyaku: Phys. Rev. B **34**, 8125 (1986)
12. T. Tsuzuki: Prog. Theor. Phys. **73**, 1352 (1985); **75**, 225 (1986)

13. T. Sakaguchi, K. Kubo and S. Takada: J. Phys. Soc. Jpn **54**, 861 (1985);
K. Kubo and T. Takada: *ibid.* **55**, 438 (1986);
T. Takada and K. Kubo: *ibid.* **55**, 1671 (1986)
14. M. Suzuki: Prog. Theor. Phys. **56**, 1454 (1976)
15. M. Suzuki: J. Stat. Phys. **43**, 883 (1986), and references cited therein.
16. M. Barma and B.S. Shastry: Phys. Rev. B **18**, 3351 (1978)
17. H. De Raedt and B. D. Raedt: Phys. Rev. A **28**, 3575 (1983)
18. H. De Raedt, A. Lagendijk, and J. Fivez: Z. Phys. B **46**, 261 (1982)
19. R.M. Fye: Phys. Rev. B **33**, 6271 (1986)
20. S. Katsura: Phys. Rev. **127**, B1508 (1962)
21. J.J. Cullen and D.P. Landau: Phys. Rev. B **27**, 297 (1983)
22. M. Suzuki: Phys. Lett. **113A**, 299 (1985)
23. J.C. Bonner and M.E. Fisher: Phys. Rev. **135**, A640 (1964)
24. M. Suzuki: J. Stat. Phys. **42**, 1047 (1986), and references cited therein.

Part II

Quantum Spin Systems and Quantum Spin Glasses

Investigation of One- and Two-Dimensional Quantum Spin Systems by Monte Carlo Simulations

M. Marcu

II. Institut für Theoretische Physik, Universität Hamburg,
Luruper Chaussee 149, D-2000 Hamburg 50, Fed. Rep. of Germany

Abstract: A method for simulating quantum spin systems with nearest neighbour interaction is reviewed. In the framework of a path-integral approach based on the Trotter formula, the quantum spin system is mapped into the limit of a sequence of classical spin systems with one extra dimension. Numerical questions connected to this limit are discussed. The role of particle and winding numbers is clarified. An overview of the problems investigated by our group is given. In particular, two new results are presented. A simulation of the one-dimensional spin S isotropic Heisenberg antiferromagnet, S=1/2, 1, 3/2, 2, supports Haldane's conjecture that the mass gap is zero for half-integer S and nonzero for integer S. For the two-dimensional spin 1/2 xy model we discuss evidence for a Kosterlitz-Thouless transition. In particular, we show that the helicity modulus can be computed directly from our simulaton.

1. Introduction

Since the pioneering work of Suzuki [1] there have been numerous attempts to simulate quantum spin systems using a path-integral approach (see [2,3] for reviews).

We consider for simplicity the Hamiltonian

$$H_\Lambda = -\sum_{<j,k>} (J_x S_j^x S_k^x + J_y S_j^y S_k^y + J_z S_j^z S_k^z + \vec{h}\cdot\vec{S}). \qquad (1.1)$$

Here $<j,k>$ is a nearest neighbour pair of lattice sites, S_j^α (α=x,y,z) are spin S generators of SU(2) at the site j, J_α are couplings and $\vec{h}$ is an external magnetic field. The lattice will be denoted by Λ, $|\Lambda|=L$ in d=1 dimensions, $|\Lambda|=L_1 \times L_2$ in d=2, L, L_1 and L_2 are even. Periodic boundary conditions are assumed.

The expectation value of an operator O is defined by taking the thermodynamic limit:

$$<O_\Lambda>_\Lambda = \frac{\mathrm{Tr}\, O_\Lambda \exp(-\beta H_\Lambda)}{\mathrm{Tr} \exp(-\beta H_\Lambda)} \xrightarrow[\Lambda\to\infty]{} <O> \qquad (1.2)$$

(β is the inverse of the temperature T). Notice that O_Λ is not necessarily equal to O, only $O_\Lambda \to O$ is required. Now we decompose H_Λ into sums over commuting local operators K(r)

$$H_\Lambda = -A_\Lambda - B_\Lambda, \quad A_\Lambda = \sum_{r\varepsilon S_1} K(r), \quad B_\Lambda = \sum_{r\varepsilon S_2} K(r), \qquad (1.3)$$

$S_{1,2}$ being two sublattices, and we use the Trotter formula [4]

$$\exp(-\beta H_\Lambda) = \lim (T_{\Lambda,M})^{M/2} \ , \quad T_{\Lambda,M} = \exp(\bar\beta A_\Lambda)\, \exp(2\bar\beta B_\Lambda)\, \exp(\bar\beta A_\Lambda) \qquad (1.4)$$

with $\bar\beta=\bar\beta/M$, M=even. The expectation value $<0_\Lambda>_\Lambda$ is obtained as the limit

$$<0_{\Lambda,M}>_{\Lambda,M} = \frac{\mathrm{Tr}\ 0_{\Lambda,M}(T_{\Lambda,M})^{M/2}}{\mathrm{Tr}\ (T_{\Lambda,M})^{M/2}} \xrightarrow[M\to\infty]{} <0_\Lambda>_\Lambda \ . \qquad (1.5)$$

Again, $0_{\Lambda,M}$ is in general different from 0_Λ, but $0_{\Lambda,M} \to 0_\Lambda$ as $M\to\infty$. By inserting complete sets of eigenstates of the S_j^α operators, α = z or x, a classical spin system on the lattice ΛxM is obtained. For $\vec{h}=0$, this system will have constraints [5]. In particular, if we use z-spins as intermediate states the configurations can be interpreted as sets of particle world lines [6-8] (in [6] a preliminary account of our work was given).

In section 2 the classical spin system in d+1 dimensions (d=1,2) is discussed. The Monte Carlo procedure used to simulate it can update also the particle and winding numbers of the world lines. Section 3 deals with the numerical problems encountered when taking the $M\to\infty$ limit and the thermodynamic limit. In section 4 questions related to the particle and winding numbers are discussed. The conclusion is that the finite size effects are smaller if the particle number is allowed to vary but the winding number is fixed to zero. If we allow the winding number to vary, then in d=1 it turns out that in the thermodynamic limit the winding number only takes the value zero. In d=2 this is true at high temperatures only. At low temperatures the winding number has a smooth distribution.

We have used our quantum Monte Carlo method to investigate a variety of models. In [8] we started to develop the method. For the first time a model with S>1/2 was simulated. In [9] the continuum limit of the Thirring model with Susskind Fermions was investigated. The first high-precision quantum Monte Carlo simulation was performed in [10] for the d=1 S=1/2 isotropic ferromagnet and antiferromagnet. Our original analysis of the data in [10] was incorrect, as shown by comparison to Bethe-Ansatz results for the ferromagnet [11]. The reason was, paradoxically, that our low-temperature data had too small statistical errors, so that the method we had previously used to extrapolate to the $M=\infty$ limit no longer worked. In [12] we showed that the discrepancies to [11] disappear when the extrapolation is done more carefully.

In order to perform high-precision calculations, we wrote fully vectorized programs for general spin S in d=1 [13] and for S=1/2 in d=2. Further savings in computer time can be achieved using a variance reduction technique for quantum spin systems [14].

One-dimensional quantum spin systems have been studied in numerous experiments [15]. Typically the spin and coupling constants of some quasi-one-dimensional material are determined by some fairly reliable method (e.g. ESR) and then neutron scattering is studied. One difficulty in comparing theory with experiment is the lack of a reliable low-temperature approximation method. In [16] we showed for several S=1/2 and S=1 examples that quantum Monte Carlo results compare well to experiment in cases where other theoretical methods have failed.

Haldane [17] gave arguments that the $S=\infty$ limit of d=1 xxz models is approached in a different way by integer and half-integer values of the spin S. In particular, he predicted that for the isotropic antiferromagnet the mass gap is zero for half-integer S and nonzero for integer S. This "Haldane conjecture" gave rise to a considerable controversy in the literature[18]. On the balance, S>1/2 finite chain diagonalization results seem to confirm the Haldane conjecture. Doubts remain, however, since the size of the chains for which exact diagonalization is still feasible is quite small. In section 5 we present results for S=1/2, 1, 3/2 and 2. They were obtained at finite temperatures for chains large enough to be effectively in the thermodynamic limit. The $T\to 0$ behaviour strongly supports the Haldane conjecture.

In two dimensions, the S=1/2 xy model ($J_z=0$) has also been subject to some controversy. The results of two calculations using different quantum Monte Carlo methods [19,20] do not agree [21]. In particular, the procedure of [19] is criticized for not being ergodic [21]. In section 6 we describe our preliminary results for the d=2 S=1/2 xy model. We establish a connection between the winding number distribution and the helicity modulus. At low temperatures our simulation yields a well-defined value for the helicity modulus. This is strong evidence for a massless phase [22] (i.e. a phase with power law decay of correlations). The possible interpretation of this result as evidence for a Kosterlitz-Thouless transition [23] is discussed. The acceptance rate for the update of the winding numbers is extremely small, and it decreases with increasing lattice size. Moreover, a simulation with variable winding numbers produces larger finite size effects than for fixed winding numbers. We indicate a way to compute the helicity modulus from fixed-winding-number simulations.

Section 7 contains our conclusions and outlook.

2. The Classical Spin System and the Monte Carlo Algorithm

In order to map the d-dimensional quantum spin system into a d+1-dimensional classical spin system, we used the following break-up of the Hamiltonian (see (1.3)):

$$d=1: \quad A_\Lambda = \sum_{j=\mathrm{odd}} H(j,j+1)\ , \quad B_\Lambda = \sum_{j=\mathrm{even}} H(j,j+1)\ , \tag{2.1.a}$$

$$d=2: \quad A_\Lambda = \sum_{j_1,j_2=\mathrm{odd}} K(p_j)\ , \quad B_\Lambda = \sum_{j_1,j_2=\mathrm{even}} K(p_j)\ , \quad j=(j_1,j_2)$$

$$\text{with } K(p_j) = H(j,j+\hat{1})+H(j+\hat{1},j+\hat{1}+\hat{2})+H(j+\hat{2},j+\hat{1}+\hat{2})+H(j,j+\hat{2})\ . \tag{2.1.b}$$

Here $H(j,k)$ is the link term in (1.1), $\hat{1}$, $\hat{2}$ are lattice unit vectors in direction 1, 2 (the lattice spacing is one), and $K(p_j)$ is the sum over the 4 link terms around the plaquette p_j (with lower left corner j). By inserting complete sets of eigenstates of S_j^z in (1.5), a classical spin system on the lattice $\Lambda \times M$ is obtained. A classical spin configuration s consists of spins $s(j,r)$, $j\in\Lambda$, $r=1,\ldots,M$ and $s(j,r)\in\{-S,-S+1,\ldots,S-1,S\}$. The partition function is [1, 5, 7,6,8,20]:

$$d=1: \quad Z_{\Lambda,M} = \sum_s \prod_{p=\mathrm{int}} f(p)\ , \tag{2.2.a}$$

$$d=2: \quad Z_{\Lambda,M} = \sum_s \prod_{c=\mathrm{int}} f(c)\ . \tag{2.2.b}$$

Here p=int means an interacting plaquette in 1+1 dimensions, c=int an interacting cube in 2+1 dimensions. For p=int the coordinates of the lower-left corner are both even or both odd. For c=int the coordinates of the lower-left-forward corner are all three even or all odd (there are LM/2 interacting plaquettes and $L_1L_2M/4$ interacting cubes). With the notation introduced in fig.1, the contribution of an interacting plaquette or cube to the Boltzmann weight of a configuration is:

$$\text{d=1:} \quad f(p) = \langle s_1,s_2|\exp\{-2\bar{\beta}H(j,j+1)\}|s_1',s_2'\rangle\,, \tag{2.3.a}$$

$$\text{d=2:} \quad f(c) = \langle s_1,s_2,s_3,s_4|\exp\{-2\bar{\beta}K(p_j)\}|s_1',s_2',s_3',s_4'\rangle\,. \tag{2.3.b}$$

As a consequence of the symmetries of (1.1), the configurations of the classical spin system may be subject to constraints. For the xxz model, defined by $J_x=J_y$, $\vec{h}=0$ (or $\vec{h}$ in z-direction) the total magnetization in z-direction $S^z_{tot} = \Sigma\, S^z_j$ is conserved and f is zero unless (xxz constraint):

$$\text{d=1:} \quad s_1+s_2 = s_1'+s_2'\,, \tag{2.4.a}$$

$$\text{d=2:} \quad s_1+s_2+s_3+s_4 = s_1'+s_2'+s_3'+s_4'\,. \tag{2.4.b}$$

For the xyz model, defined by $\vec{h}=0$ (or $\vec{h}$ in z-direction) the total magnetization is conserved, only modulo 2, and eqs.(2.4) hold modulo 2 (xyz constraint). The other symmetries do not lead to constraints, but to symmetries of f. Thus in d=2 f is invariant under the group D_{4h}.

For the xxz model (2.4) implies:

$$n_p = \sum_{j\in\Lambda} s(j,r) = \text{constant in } r\,. \tag{2.5}$$

A configuration can be seen as a set of particle lines. There are $s(j,r)+S$ world lines of particles passing through the point j,r of $\Lambda \times M$. The particle lines wind around the torus both in Trotter (Euclidean time) and in space directions [7,6,8]. The conserved total number of particle lines is $n_p+|\Lambda|S$. Eqs.(2.4) also imply conservation of the winding number(s) n_w [8,25]:

$$\text{d=1:} \quad n_w = \sum_r (-1)^{j+r} s(j,r) = \text{constant in } j\,, \tag{2.6.a}$$

$$\text{d=2:} \quad n_{w_1} = \sum_{j_2,r} (-1)^{j_1+r} s(j,r) = \text{constant in } j_1\,, \quad j=(j_1,j_2) \tag{2.6.b}$$

and n_{w_2} is defined similarly, by interchanging j_1 and j_2.

Fig.1: An interacting plaquette (j,r = even or j,r = odd) and an interacting cube (j_1,j_2,r = even or j_1,j_2,r=odd).

For the xyz model the particle and winding numbers are conserved only modulo 2.

The existence of constraints restricts the class of expectation values (1.2) that can be computed using (1.5) [8-10]. The general rule is that $O_{\Lambda,M}$

is a product of S_j^z operators or is obtained by taking derivatives of the free energy of the classical system on $\Lambda \times M$. Sometimes other quantities are also accessible. In order to compute x-x correlations however, a new simulation with J_x and J_z interchanged has to be performed. As will be discussed later, all these quantities can be computed both for fixed and for variable particle and winding numbers.

Let us now describe the Monte Carlo algorithm used to simulate the xxz model. There are three different updating procedures: a local update that preserves the values of n_p and n_w, an update of n_p and an update of n_w.

In 1+1 dimensions the local update changes the values of the 4 spins at the corners of a noninteracting plaquette. In the notation of fig.2:

$$s_1 \to s_1+\sigma \ , \quad s_2 \to s_2-\sigma \ , \quad s_3 \to s_3+\sigma \ , \quad s_4 \to s_4-\sigma \ , \ \sigma = \pm 1 \ . \qquad (2.7)$$

The constraint (2.4.a) is not violated for any of the 4 neighbouring interacting plaquettes. In 2+1 dimensions the local update involves 4-spin changes for the plaquettes that are not on the boundary of some interacting cube. If such a plaquette is time-like, then (2.7) is used for the update; if it is space-like, then the 4-spin change is:

$$s_1 \to s_1+\sigma \ , \quad s_2 \to s_2-\sigma \ , \quad s_3 \to s_3-\sigma \ , \quad s_4 \to s_4+\sigma \ , \ \sigma = \pm 1 \ . \qquad (2.8)$$

The constraint (2.4.b) is not violated for any of the 4 neighbouring interacting cubes. In 1+1 dimensions (2.7) can "deform" a particle line locally in all directions. This is not true however in 2+1 dimensions. In this case we supplement the local update procedure by offering changes that result from the formal product of two one-plaquette changes (2.7)-(2.8), such that the two plaquettes are faces of the same noninteracting cube. It may happen that a single plaquette-change leads to a zero weight (i.e. not allowed) configuration, but a formal product of two such changes again produces a configuration with nonzero weight. Using this local update procedure, a particle line in 2+1 dimensions can be also locally deformed in all directions [25].

Notice that for S=1/2 eqs.(2.7)-(2.8) can be replaced by a simple flip of all 4 spins.

The particle number update is performed by offering:

$$s(j,r) \to s(j,r)+\sigma \ , \ \sigma = \pm 1 \ , \quad \text{for all } r=1,\ldots,M \ . \qquad (2.9)$$

This amounts to inserting or removing a straight particle line. The winding number is updated by (update of n_w and n_{w_1}):

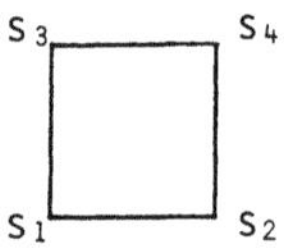

Fig.2: Notation for the 4-spin changes (2.7)-(2.8)

$$d=1: \quad s(j,r) \to s(j,r)+(-1)^{j+r}, \ \sigma = \pm 1 \ , \quad \text{for all } j=1,\ldots,L \qquad (2.10.a)$$

$$d=2: \quad s(j,r) \to s(j,r)+(-1)^{j_1+r}, \ \sigma = \pm 1 \ , \quad j = (j_1,j_2) \ , \quad \text{for all } j_1=1,\ldots,L_1 \ . \qquad (2.10.b)$$

Interchanging j_1 and j_2 in (2.10.b) we get the update of n_{w_2}.

For the xyz model, the local update is supplemented by offering a one-spin change:

$$s(j,r) \to s(j,r)+\sigma \ , \quad \sigma = \pm 2 \ . \tag{2.11}$$

For S=1/2 this is not necessary.

In 1+1 dimensions (2.7)-(2.10) define an ergodic Markov process for the xxz model. Unfortunately we could not prove that this is true also in 2+1 dimensions. However, since the particle lines can be deformed locally in all directions (if 2-plaquette changes are included), we are confident that (2.7)-(2.10) define an ergodic process in this case too.

Eqs. (2.9)-(2.11) define an ergodic Markov process for the xyz model, both in 1+1 and in 2+1 dimensions.

In section 4 it will be argued that the nonlocal parts (2.9)-(2.10) of the Monte Carlo algorithm can be dropped sometimes. Furthermore it will be shown that an optimal procedure is to update n_p but to keep n_w=0. This is extremely important in view of the typically very small acceptance rate for (2.10) [8, 25 ,26]. The acceptance rate for the local and the particle number update is usually reasonable.

3. Problems Concerning the $M \to \infty$ and the Thermodynamic Limits

It was noticed by several workers that the linear correction in 1/M to the $M \to \infty$ limit vanishes (see [2,3] for reviews). In particular, in refs. [8-10] we observed that

$$\langle O_{\Lambda,M}\rangle_{\Lambda,M} = \langle O_\Lambda\rangle_\Lambda + \bar{\beta}^2 \beta \sum_{j\epsilon\Lambda} (\langle OB_j\rangle_\Lambda - \langle O\rangle_\Lambda \langle B_j\rangle_\Lambda) + \dots , \tag{3.1}$$

where B_j are local operators obtained from double commutators of A_Λ and B_Λ.

Suzuki [27] gave a one-line proof that the r.h.s of (3.1), considered as a function of $\bar{\beta}$ ($\bar{\beta}=\beta/M$) and β, is even in $\bar{\beta}$ (one must be careful not to introduce odd powers of $\bar{\beta}$ by an unhappy choice of $O_{\Lambda,M}$).

As $T \to 0$ one has to take larger values of M in order to be numerically close to the $M \to \infty$ limit. In [7] it was argued that the size of the corrections to the $M \to \infty$ limit only depends on the value of $\bar{\beta}$. Our investigations however [8-10,12,16,25,26,28] do not confirm this as a general rule. In the important case that T=0 is a critical point, the coefficients in front of the even powers of $\bar{\beta}$ (see (3.1)) sometimes have a nonnegligible dependence on β, and they may behave differently for different physical quantities.

At the beginning [8,10] we extrapolated to the $M \to \infty$ limit by taking values of M high enough such that the r.h.s. of (3.1) was effectively linear in $\bar{\beta}$. This is however very dangerous, as discussed in [12,26]. A correct procedure is to fit (3.1) with polynomials in $\bar{\beta}^2$ whose order is increased until the extrapolated result stabilizes. The fits must be done for every quantity for every temperature and for every set of coupling constants separately, and goodness-of-fit tests should not be ommited.

Suzuki argued that the thermodynamic limit $\Lambda \to \infty$ and the "Trotter limit" $M \to \infty$ can be interchanged [29]. His proof however does not seem to hold for the systems (2.2) that interest us here. In [12,26] we checked numerically that the two limits can indeed be interchanged. For various one-dimensional xxz models we compared the results obtained by first taking the $M \to \infty$ limit and then the $L \to \infty$ limit with those obtained by taking the $M = L \to \infty$ limit. Within errorbars they were always the same.

4. Role of Particle and Winding Numbers in xxz Models

In [7] it was argued that, similar to the equivalence of canonical and grand-canonical ensembles in the thermodynamic limit, the expectation values of local oprators should not depend on particle and winding numbers:

$$\lim_{\Lambda\to\infty} \lim_{M\to\infty} \langle O_{\Lambda,M} \rangle_{\Lambda,M,n_p=n,n_w=w} = \text{independent of } n \text{ and } w \,. \qquad (4.1)$$

As far as particle number is concerned, the idea for a proof can be found in many Statistical Mechanics textbooks. Let us discuss the reason why we do not expect (4.1) to depend on w . The winding number is defined by counting the particle lines that go around the torus in space directions. Therefore, fixed and variable winding numbers should correspond to different types of boundary conditions. In the thermodynamic limit the boundary conditions should play no role.

In a Monte Carlo simulation the problem of finite size effects is very important. In general, finite size effects are smaller if we use a grand-canonical ensemble. Thus in a simulation with variable particle number the thermodynamic limit is approached faster than in the case of fixed particle number. Moreover, we are often interested in nonlocal quantities (e.g. susceptibility) that cannot be computed if the particle number is not allowed to vary [8,10,21].

Let us discuss the shape of the particle and winding number distributions

$$\rho_p(n) = \langle \delta(n_p-n) \rangle \,, \quad \rho_w(w) = \langle \delta(n_w-w) \rangle \qquad (4.2)$$

(δ is the Kronecker symbol). In all the models we have studied, $\rho_p(n)$ is a smooth curve, even in the thermodynamic limit. If at T=0 there is no z-magnetization, the peak at n=0 becomes narrower with decreasing temperature. If we are interested in obtaining T=0 results, it might be useful to keep the particle number $n_p=0$ [9,10]. If at T=0 the spontaneous magnetization is nonzero, then $\rho_p(n)$ broadens and tends to a constant as $T \to 0$ [8]. The winding number distribution behaves differently. In one-dimensional systems we find the following behaviour [12,26]. In a finite volume, $\rho_w(w)$ will be of course a smooth curve. As the volume is increased however, $\rho_w(w) \to \delta(w)$. Thus it is to be expected that simulations with $n_w=0$ give smaller finite size effects than in the case of variable winding number.

For the one-dimensional xy model ($J_x=J_y$, $J_z=\vec{h}=0$), we confirmed this conclusion using exact solutions [12]. Notice that the $n_w=0$ exact solution was obtained by diagonalizing the transfer matrix in space (as opposed to the usual transfer matrix in Euclidean time) direction. The original winding number was mapped into the particle number of this new transfer matrix [26]. Monte Carlo simulations for $J_z\neq 0$ (d=1) also confirmed the fact that the finite size effects are smaller for $n_w=0$.

In the two-dimensional xy model we found a different picture for the winding number distribution. The model has a phase transition at finite temperatures. In the high-T phase we found again that $\rho_w(w) \to \delta(w)$ in the thermodynamic limit. In the low-T phase however, $\rho_w(w)$ seems to converge towards a smooth curve [25]. Nevertheless, at all temperatures the finite size effects are smaller for $n_w=0$ than for variable winding number.

In section 6 a connection between $\rho_w(w)$ and the helicity modulus will be established. This is in agreement to our general statement that different winding numbers mean different boundary conditions.

We conclude this section by restating our main conclusion: the most efficient simulation of an xxz system is with variable n_p and with $n_w=0$. Fixed n_p is useful if we study T=0 properties and there is no magnetization at T=0. Variable n_w is useful in computing the helicity modulus, although in section 6 a fixed n_w method will be also proposed.

5. Quantum Monte Carlo Test of the Haldane Conjecture

We performed a high-precision calculation for the d=1 isotropic antiferromagnet with S = 1/2, 1, 3/2, 2 [28]. In order to extrapolate to the thermodynamic and $M \to \infty$ limits, the largest lattices we took were L=M=120.

If there is a mas gap, the S^z-S^z correlation function should decay exponentially even in the limit T=0. Its Fourier transform, the static structure factor I(q), will, in a small region around the peak at $q=\pi$, converge towards a Lorentzian curve as $T \to 0$. If on the other hand at T=0 we have a power-law decay of the correlation function, then $I(\pi)$ will diverge as $T \to 0$.

In fig.3 we present our results for $I(\pi)$ as a function of temperature. The behaviour is very different for S = 1/2 and 3/2 as compared to S = 1 and 2. If $I(\pi)$ diverges we have a power-law approach to T=0

$$I(\pi) \sim T^{-\gamma_1} \tag{5.1}$$

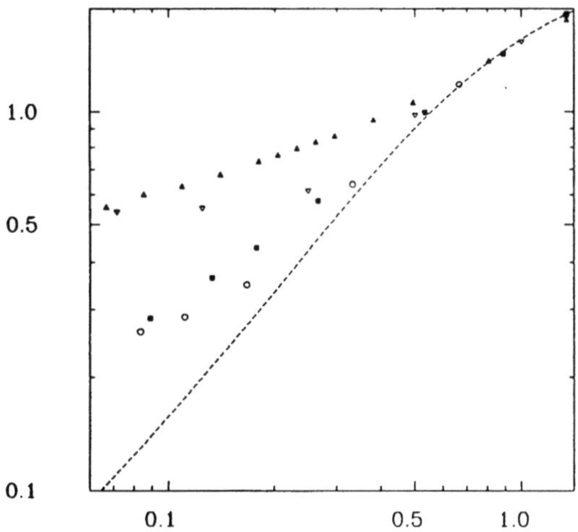

Fig.3 Log-log plot of $[I(\pi)/S(S+1)]^{-1}$ against T/S(S+1). The full triangles are S=1/2, the empty triangles are S=1, the full squares are S=3/2, and the empty circles are S=2. The dashed line is the S=∞ limit.

and in fig.3 we should get a straight line with slope γ_1. For S = 1 and 2 this is evidently not the case. For S = 1/2 and 3/2 a fit with a straight line is possible, giving the values $\gamma_1=.22\pm.03$ for S=1/2 and $\gamma_1=.70\pm.05$ for S=3/2. For S=1/2 it is known that $\gamma_1=0$ and that $I(\pi)$ diverges as the logarithm of T (see [7] and references therein). This is not inconsistent with our data since it is very difficult to distinguish between a small power and a log.

We also computed the staggered susceptibility, which in the case of zero-mass gap should diverge with the exponent $\gamma=\gamma_1+1$ [30]. An analysis similar to that for $I(\pi)$ gave the result $\gamma=1.05\pm.06$ for S=1/2 and $\gamma=1.70\pm.06$ for S=3/2.

From our data it seems that for half-integer spins the exponent γ_1 grows from 0 to 1 as S increases from S=1/2 to $S=\infty$ (γ goes from 1 to 2). In [31] however, it is argued that $\gamma_1=0$ for all finite half-integer S. For S>1/2 this would mean that at very low T there is a crossover from γ_1 close to 1 to $\gamma_1=0$. In the $S=\infty$ limit the crossover temperature would go to zero. It is not clear whether a quantum Monte Carlo simulation could decide which alternative is correct.

6. The Two-Dimensional Spin 1/2 xy Model

For this model we performed two simulations, one with variable winding number and one with $n_w=0$. The behaviour of the winding number distribution and of the finite size effects was discussed in section 4. Here a connection between $\rho_W(w)$ and the helicity modulus will be established.

In the classical case, the helicity modulus was defined in [22]. Consider the partition function $Z(\theta)$ with a twist of θ in the angle variables across the boundary in 1-direction. Then [22]

$$\frac{Z(\theta)}{Z(0)} \underset{\theta/L_1=\text{small}}{=} \exp(-\theta^2\beta\Upsilon L_2/2L_1) \; ; \tag{6.1}$$

Υ is called helicity modulus . In the low-temperature phase we expect a non-zero value of Υ, while for $T>T_c$ $\Upsilon=0$ in the thermodynamic limit. Notice that (6.1) depends very sensitively on the values of L_1 and L_2.

For the quantum xxz model we define the twisted boundary conditions by changing the Hamiltonian across the boundary as follows:

$$H(j,j+\hat{1}) \rightarrow \exp(i\theta S_j^z)\, H(j,j+\hat{1})\, \exp(-i\theta S_j^z) \;, \quad j=(j_1,j_2) \;,\; j_1=L_1 \quad j_2=1,\ldots,L_2 \,. \tag{6.2}$$

It is easily seen that for $S\rightarrow\infty$ the correct twisted boundary conditions are obtained. The path-integral method described in sections 1 and 2 immediately leads to:

$$\frac{Z(\theta)}{Z(0)} = \langle\exp(i\theta n_w)\rangle = \sum_w \exp(i\theta w)\, \rho_W(w) \,. \tag{6.3}$$

$Z(\theta)/Z(0)$ can thus be computed directly in a simulation with variable winding number, and it is the Fourier transform of $\rho_W(w)$.

In the low-temperature region we could fit our data well with the form (6.1) for 4 x 4, 4 x 8, 8 x 4 and 8 x 8 lattices (the highest value of M was 32). This strongly suggests that $\Upsilon\neq0$ for $T<T_c$ and therefore the low-T phase is massless [25].

We interpret this as evidence for a Kosterlitz-Thouless phase transition [23], i.e. a transition from a massive to a massless phase such that the correlation length blows up exponentially as $T \to T_c$, $T>T_c$, and such that there is no symmetry breaking for $T<T_c$. Other groups have also found evidence for a massless phase at $T<T_c$ [20,32]. However, in [19] it is argued that for $T>T_c$ the data show a power-law divergence of the correlation length. No model is known in the literature where this $T>T_c$ behaviour is combined with the Kosterlitz-Thouless behaviour for $T<T_c$. It is therefore very important to do further investigations.

Notice that if we take $\theta=\pi$ in (6.3), only the winding number modulo 2 is important. Then the formula holds also for the xyz model.

Because of a very low acceptance rate, it is very difficult to compute Υ for large lattices using (6.3). Manipulating (6.3) it can be shown that the difference between the energies for $n_w=1$ and $n_w=0$ is a numerically simple function of Υ. Using this method, the helicity modulus could be computed for much larger lattices.

7. Conclusions and Outlook

In this paper the quantum Monte Carlo work done by our group has been reviewed. In particular, we discussed the various limits carefully, as well as the problems connected to the particle and winding numbers. We argued that it is safe and even desirable to keep the winding number zero.

In a high-precision simulation we gave strong numerical evidence for the validity of the Haldane conjecture. For the d=2 S=1/2 xy model we computed the helicity modulus as the Fourier transform of the winding number distribution.

The work on the Haldane conjecture will be continued in order to see how the $S=\infty$ limit is approached for half-integer spins. Our main effort in the near future will be in d=2, in the investigation of the S=1/2 xxz model.

References

1. M.Suzuki: Commun.Math.Phys.51 (1976) 183
   Prog.Theor.Phys.56 (1976) 1454
   M.Suzuki, S.Miyashita and A.Kuroda: Prog.Theor.Phys.58 (1977) 1377
2. M.Suzuki: J.Stat.Phys.43 (1986) 883
3. H.de Raedt and A.Lagendijk: Phys.Repts.127 (1985) 223
4. H.F.Trotter: Proc.Am.Math.Soc.10 (1959) 545.
5. M.Barma and B.Shastry: Phys.Rev.B18 (1978) 3351
6. J.Honerkamp: In "Recent Advances in Field Theories and Statistical Mechanics", Les Houches 1982, eds. J.-B.Zuber and R.Stora (Elsevier Sciece Publishers B.V., 1984) p.248
7. J.E.Hirsch et.al.: Phys.Rev.Lett.47 (1981) 1628; Phys.Rev.B26 (1982) 5033
8. M.Marcu and A.Wiesler: J.Phys.A18 (1985) 2479
9. F.-K.Schmatzer: "Massenberechnungen am Massiven Thirringmodell mit Hilfe der Monte Carlo Methode", Diplomarbeit, Freiburg 1983
10. M.Marcu, J.Mueller and F.-K.Schmatzer: J.Phys.A18 (1985) 3189
11. P.Schlottmann: Phys.Rev.Lett.54 (1985) 2131
    M.Takahashi and M.Yamada: J.Phys.Soc.Japan 54 (1985) 2808
12. M.Marcu, J.Mueller and F.-K.Schmatzer: Phys.Lett.116A (1986) 447
13. M.Marcu, J.Mueller and F.-K.Schmatzer: "Fast Algorithm for One-Dimensional Quantum Monte Carlo Simulations", Freiburg preprint THEP 86/6, submitted to Computer Phys.Comm.

14. M.Marcu and J.Mueller: "Variance Reduction Technique for Quantum Monte Carlo Simulations", Freiburg preprint THEP 86/7, to appear in Phys.Lett.A
15. L.J.de Jongh and A.R.Miedema: Adv.Phys.23 (1974) 1
M.Steiner, J.Villain and C.G.Windsor: Adv.Phys.25 (1976) 87
16. M.Marcu and J.Mueller: "Quantum Monte Carlo versus Experimental Results for xxz Chains", Freiburg preprint THEP 86/8, submitted to Phys.Lett.A
17. F.D.M.Haldane: Phys.Rev.Lett.50 (1983) 1153; Phys.Lett.93A (1983) 464
18. R.Botet and R.Julien: Phys.Rev.B27 (1983) 613
M.Kolb, R.Botet and R.Julien: J.Phys.A16 (1983) L673
J.B.Parkinson et.al.: J.Appl.Phys.57 (1985) 3319
19. H.de Raedt, J.Fivez and A.Lagendijk: Phys.Lett.104A (1984) 430
H.de Raedt and A.Lagendijk: Z.Phys.B57 (1984) 209
20. E.Loh, D.J.Scalapino and P.M.Grant: Phys.Rev.B31 (1985) 4712
21. H.de Raedt and A.Lagendijk: Phys.Rev.B33 (1986) 5102
E.Loh, D.J.Scalapino and P.M.Grant: Phys.Rev.B33 (1986) 5104
22. M.E.Fisher, M.M.Barber and D.Jasnow, Phys.Rev.A8 (1973) 1111
23. J.M.Kosterlitz and D.J.Thouless: J.Phys.C6 (1973) 1181
J.M.Kosterlitz: J.Phys.C7 (1974) 1046
25. M.Marcu and F.-K.Schmatzer: in preparation
26. M.Marcu, J.Mueller and F.-K.Schmatzer: in preparation
27. M.Suzuki: Phys.Lett.113A (1985) 299
28. M.Marcu and J.Mueller, in preparation
29. M.Suzuki: Phys.Rev.B31 (1985) 2957
30. I thank Dr. Miyashita for teaching me this point
31. H.J.Schulz: Orsay preprint (1986)
32. T.Onogi, S.Miyashita and M.Suzuki: in preparation

Critical Properties of the Two-Dimensional Spin-1/2 XY Ferromagnet

T. Onogi, S. Miyashita, and M. Suzuki

Department of Physics, University of Tokyo,
Hongo, Bunkyo-ku, Tokyo 113, Japan

1. Introduction

Critical phenomena of the two-dimensional XY model have been arousing a great deal of interest for a couple of decades [1,2] because of the model's marginal feature that the internal degree of freedom is two, which is a critical value in two dimensions. It is proved that in two dimensions a continuous symmetry breaks the long-range order at finite temperatures [3]. Nevertheless, the possibility of the existence of a phase transition without long-range order was suggested by numerical investigations [4]. Thereafter, the nature of phase transitions of such systems, namely the XY and Heisenberg models, has been studied by various methods in classical systems. While the Heisenberg model turned out to have no phase transition at finite temperatures [5], it has been found that a marginal type of phase transition, known as the Kosterlitz-Thouless transition, takes place where a vortex, namely a topological singularity of this system, plays an important role [6-9].

The way the quantum effect affects the nature of a system's phase transition and ordered state is also an interesting problem. It is believed that this effect does not cause any essential change for three-dimensional spin systems. On the other hand, for the two-dimensional spin-1/2 XY model

$$H = - J \Sigma_{<ij>} (\sigma_i^x \sigma_j^x + \sigma_i^y \sigma_j^y) , \tag{1.1}$$

(hereafter we often put J = 1), we may expect some qualitative changes due to the quantum effect (spin = 1/2) because of its marginal nature. The result of high-temperature expansions for the quantum system (1.1) does not necessarily support the possibility that it shows a type of phase transition found in the corresponding classical system [10]. It is also suggested that the nature of the ground state might be qualitatively different from the classical one (spin = ∞) [11,12]. It is, therefore, interesting to study the phase transition and the nature of the low-temperature phase of the two-dimensional spin-1/2 XY model and to compare them with the classical cases.

For the study of properties of quantum systems at finite temperatures, there are many methods, such as the spin wave theory, the thermal Green's function and a method with direct diagonalization. Complementary to these methods, a general quantum Monte Carlo method for general short-range interacting systems was introduced by SUZUKI [13] using the Suzuki-Trotter formula

$$\exp(-\beta H) = \lim_{n\to\infty} [\ \Pi_i \exp(-\beta H_i/n)]^n , \tag{1.2}$$

where $H = \Sigma_i H_i$. A preliminary study of the present model was done with this method, and the critical temperature was estimated by investigating the temperature and size dependences of the square of magnetization $<M_x^2>$ and specific heat [14]. This simulation was accomplished only for the n=1 approximant based upon the real space pair decomposition. After this work, several more advanced studies have been done using various decompositions, representations, and algorithms. LOH et al. [15] studied the present model by using the cell decomposition and diagonalizing the z-axis of spin space. The states updated by their Monte Carlo algorithm were confined within a space with total magnetization $M_z = 0$ and with a topological constraint, the so-called winding number, w = 0. They measured the thermodynamic quantities at each temperature by changing the value of the Trotter number to fix $\Delta\tau = \beta/n$ at a small value ($\simeq 0.25$). This procedure takes into account the quantum effect properly at each temperature. DE RAEDT and LAGENDIJK [16,17] studied the same model by using the real space pair decomposition, and diagonalizing the x-axis of spin space. They developed an algorithm to ensure the ergodicity of Monte Carlo sampling, namely, they allowed the fluctuation of both values of M_z and w in principle. There are some controversies between both the groups, and therein several fundamental problems of the quantum Monte Carlo method are raised [18]: (1) The effect of conservation of physical quantities on thermodynamic properties. If we change a configuration of the system only locally, the new state still stays within a space characterized by quantum numbers of the old configuration. For example, M_z is a good quantum number in the XY model. Without any external field, this constraint does not give any change in the thermodynamic limit. It would be, however, a practical problem in finite-size lattices. (2) The effect of boundary conditions. Because of the trace procedure, we have to use the periodic boundary condition for the additional direction due to decomposition of the Boltzmann factor (Trotter direction). For the original space we also usually adopt the periodic boundary condition. Thus the space is a torus, where topologically different classes of line configurations exist. These classes are specified by the winding number. This effect again becomes irrelevant in the thermodynamic limit but is an important problem in finite lattices. It is rather difficult to introduce an efficient algorithm for transitions between different winding numbers. In order to overcome this problem, DE RAEDT et al. [17] proposed a method mapping the (d+1)-dimensional lattice (ZX-model) to another lattice interacting with a long-range interaction. Recently MARCU [19] introduced an algorithm for a (d+1)-dimensional lattice (XY-model), constructed by the cell decomposition, to allow transitions between states with different total magnetizations and different winding numbers, and he studied the system with a representation diagonalizing the z-axis of spin space.

In the present study, fluctuation of the value of M_x is allowed, but any transition to a space with odd winding numbers is not included. Some details of our Monte Carlo procedure are illustrated in the next section. Sections 3 and 4 give the results for the specific heat and the order parameter, respectively. Section 5 is devoted to summary and discussion.

2. Method

The Boltzmann factor of the present system is decomposed into four kinds of pair-interacting parts. The order of decomposition and a transformed (2+1)-dimensional lattice are given in Figs. 1 and 2, respectively. The case with n = 1 has been studied analytically [20] and some exact results for thermodynamic properties are obtained. In particular, the specific heat shows a logarithmic divergence, and $\langle M_x^2 + M_y^2 \rangle$ gives a divergence at T_c with exponents: $\gamma = 3/2$ and $\eta = 1/2$, although it remains finite except at $T = T_c$.

For the properties of the original quantum system we have to study the systems with $n \geq 2$. A Monte Carlo simulation is implemented for these cases. Here we take the σ^x-representation. Three types of flips (see Fig. 2) are used to ensure the ergodicity at least in the space with $w = 0$: (1) "global flip", (2) "four spin flip" and (3) "eight spin flip", which may vary the value of M_x. With the choice of this representation, one is able to measure the order parameter, simultaneously with the energy and specific heat. In the present simulation, the statistical average of physical quantities was taken over 10000 Monte Carlo steps (MCS) after the initial 5000 MCS were discarded.

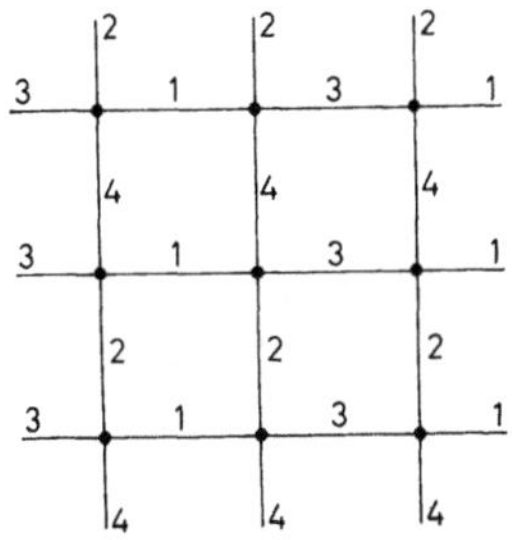

Fig. 1 The original two-dimensional lattice. The attached numbers correspond to the order in the decomposition (1.2).

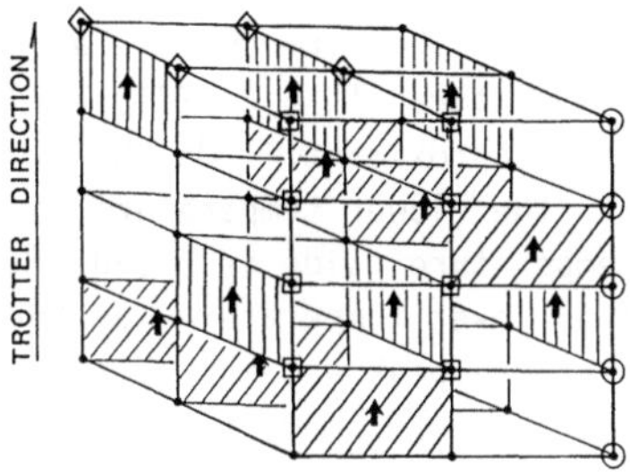

Fig. 2 The transformed (2+1)-dimensional lattice. Shaded squares denote elementary four-spin interactions on this lattice. The symbols ○, ◇ and □ denote "global flip", "four spin flip" and "eight spin flip", respectively.

It is plausible that the specific heat remains finite in the original quantum system. The logarithmic divergence which appears in the case n = 1 is apparently due to an energy gap introduced by the decomposition procedure. Thus it is expected that such divergence will disappear as n increases. There are two possible routes for the disappearance: (1) The critical amplitude of the divergence may decrease as a function of n, for example,

$$C_n \propto (1/n^2)\log|T-T_c(n)| \ , \qquad (2.1)$$

or (2) the universality class of the n-th approximant may vary according to the value of n. The symmetry of the original Hamiltonian remains even in the decomposed systems, but the energy spectrum with a gap may change the nature. One might expect such changes according to n, analogously to those which are found in the q-state Potts model [21] and in the n-state clock vector model [7,22,23]. In order to study how such changes are realized, we investigate thermodynamic properties separately for the system with each value of n.

3. The Specific Heat

The specific heat of the n-th approximant was measured using the formula in [14] obtained from the definition

$$C_n = k_B \beta^2 \partial^2 \ln Z_n / \partial\beta^2 \quad , \quad \beta = 1/k_BT \ . \qquad (3.1)$$

The data of the specific heat thus obtained were consistent with those of the numerical differentiation of the energy. Figure 3 shows the temperature-dependences of the data (per one quantum spin) with n = 1, 2, 3 and 4, where L denotes the linear dimension of the quantum system. We observe here a qualitative difference between the data of n = 1 and those of n = 2, 3 and 4. The curve $C_1(T)$ has a single sharp peak around the Onsager critical temperature ($T_c \simeq 2.27J$), as is expected from the analytic investigation of this approximant. On the other hand, $C_n(T)$ ($n \geq 2$) has two peaks. The peak at the lower temperature side is very sharp and the other at the higher temperature side is broad. Note that the lower temperature peak shifts to the lower temperatures as the value of n increases, while the higher temperature peak remains near T = 2J. We believe that the higher temperature peak corresponds to the ordering of the XY model and the lower temperature peak is caused by the energy gap due to the decomposition procedure. Figure 4 shows the size dependence of the value of the higher temperature peak. The n = 1 data clearly show the logarithmic divergence. On the other hand, the peak values for $n \geq 2$ show slight L-dependence, but it is too faint for a divergence to be concluded. Indeed, a similar increase with the size is observed for the classical XY system where we believe the specific heat does not diverge. Thus we conclude the second route of disappearance of diver-

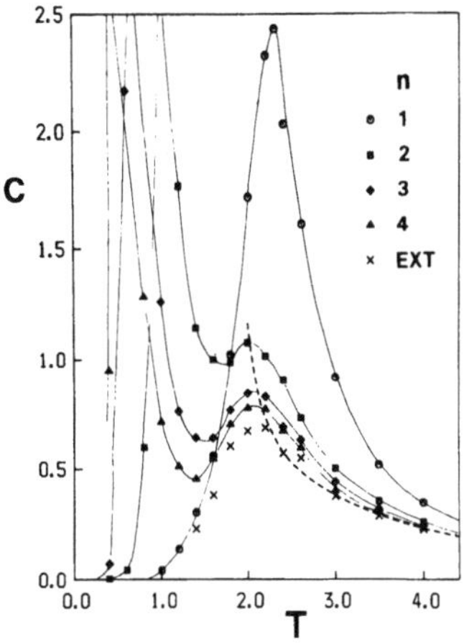

Fig. 3 Temperature and n dependences of the specific heat(L = 8). The broken line corresponds to the result of the high-temperature expansion. "EXT" denotes the extrapolation with n, as given in the text.

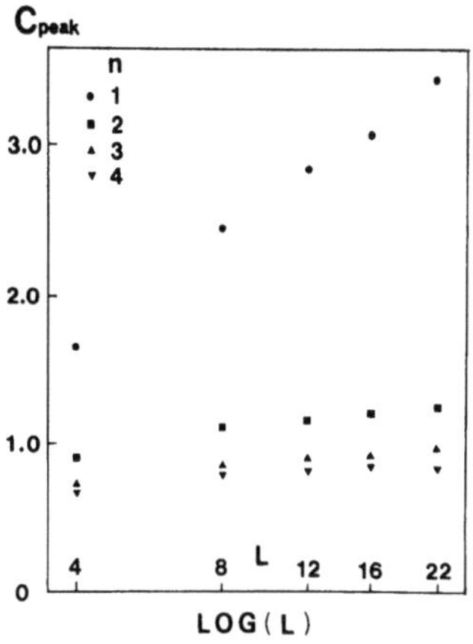

Fig. 4 Size dependence of peak values of the specific heat.

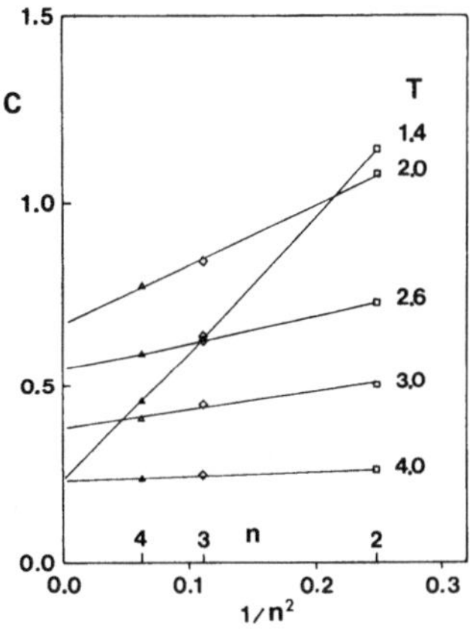

Fig. 5 Extrapolation $n \to \infty$ of the specific heat (L = 8).

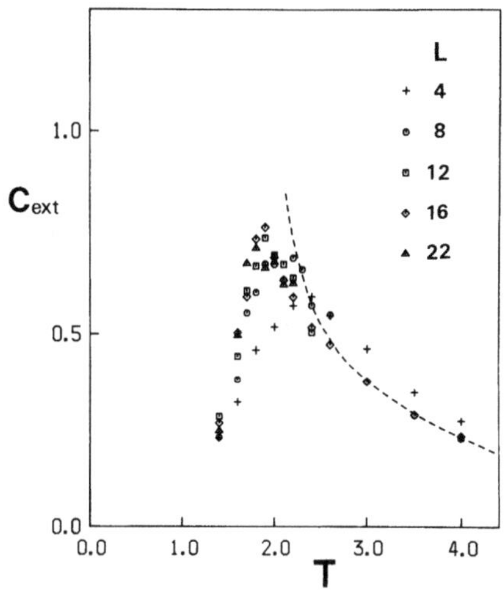

Fig. 6 Size and temperature dependences of the specific heat of the two-dimensional spin-1/2 XY model.

gence, although of course we cannot exclude the existence of a very weak amplitude of divergence.

Now, in order to discuss the quantitative properties of the original quantum system, we should take the limit $n \to \infty$. Figure 5 shows the n-dependence of the data. It is generally stated that the n-dependence of the data should be given [24] by an even function of n such as

$$A(n) = A(\infty) + a/n^2 + b/n^4 + c/n^6 + \cdots , \qquad (3.2)$$

where A is an observable. Fortunately, the present data in Fig. 5 show good behavior for estimation of the limiting values by $1/n^2$ extrapolation of the data with n = 2, 3 and 4. In Fig. 3 we plot the extrapolated values of the specific heat which are denoted by "EXT". At high temperatures the present results are in good agreement with those of the high-temperature expansion. Finally, the extrapolated specific heat, i.e. the specific heat of the quantum XY model, is shown in Fig. 6. From this figure, we may conclude that the specific heat is non-divergent and the peak value is estimated as 0.70 ± $0.05k_B$ at T_p = 1.8-2.0J from the data of the size up to L = 22.

4. The Order Parameter

In the present investigation we take the x-axis for the diagonalization axis so that the fluctuation in the xy plane can be studied. Because of the non-commutativity of the Hamiltonian and the order parameter M_x or M_y, $[H, M_x] \neq 0$, the square of in-plane magnetization and the in-plane susceptibility are different from each other. We define their approximants as

$$\langle M_x{}^2\rangle = \langle (\Sigma_i \sigma_i{}^x)^2\rangle \simeq \langle (\Sigma_r \sigma(r,1))^2\rangle_{MC} \qquad (4.1)$$

and

$$\chi^{xx} = \mathrm{Tr} \int_0^\beta e^{(\lambda - \beta)H} M_x e^{-\lambda H} M_x \, d\lambda / Z$$

$$\simeq (\beta/n^2)\langle (\Sigma_{r,1} \sigma(r,1))^2\rangle_{MC} , \qquad (4.2)$$

where $\sigma(r,1)$ denotes an Ising variable (±1) on the (d+1)-dimensional lattice and r and 1 denote positions in the original d-dimensional space and in the Trotter direction, respectively. Here $\langle\cdots\rangle$ and $\langle\cdots\rangle_{MC}$ denote the canonical average and an average over Monte Carlo samples, respectively, and Z denotes the partition function.

Figure 7 shows the n dependence and temperature dependence of $\langle M_x{}^2\rangle/N$ where $N = L^2$. The data with n=1 are well understood from the analytic result of the n=1 approximant, where both low- and high-temperature phases are paramagnetic and there exists a critical point only at the Onsager critical temperature. Numerical analysis of the size dependence of this quantity reproduces the previous exact result that $\eta = 1/2$. Next we proceed to the case $n \geq 2$. Therein we observe a certain intermediate temperature region where the spin correlation seems to be well developed, while outside the region such ordering is suppressed. Recall here the remarkable feature of the specific heat, i.e. the double peak, discussed in the preceding section. Studying the behavior of the specific heat and the squared magnetization, we are tempted to propose the following picture about the nature of the n-th approximant ($n \geq 2$). The systems with $n \geq 2$ have two phase transitions T_{c1} and T_{c2}: For $T > T_{c2}$ the systems are paramagnetic, for T_{c2}

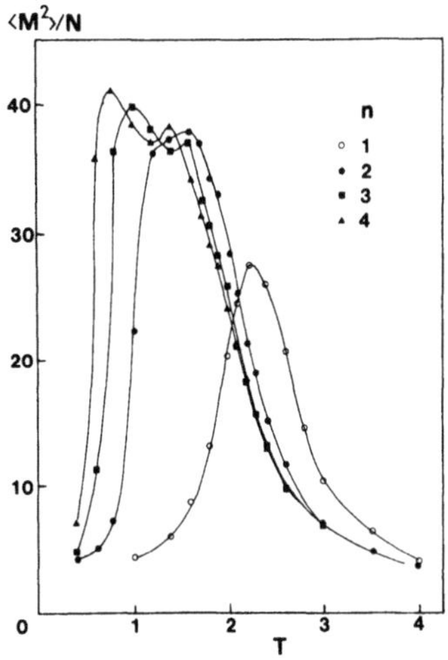

Fig. 7 Temperature and n dependences of the squared magnetization (L = 8).

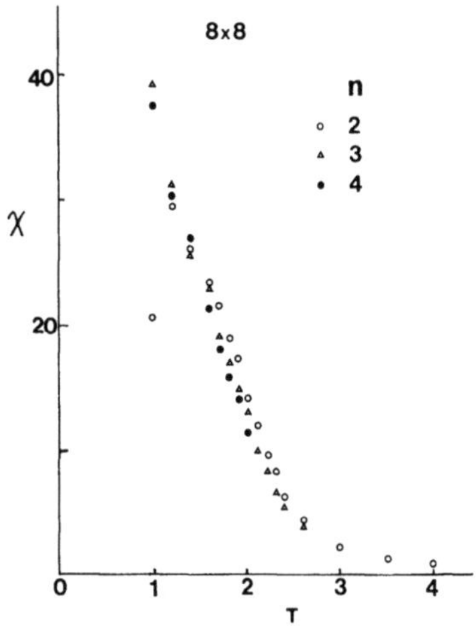

Fig. 8 Temperature and n dependences of the susceptibility.

> T > T_{c1} they show a massless phase corresponding to the XY model, and for $T < T_{c1}$ they show a low-temperature phase analogous to the paramagnetic phase derived in the case n = 1. Although it is thus interesting to study not only the nature of the massless phase but also that of the low-temperature phase, here in the present investigation we focus upon the high-temperature transition-point T_{c2} from the paramagnetic phase to the massless phase.

Figure 8 shows the n dependence ($n \geq 2$) of susceptibility data. Contrary to the case of the specific heat, the n dependence is rather small. Thus we use the data with n = 4 for the limiting value. In Fig. 9, the size dependence of data is plotted. It is very difficult to determine the critical temperature from the data in Fig. 9. In the classical case one may use the Kosterlitz-Thouless criterion: $\eta(T_c) = 1/4$, which has been supported by a Monte Carlo renormalization analysis [9]. In the quantum case we cannot use this criterion a priori. One of the methods to analyze the data is to try to use finite-size scaling. To perform the finite-size scaling plot, the values of T_c and exponents are necessary. We adopt, then, the data of high-temperature expansion [10]: $T_c = 1.57$, $\gamma = 2.5$ and $\nu = 1.43$. In Fig. 10, the data are plotted in $(L^{1/\nu}\,\varepsilon,\ L^{-\gamma/\nu})$. This plot yields a scaling form, which suggests that our Monte Carlo data are consistent with the result of high-temperature expansion. This means that both the results agree qualitatively with each other concerning the global shape of the susceptibility in the thermodynamic limit, so we may conclude that the critical temperature is close to 1.57. It does not, necessarily, exclude the existence of a Kosterlitz-Thouless type transition.

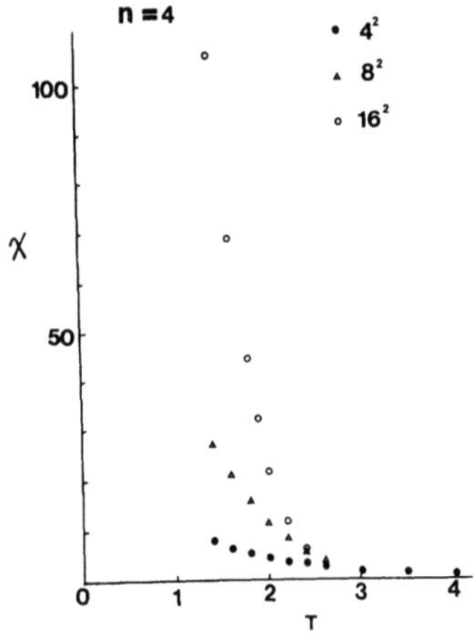

Fig. 9 Size and temperature dependences of the susceptibility (n = 4).

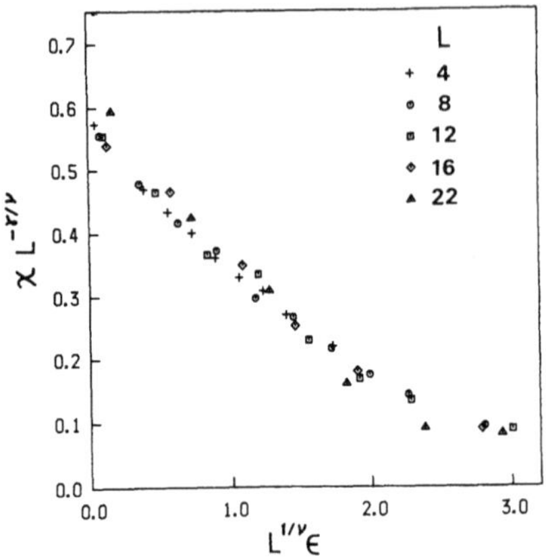

Fig. 10 Scaling plot of the susceptibility. The values of T_c and exponents used in ref. 10 are used, and $\varepsilon = (T - T_c)/T_c$.

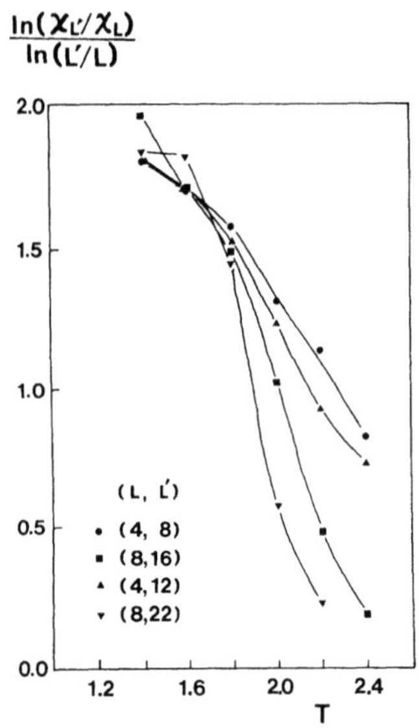

Fig.11 A plot of data in Fig. 9 according to the phenomenological renormalization group method[23].

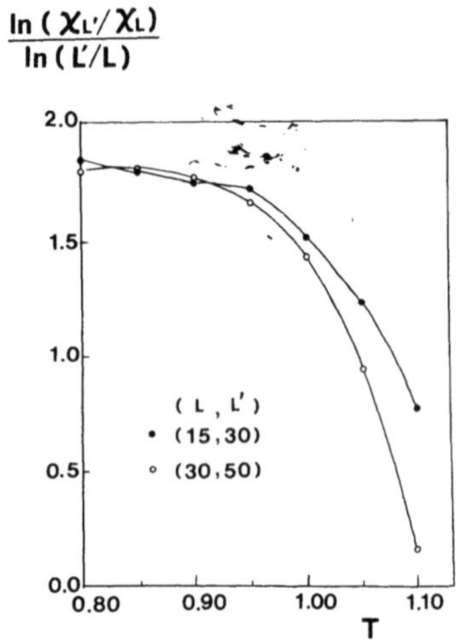

Fig. 12 A plot similar to Fig. 11 for the corresponding classical model (from the data in ref. 8).

Without any a priori assumption, we have also analyzed the data based on the the phenomenological renormalization group method [23] (see Fig. 11). In this plot a cross point gives the critical temperature and the value of $\eta(T_c)$. From Fig. 11, we estimate that

$$T_c \simeq 1.6 \quad \text{and} \quad \eta \simeq 0.3 \,, \tag{4.3}$$

which are consistent with the previous results.

For reference, in Fig. 12, we give the corresponding classical data where we find a similar behavior at the higher-temperature side and we obtain a critical line below the critical temperature. In Fig. 11, scattering of data at the lower-temperature side is large, and it is difficult to determine the behavior of the critical line accurately. Both the figures are, however, qualitatively similar. This suggests some similarity of the quantum XY transition to the classical one. Much more detailed analyses for the size-dependence of the susceptibility are necessary to determine the type of divergence.

5. Summary and Discussion

In the present article thermodynamic properties of the two-dimensional spin-1/2 XY model are studied by the quantum Monte Carlo method introduced by Suzuki [13] with pair decoupling. The specific heat and susceptibility are given in Figs. 6 and 9, respectively. The critical temperature is estimated to be close to 1.6 and non-divergence of the specific heat is concluded. The nature of phase transition, namely whether a type of Kosterlitz-Thouless transition or an ordinary power law type of transition occurs, is still an open problem.

So far we have investigated mainly the nature of the system at the higher temperature side. On the other hand, as was mentioned in the introduction, the nature of lower temperature phase of the present quantum system is also a very interesting problem. FUJIKI and BETTS [12] have proposed that the value of the in-plane correlation exponent η remains nonzero ($\simeq 0.12$) even in the ground state. Thus it is interesting to investigate temperature dependence of η at low temperatures by the present method. We may expect a kind of reentrant behavior, in other words, non-monotonic behavior in $\eta(T)$ due to the quantum fluctuation which has been emphasized in excess for the case with $n = 1$ [20]. With respect to this point, the nature of phases at low temperatures of systems with $n \geq 2$ is very interesting for future study.

Acknowledgements

Numerical calculations were done by HITAC M680H at University of Tokyo and FACOM M190 of the Laboratory of the International Collaboration on Elementary Particle Physics of University of Tokyo, which should be sincerely acknowledged. One of the authors (S. M.) thanks Professor D. D. Betts and Dr. S. Fujiki for stimulating discussions and for providing data prior to publication. The present work is partially supported by a Grant-in-Aid from the Ministry of Education, Science and Culture.

References

1. D.D. Betts: in Phase Transitions and Critical Phenomena, Vol.3, ed. by C. Domb and M.S. Green (Academic Press, N.Y. 1974)
   D.D. Betts: Physica 86-88B, 556 (1977)
2. D.R. Nelson: in Phase Transitions and Critical Phenomena, Vol.7 ed. by C. Domb and J.L. Lebowitz (Academic Press, London, 1983) p.1
3. R.E. Peierls: Helv. Phys. Acta 7, 81 (1934)
   F. Bloch: Z. Phys. 61, 206 (1930)
   N.D. Mermin and H. Wagner: Phys. Rev. Lett. 17, 1133 (1966)
   P.C. Hohenberg: Phys. Rev. 158, 383 (1967)
4. H.E. Stanley and T.A. Kaplan: Phys. Rev. Lett. 17, 913 (1966)
   G.S. Rushbrooke and P.J. Wood: Mol. Phys. 1, 257 (1958)
5. K. Yamaji and J. Kondo: J. Phys. Soc. Jpn. 35, 25 (1973) and Phys. Lett. 45A, 317 (1973)
   A.A. Belavin and A. M. Polyakov: JETP Lett. 22, 245 (1975)
   A.A. Migdal: Sov. Phys.-JETP 42, 743 (1976)
   D.S. Fisher and D.R. Nelson: Phys. Rev. B16, 2300 (1977)
   S.H. Shenker and J. Tobochnik: Phys. Rev. B22, 4462 (1980)
6. J.M. Kosterlitz and D.J. Thouless: J. Phys. C6, 1181 (1973)
   J.M. Kosterlitz: J. Phys. C7, 1046 (1974)
7. J.V. Jose, L.P. Kadanoff, S. Kirkpatrick and D.R. Nelson:Phys. Rev. B16, 1217 (1977)
8. S. Miyashita, H. Nishimori, A. Kuroda and M. Suzuki: Prog. Theor. Phys. 60, 1669 (1978); 63, 797 (1980)
   J. Tobochnik and G.V. Chester: Phys. Rev. B20, 3761 (1979)
9. S. Miyashita: Prog. Theor. Phys. 65, 1181 (1981)
10. J. Rogiers, E.W. Grunke and D.D. Betts: Can. J. Phys. 57, 1719 (1978)
11. J. Oitmaa and D.D. Betts: Can. J. Phys. 56, 897 (1978)
    M. Suzuki and S. Miyashita: Can. J. Phys. 56, 902 (1978)
12. S. Fujiki and D.D. Betts: Can. J. Phys. 64, 876 (1986)
13. M. Suzuki: Commun. Math. Phys. 51, 183 (1976) and 57, 193 (1977)
    M. Suzuki: Prog. Theor. Phys. 56, 1454 (1976)
    M. Suzuki: Phys. Rev. B31, 2985 (1985)
    M. Suzuki: J. Stat. Phys. 43, 883 (1986) [review article]
    M. Suzuki: article in this volume
14. M. Suzuki, S. Miyashita and A. Kuroda: Prog. Theor. Phys. 58, 1377 (1977)
15. E. Loh, Jr., D.J. Scalapino and P.M. Grant: Phys. Rev. B31, 4712 (1985)
16. H. De Raedt and A. Lagendijk: Phys. Rep. 127, 233 (1985)
17. H. De Raedt, B. De Raedt and A. Lagendijk: Z. Phys. B57, 209 (1984)
18. H. De Raedt and A. Lagendijk: Phys. Rev. B33, 5102 (1986)
    E. Loh, Jr., D.J. Scalapino and P.M. Grant: Phys. Rev. B33, 5104 (1986)

19. M. Marcu: article in this volume
20. T. Onogi, S. Miyashita and M. Suzuki: J. Stat. Phys. 45, 183 (1986)
21. H.N.W. Temperley and E.H. Lieb: Proc. Roy. Soc. (London) A222, 251 (1971)
R. J. Baxter: J. Phys. C6, L445 (1973)
22. S. Elitzur, R.B. Pearson and J. Shigemitsu: Phys. Rev. D19, 3698 (1978)
J.L. Cardy: J. Phys. A13, 1507 (1982)
23. M.N. Barber: in Phase Transitions and Critical Phenomena, Vol.8, ed. C. Domb and J.L. Lebowitz (Academic Press, London, 1983) p.145
24. M. Suzuki: Phys. Lett. 113A, 299 (1985)

Critical Properties of Spin Chains

S. Takada

Institute of Physics, University of Tsukuba, Sakura-mura,
Niihari-gun, Ibaraki 305, Japan

The S^x-S^x correlation functions of S=1/2 and 1 XXZ chains are numerically investigated at low temperatures by means of the transfer matrix with use of the Trotter approximation. It is shown that extrapolation to $N=\infty$ works quite effectively and makes the estimation of the critical indices ν, γ and η possible. In the case of S=1/2 the obtained values of ν, γ and η agree quite well with the exact values obtained by Luther and Peshel. Haldane's conjecture for integer spins is confirmed in the case of S=1.

1. INTRODUCTION

The numerical method of quantum statistical mechanics, which was proposed by Suzuki based on the Trotter formula[1], has been successfully applied to many quantum systems[2]. In this talk I would like to show the effectiveness of this method, taking a one-dimensional (1-D) XXZ model as an interesting example [3 - 5]. The Hamiltonian is described by

$$H = -\frac{1}{S^2}\Sigma\ (S_i^x S_{i+1}^x + S_i^y S_{i+1}^y + \Delta S_i^z S_{i+1}^z) \quad . \tag{1}$$

The reasons why we find this model interesting and important in the application of Suzuki's method are the following.

1) The problem in 1-D is naturally the simplest case to which the numerical method can be applied and hence this method enables us to treat this model most precisely. In addition there exists the exact solution for S=1/2 with which we can compare our results.

2) In this model the absolute zero temperature can be regarded as the critical point for the X-Y phase around $\Delta=0$ and hence the critical phenomenon appearing here is of fully quantum nature: The critical indices $(\sigma, \eta, \nu, \gamma)$ are defined as follows:

$$f(r) = \langle S_r^x S_0^x \rangle = \left(\frac{T}{T_0}\right)^{\sigma} \phi\,(r/\xi)$$

$$\propto \begin{cases} e^{-r/\xi} & r \gg \xi \\ r^{-\eta} & r \ll \xi\ , \end{cases} \tag{2}$$

$$\xi \propto T^{-\nu}\ , \tag{3}$$

and

$$\chi = \sum_{r=\infty}^{\infty} \int_0^{\beta} \langle S_r^x(\tau)\ S_0^x \rangle\, d\tau \propto T^{-\gamma}\ . \tag{4}$$

They are related to the scaling laws

$$\sigma = \nu\eta, \quad (5)$$

$$\gamma = \nu(1 - \eta) + 1 . \quad (6)$$

The second equation corresponds to the two-dimensional scaling law in usual critical phenomena. This is because the imaginary time direction or the Trotter direction appears in the infinite size at OK. As is well known, determination of the critical indices is the most delicate problem in the investigation of critical phenomena. Therefore it seems to be interesting to test the numerical method in the fully quantum critical phenomena by calculating critical indices.

3) This model is physically interesting regarding the nature of ground states: Haldane conjectured that the phase diagram of the ground states for integer spins is quite different from that for half integer spins. In the case of half integer spins there exists the X-Y singlet phase or Kosterlitz-Thouless (KT) phase with gapless excitations and the S^x-S^x correlation behaves as a power law:

$$f(r) = \langle S_r^x S_0^x \rangle \sim r^{-\eta} \quad (7)$$

for larger r. On the other hand, Haldane conjectured on the basis of the equivalence of the antiferromagnet (Δ=-1) to the nonlinear σ model in the limit of infinite S that a novel singlet ground state should appear around the antiferromagnetic point for integer S. His conjecture is as follows: For $1 > \Delta > \Delta_1$ ($0 > \Delta_1 > -1$) the ground state is the XY singlet with a gapless excitation while for $\Delta_1 > \Delta > \Delta_2$ ($-1 > \Delta_2$) the ground state is a singlet with an excitation gap which never exists for half integer spins. At $\Delta = \Delta_2$ the excitation gap opens again from the Néel doublet ground state for $\Delta < \Delta_2$.

We calculate the S^x-S^x correlation function and the susceptibility for for S=1/2[3] and 1[4]. We employ transfer matrix in the spatial direction to obtain the trotter approximations[3],[5]. We are therefore free from statistical errors which necessarily accompany the Monte Carlo calculations, and thus are able to calculate the approximants numerically but exactly. Another merit of this method is that we can treat sufficiently large systems which are practically regarded as in the thermodynamic limit except at very low temperatures. On the other hand this method has a demerit that a large memory size is needed in the calculation since the vector space increases as $(2S+1)^{2N}$ with the Trotter size N. We are able to calculate only up to N=9 for S=1/2, N=6 for S=1 and N=5 for S=3/2 with our computer facilities (HITAC S810).

2. METHOD

We divide the Hamiltonian (1) into H_1 and H_2

$$H_1 = \sum_{i=\mathrm{odd}} h_i , \quad H_2 = \sum_{i=\mathrm{even}} h_i , \quad (8)$$

$$h_i = -\frac{1}{S^2}(S_i^x S_{i+1}^x + S_i^y S_{i+1}^y + \Delta S_i^z S_{i+1}^z) . \quad (9)$$

Then we approximate a thermal quantity by replacing

$$e^{-\beta H} \to (e^{-\frac{\beta H_1}{N}} e^{-\frac{\beta H_2}{N}})^N . \quad (10)$$

Then a chain with L quantum spins is mapped onto a 2D system with 2NL classical spins $\{S_{ij}\}$, which is represented as a checker board lattice[3]. We adopt S^x-representation for convenience of calculation of the S^x-S^x correlation function.

Now we introduce the transfer matrix T_i in the spacial direction which transforms $|\psi_i\rangle$ to $|\psi_{i+1}\rangle$ where $|\psi_i\rangle$ denotes a vector of $(2S+1)^{2N}$ components specified by

$$S_i = (S_{i,1}, \dots S_{i,2N}):$$

$$\langle S_i|T_i|S_{i+1}\rangle = \prod_j \langle S_{i,j+1}\ S_{i+1,j+1}|e^{-\frac{\beta}{N}h_i}|S_{i,j}\ S_{i+1,j}\rangle , \qquad (11)$$

where $T_i = T_1$ for i=odd and $T_i = T_2$ for i=even. Thus the Nth approximent of the partition function is given by

$$Z_N = \langle\psi_1|\prod_{i=1}^{L-1} T_i|\psi_L\rangle , \qquad (12)$$

where $|\psi_1\rangle$ and $|\psi_L\rangle$ are determined by the free boundary condition[3]. Similarly, the correlation function $f_N(i,j)$ and the canonical correlation function $g_N(i,j)$ are calculated by

$$f_N(i,j) = \frac{1}{2N}\sum_{\{S_{k,\ell}\}}\{\langle\psi_1|T_1|S_2\rangle\prod_{i=2}^{L-2}\langle S_i|T_i|S_{i+1}\rangle\langle S_{L-1}|T_{L-1}|\psi_L\rangle$$
$$\times\ (\frac{1}{4N^2}\sum_{m,n=1}^{2N} S_{i,m}S_{j,m})\}, \qquad (13)$$

$$g_N(i,j) = \frac{1}{2N}\sum_{\{S_{k,\ell}\}}\{\langle\psi_1|T_1|S_2\rangle\prod_{i=1}^{L-2}\langle S_i|T_i|S_{i+1}\rangle\langle S_{L-1}|T_{L-1}|\psi_L\rangle$$
$$\times\ (\frac{1}{4N^2}\sum_{m,n=1}^{2N} S_{i,m}\ S_{j,m})\}. \qquad (14)$$

Employing the transfer matrix we can calculate correlation functions of sufficiently long chains as the computing time only increases linearly with the size. In practice we have calculated $f_N(\ell, \ell+r)$ and $g_N(\ell, \ell+r)$ of a chain with the size $L=2\ell+r$ where the size of the margins ℓ has been chosen to be larger than $4\xi(T)$ ($\xi(T)$ is the coherence length). Accordingly, the correlation functions are regarded as the those of infinite systems.

Next we estimate $\xi_N(T)$ and $C_N(T)$ and $\chi_N(T)$ as

$$f_N(r) = \exp[-C_N(T) - r/\xi_N(T)] , \qquad (15)$$

$$\chi_N(T) = g_N(0) + 2\sum_{r=1}^{r_0} g_N(r) + \Delta\chi_N , \qquad (16)$$

where r_0 is the maximum distance where $g_N(r)$ is calculated and $\Delta\chi_N$ is contribution from $r > r_0$, which we estimate using the scaling form:

$$\Delta\chi_N = 2g_N(r_0)[e^{\frac{1}{\xi(T)}} - 1]^{-1} . \qquad (17)$$

Next we estimate the extrapolated value of $\xi_N(T)$, $C_N(T)$ and $\chi_N(T)$ for the infinite N fitting ℓ_0 approximants: For example

$$\xi_N(T) = \xi(T) + \sum_{\ell=1}^{\ell_0-1} a_\ell \frac{1}{N^{2\ell}} \quad . \tag{18}$$

In practice we utilize expansions up to N^{-6}. Then critical indices are determined according to

$$\gamma = \frac{d \ln \chi}{d \ln \beta}, \quad \nu = \frac{d \ln \xi}{d \ln \beta} \text{ and } \eta = \frac{d \ln C}{d \ln \xi} \quad , \tag{19}$$

using a least squares fit.

3. RESULT FOR S=1/2

3.1. Scaling Property

Since we use the scaling form of the correlation function $f_N(r)$ given by (14), we first examine the scaling property expressed by

$$\tilde{\phi}_N(r/\xi_1) = e^{C_N + r/\xi_N} f_N(r) \quad . \tag{20}$$

We show the scaling function $\tilde{\phi}_9(x)$ for Δ=0 and N=9 in Fig.1. Data of six β from β=3 to β=8 are ploted. We find in this figure that $f_N(r)$ beautifully shows the assumed scaling behaviour given by (20). In particular the long range behaviour shows the simple exponential decay.

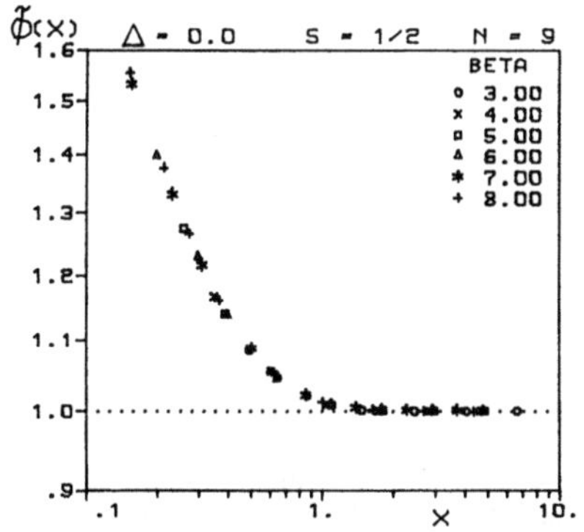

Fig.1. Data of the scaling function

3.2 Susceptibility χ and Correlation Length ξ

Data of χ at Δ=0 are plotted in Fig.2. In this figure EXT $N_1, N_2, \ldots N_m$ denotes the extrapolation obtained from N_1, N_2 ... and N_m approximants by fitting the original data up to $N^{-2(m-1)}$. Generally speaking, the temperatures can be classified into three regions according to the convergence of data. At sufficiently high temperatures ($\beta < \beta_1$) all data coincide with each other very well in region I. We can regard original data with N=9 as well as N=7,8 in the limit $N \to \infty$ without extrapolation. In intermediate temperatures ($\beta_1 < \beta < \beta_2$) all extrapolated results agree with each other, although original data deviate from each other as well as from extrapolated ones. As the temperature is further decreased ($\beta > \beta_2$), finally all data deviate from each other, suggesting that the extrapolation to $N=\infty$ is not reliable any more within the present Trotter stage ($N \leqq 9$).

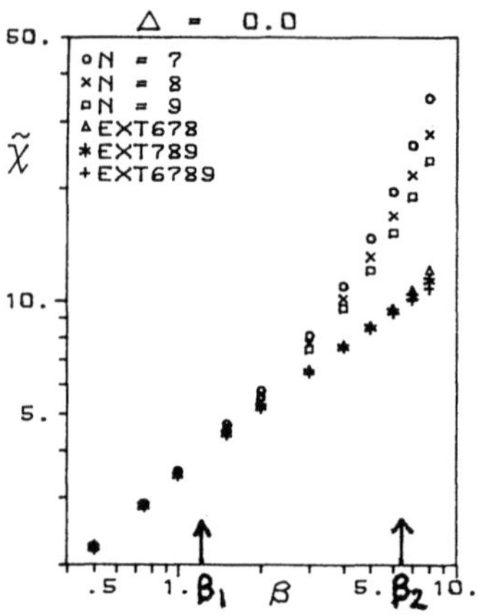

Fig.2. Data of χ at Δ=0.

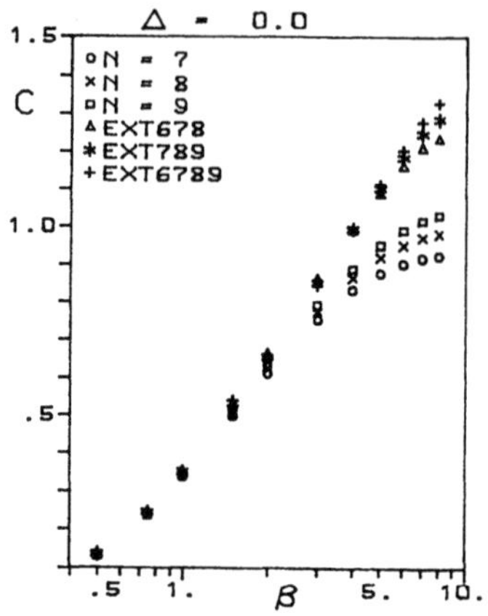

Fig.3. Data of C at Δ=0.

It should be noted that even in the case where the difference between original and extrapolated data is rather large, the convergence of the extrapolations is fairly good in region II. For example, we find in Fig.2 that the original data are about twice the extrapolations at $\beta \approx \beta_2$. This means that the extrapolation works very well.

Data for ξ show similar behaviour, while on the other hand C increases with increasing N, contrary to the cases of χ and ξ as is shown in Fig.3. Thus, the quantum effect increases C. In addition the convergence gets worse as β increases compared with data of χ and ξ.

3.3 Determination of Critical Indices ν, γ and η

Data of $d\log\xi/d\log\beta$ at Δ=0 are plotted in Fig.4 [3].

In principle critical indices should be determined in the limit $\beta \to \infty$ ($T \to 0$). Practically, however, we must determine those for $\beta \leq \beta_2$ where different extrapolations agree with each other. Thus, if data shows a stationary behaviour around $\beta = \beta_2$, we estimate it as the critical index. It should be noted in Fig.4 that extrapolated data (EXT 89) using only N=8 and N=9 deviate from other extrapolations around β=4 where other extrapolations still converge with each other. This shows that EXT 89 by fitting data of N=8 and N=9 up to N^{-2} is not sufficient and higher fitting beyond N^{-2} works effectively. Thus, excluding EXT 89, we find $\beta_2 \simeq 5$ and estimate ν as ~ 1 from the stationary value around $\beta_2 \simeq 5$. It is interesting to note that only EXT 6789 remains at the stationary value around β=6 where other extrapolations deviate from the stationary value. It suggests that higher-order fitting up to N^{-6} works effectively.

In this way we can estimate ν for $-1.0 \lesssim \Delta \lesssim 0.4$. However, the stationary region changes to the low temperature side and the convergence between extrapolations gets worse as Δ increases. Thus at Δ=0.5 we cannot get a stationary behaviour for $\beta \leq \beta_2$. In this case only EXT 6789 up to N^{-6} fitting shows stationary behaviour for $\beta > \beta_2$ and hence we estimate ν from the stationary value, giving $\nu \approx 1$. For $\Delta \gtrsim 0.6$ we cannot estimate ν since no extrapolation shows a stationary behaviour any more.

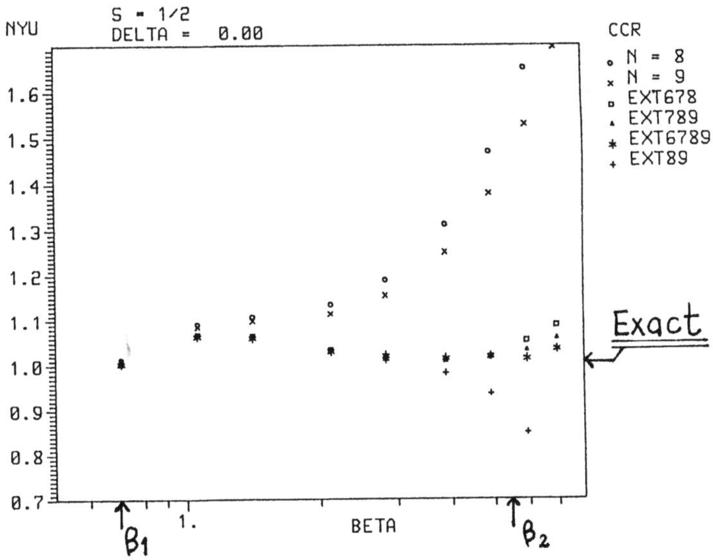

Fig.4. Data of $d\log\xi/d\log\beta$ ($=\nu$): This graph shows that EXP 89 up to N^{-2} is not sufficient.

Data of $d\ln\chi/d\ln\beta$ at $\Delta=0$ are plotted in Fig.5. In this graph we find that even for $\beta>\beta_2\approx 4$,EXT 6789, which is fitted up to N^{-6}, shows stationary behaviour for the region between $\beta\approx\beta_2$ and $\beta=8$ where the other two extrapolations deviate rapidly from stationary behaviour. This stationary value gives $\gamma\approx 1.5$ which is the exact value. In this way we can determine γ for $-0.75\lesssim\Delta\lesssim 0.5$. For $\Delta<-0.75$ the difference between original and extrapolated data becomes large and we cannot get the stationary value for $\beta<\beta_2$ since β_2 decreases with decreasing Δ. The region of β where data show stationary behaviour moves to the low temperature side so that we cannot get the stationary behaviour for $\Delta\gtrsim 0.6$.

The determination of η is more subtle than that of ν and γ, because fluctuations due to the least squares fit are rather large [3]. In this case we estimate η in the following two ways. First we get $\bar{\eta}$ by averaging $dC/d\ln\xi$ over the nearly stationary region. Secondly we simply determine it using the scaling law $\eta_S = 1-(\gamma-1)/\nu$ with the aid of estimated values of ν and γ.

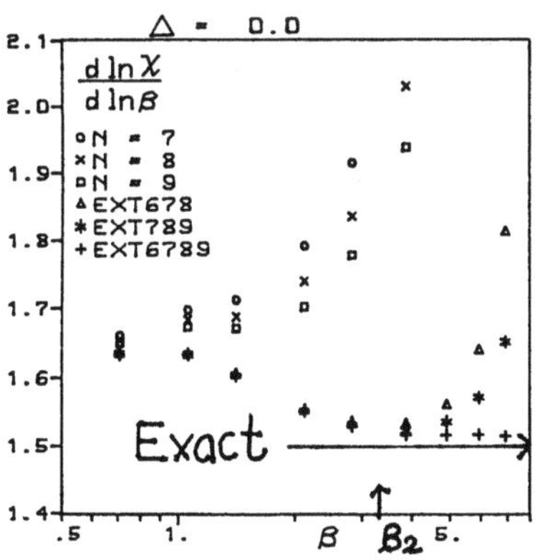

Fig.5. Data of $d\log\chi/d\log\beta=\gamma$: EXP 6789 up to N^{-6} shows the stationary behaviour.

3.4 Results of Critical Indices

In Fig.6 we plot indices ν, γ, η_s and $\bar{\eta}$ together with Luther-Peshels exact results which are drawn by dotted lines. Agreement of our results with the exact ones is quite good, and $\bar{\eta}$ and η_s conincide with each other within errors.

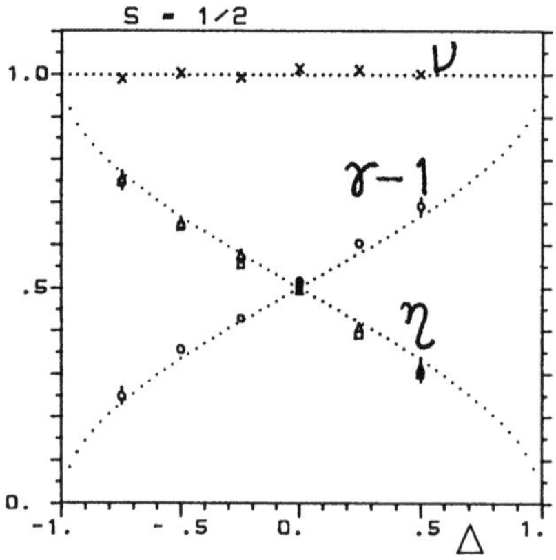

Fig.6. Results of γ, ν, $\bar{\eta}$ and η_s.

4. RESULTS FOR S=1 (HALDANE GAP)

In this section I shall discuss our results for S=1, concentrating on Haldane's conjecture [4].

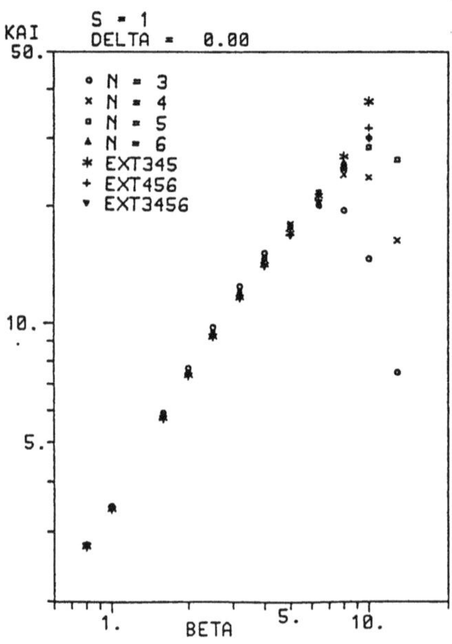

Fig.7. Data of χ at Δ=0.

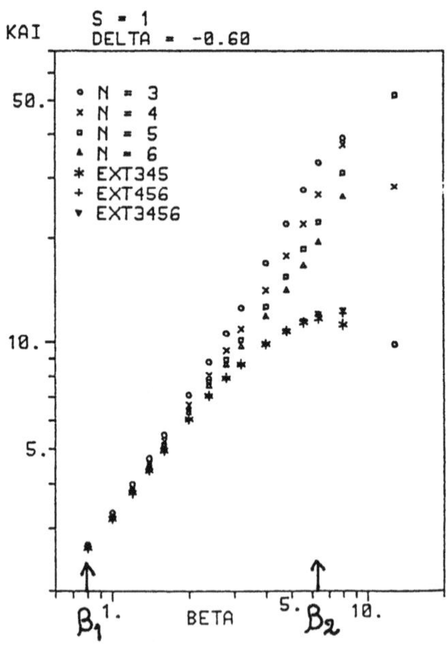

Fig.8. Data of χ at Δ=-0.6.

In Fig.7 we plot data of χ at $\Delta=0$. Compared with the case of S=1/2, fluctuations of original data suddenly begin at $\beta\approx 8$ as β increases as well as the extrapolations. The behaviour of χ at low temperatures changes for $\Delta\gtrsim -0.4$ as Δ decreases. For example, it is found in Fig.8 that at $\Delta=-0.6$ χ shows stationary behaviour at $\beta\approx\beta_2$. This means that χ does not diverge and hence the excitation gap remains at T=0. This is in striking contrast to the case of S=1/2 where χ diverges at T=0 for $|\Delta|\leqq 1$.

In Fig.9 we plot ξ^{-1} m at $\Delta=-1.0$. The mass $m_0=\xi^{-1}$ (T=0) is obtained by extrapolating ξ^{-1} for $\beta\leqq\beta_2$ to T=0 in this graph. We obtain $m_0\approx 0.12$ in this way. As Δ increases, m_0 decreases. Finally, it vanishes at $\Delta_1\approx -0.3$ although the data is not accurate enough to determine Δ_1 at which m_0 vanishes [4].

Critical indices are also roughly estimated as in the case of S=1/2. The final results are shown in Fig.10. η takes $\sim 1/4$ around $\Delta_1=-0.3$ at which m_0 vanishes. This result is consistent with the finite size scaling analysis done by Glaus and Schneider [6]. It also agrees with the theoretical prediction given by den Nijs stating that the K-T transition should occur at $\eta=1/4$ for S=1 [7].

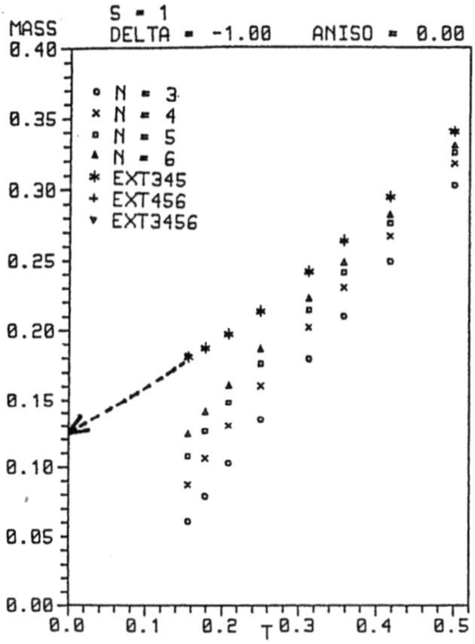

Fig.9. Data of $m_0=\xi^{-1}(0)$ at $\Delta=-1.0$. The extrapolations show the non-vanishing mass.

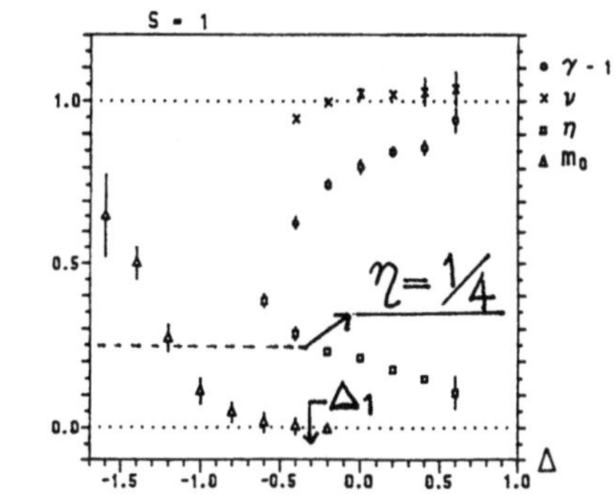

Fig.10. Results of γ, ν, η and m_0

5. CONCLUDING REMARKS

I have shown that the critical properties of the XXZ model can be investigated by Suzuki's method with the aid of the transfer matrix. I will add the following remarks before closing my talk.

1) Even critical indices can be determined by our method inspite of fully quantum nature.

2) Extrapolations to $N=\infty$ effectively work well. It is very effective to use higher order expansions beyond the N^{-2} term in the extrapolation.

3) Haldane's conjecture is confirmed.

However, the transfer matrix method has a demerit that a large memory size is needed in calculation. In fact critical indices are hardly determined around $|\Delta| \approx 1$ due to this demerit although χ and ξ can be determined rather precisely. In those cases it is desirable to obtain critical indices through Monte Carlo calculation at higher Trotter stages.

We have applied the present method to the investigation of the completely integrable model for S=1 and of XXZ model with S=3/2 as well. The detailed account will be published elsewhere.

ACKNOWLEDGEMENTS

The author is very grateful to K. Kubo and K. Saito for collaboration and useful discussions.

REFERENCES

1. M. Suzuki: Commun. Math. Phys. 51, 183 (1976); Phys. Rev. B31, 2957 (1985).
2. M. Suzuki: J. Stat. Phys. 43, 883 (1986) and references therein.
3. S. Takada and K. Kubo: J. Phys. Soc. Jpn. 55, 1671 (1986).
4. K. Kubo and S. Takada: J. Phys. Soc. Jpn. 55, 438 (1986).
5. T. Sakaguchi, K. Kubo and S. Takada: J. Phys. Soc. Jpn. 54, 861 (1985).
6. U. Glaus and T. Schneider: Phys. Rev. B30, 215 (1984).
7. M. den Nijs: Physica 111A, 273 (1982).

Monte Carlo Simulations of Three-Dimensional Heisenberg and Transverse-Ising Magnets

O. Nagai, Y. Yamada, and Y. Miyatake

Department of Physics, Faculty of Science, Kobe University, Rokkodai, Kobe 657, Japan

The three-dimensional quantum magnets are studied by use of the Monte Carlo method based on Suzuki's theory. In the calculation for a Heisenberg model, we adopted Cullen and Landau's algorithms. We find that the Monte Carlo result for a simple cubic ferromagnet is in reasonable agreement with the previous result. Also, the Monte Carlo simulations for a transverse-Ising magnet are performed by using a multispin flip or spin cluster flip method. The computed results for a ferromagnetic transverse-Ising model are satisfactory. In the case of the antiferromagnetic triangular lattice, a phase-transition-like behavior is observed at a finite value of the transverse field.

1. Introduction

The theory of quantum magnetism has been extensively developed [1] by using various mathematical techniques. Although Monte Carlo (MC) methods have been widely used [2] in the study of classical magnetism, they were not common for quantum spin models. However, since SUZUKI [3,4] proved that the Trotter expansion formula [5]

$$\exp(A+B) = \lim_{n\to\infty} [\exp(A/n)\cdot\exp(B/n)]^n$$

enables us to transform a d-dimensional quantum spin lattice into the effective (d+1)-dimensional Ising lattice, many authors tried to perform the quantum MC simulations for various spin models [6]. Thus, the MC techniques for quantum magnetism are still now being developed [7]. In this article, we describe the MC simulation results, which were obtained by ourselves, for three-dimensional spin systems, Heisenberg model and transverse-Ising (TI) model.

A reliable theoretical result for the Heisenberg chain was presented by BONNER and FISHER [8]. A comprehensive MC study for this model has been reported by CULLEN and LANDAU [9]. They proposed the algorithms of a multispin flip. That is, all spins on a string in the (1+1)-dimensional effective lattice are flipped at each MC trial. We apply Cullen and Landau's algorithms to the three-dimensional lattice and compare the result with the previous one [10]. On the other hand, the exact results for the TI linear chain have been presented by several authors [11-15]. The MC simulations based on Suzuki's theory were carried out by WIESLER [16] for the ferromagnetic TI chain. He found that the MC result does not agree with the exact one if we use the conventional single spin flip MC method for the effective (1+1)-dimensional Ising lattice. In this study, we perform the MC simulations by using a multispin flip or a spin cluster flip method and show that the MC result reasonably agrees with the exact one. In the present spin flip method, the spins on a Trotter axis compose a cluster and all possible spin states of this cluster are taken into account at each MC judgement. We apply our MC method to the two- and three-dimensional TI models [17] and compare the MC results with the high-temperature expansion results of ELLIOTT et al. [18,19].

2. Heisenberg model

The Heisenberg Hamiltonian is given by

$$H = \Sigma_{ij} H_{ij} = -\Sigma_{ij}[J_x S_i^x S_j^x + J_y S_i^y S_j^y + J_z S_i^z S_j^z + (H/2d)(S_i^z + S_j^z)], \qquad (1)$$

where H and d denote the external field and the dimension, respectively, and Σ_{ij} stands for the sum over the nearest-neighbor pairs. The magnitude of the spin is 1/2. The lattice considered in this section is confined to the hypercubic crystal. The partition function Z is written as

$$Z = \mathrm{Tr}[\exp(-H/kT)], \qquad (2)$$

where k and T denote the Boltzmann constant and temperature, respectively. Following SUZUKI [3], we divide the Hamiltonian into the "even" and "odd" parts, and apply the generalized Trotter formula to the density matrix in the partition function. Further, we adopt the so-called checkerboard decomposition [20]. Then, the n-th approximant of the partition function Z_n is written as

$$Z_n = \mathrm{Tr}[\Pi_{ij} \exp(-H_{ij}/nkT)]. \qquad (3)$$

The matrix representation of the local density matrix, $\exp(-H_{ij}/nkT)$, is given by

$$\exp(-H_{ij}/nkT) = \begin{pmatrix} P(1) & 0 & 0 & P(7) \\ 0 & P(3) & P(5) & 0 \\ 0 & P(6) & P(4) & 0 \\ P(8) & 0 & 0 & P(2) \end{pmatrix}. \qquad (4)$$

Here,

$$P(1) = \exp(K_z)\cdot[\cosh R + (2h/R)\cdot\sinh R] \qquad (5)$$

$$P(2) = \exp(K_z)\cdot[\cosh R - (2h/R)\cdot\sinh R] \qquad (6)$$

$$P(3) = P(4) = \exp(-K_z)\cdot\cosh K_+ \qquad (7)$$

$$P(5) = P(6) = \exp(-K_z)\cdot\sinh K_+ \qquad (8)$$

$$P(7) = P(8) = (K_-/R)\cdot\exp(K_z)\cdot\sinh R \qquad (9)$$

where R is defined by

$$R = [(2h)^2 + (K_-)^2]^{1/2} \qquad (10)$$

and $K_z = J_z/4nkT$, $K_\pm = (J_x \pm J_y)/4nkT$, and $h = H/4nkT$.

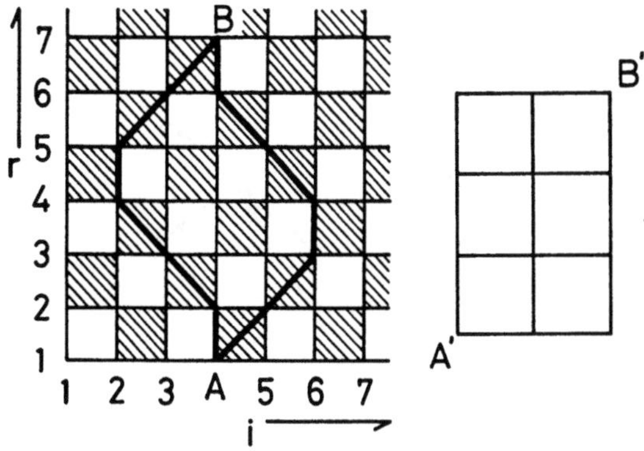

Fig.1 The (1+1)-dimensional effective lattice for the Trotter number n=3 (left figure). The number of the possible strings from A to B is equal to the number of the shortest paths from A' to B' on the right figure.

We show the effective (1+1)-dimensional lattice in Fig.1 (left figure). Here r ($r=1,2,\cdots,2n$) denotes the index of the Trotter axis and i denotes the index of the real axis. The shaded square represents the 4-spin interaction. The effective (2+1)-dimensional lattice is given by SUZUKI [3]. One may easily imagine the effective (3+1)-dimensional lattice. CULLEN and LANDAU [9] considered "zigzag" strings (bold solid lines in Fig.1) in the effective lattice and found that the MC process in which all spins on the string flip by the MC judgement gives non-zero matrix element. A zigzag string from A to B in Fig.1 is "closed" because the periodic boundary condition is assigned in the Trotter direction. The number of possible zigzag strings from A to B is equal to the number of the shortest paths from A' to B' on the right figure and the number is equal to ${}_{(2n-1)}C_n$. Thus, the number of the possible zigzag strings in the (d+1)-dimensional lattice is given by $[{}_{(2n-1)}C_n]^d$ per spin.

In order to see the accuracy of our calculation, we performed the MC simulations for the XY chain ($J_x=J_y=J$, $J_z=0$, and $H\neq0$). The calculation was done for the lattice with N=24 spins. In the present calculation for the linear chain, the Trotter size is restricted to n=4. The MC average is taken over 5000 MC steps/spin after the discard of 5000 MC steps/spin. Though our calculation includes the case $H\neq0$, the MC result for H=0 agrees with the previous calculation of CULLEN and LANDAU [9]. In Fig.2a, Fig.2b, and Fig.2c, we show the computed result for the energy, specific heat, and susceptibility, respectively. The agreement between the MC result and the exact result [11] is satisfactory.

Let us describe the MC results for the three-dimensional Heisenberg magnet ($J_x=J_y=J_z=J$ and H=0). We selected simple cubic crystal with $N=12\times12\times12$ spins. The MC calculations are performed for the cases n=1, 2, and 3. For the case n=2, the number of the possible zigzag strings is $3^3=27$ per spin. The MC average for this case is taken over 10000 MC steps/spin after the discard of 5000 MC steps/spin. The MC simulations are made with lowering the temperature gradually. We show in Fig.3a the MC results of the energy as a function of temperature for the case n=2. Also, we show the MC data of the specific heat and spin average ($\bar{s}$) for n=2 in Fig.3b. The MC data of the susceptibility for n=2 are shown in Fig.3c. The critical temperature is deduced to be $kT_c/J=0.85$ for the case n=2. In Fig.4, we show the sketch of the specific heat data (solid lines) of the ferromagnetic crystal for n=1, 2, and 3. The deduced critical temperatures are $kT_c/J=0.75$ and 0.88 for n=1 and 3, respectively. If we assume that the critical temperature is in approximation given by [21]

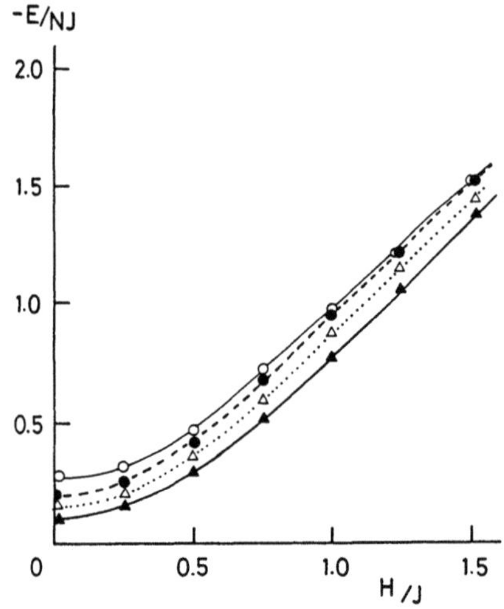

Fig.2a Magnetic field dependence of the energy in the XY chain for the case n=4. The open circle, solid circle, open triangle, and solid triangle denote MC data for kT/J=0.25, 0.50, 0.75, and 1.00, respectively. The solid, broken, dotted, and bold solid lines denote the exact results for the corresponding temperatures.

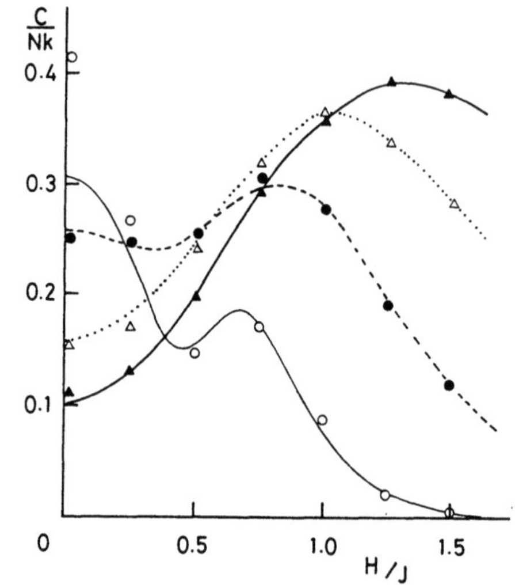

Fig.2b Specific heat data for the XY chain. For the symbols in the figure, see the caption of Fig.2a.

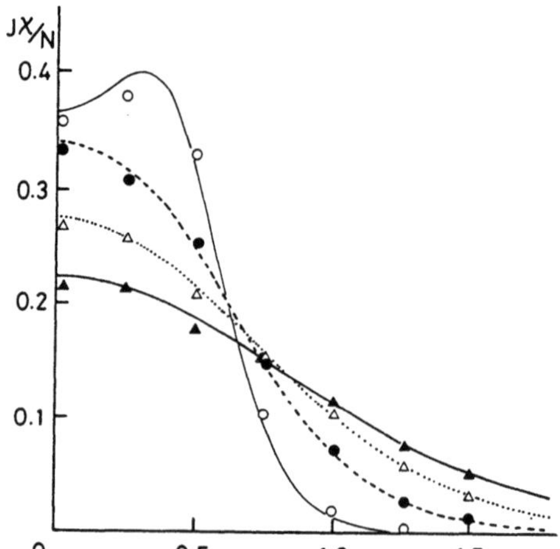

Fig.2c Susceptibility data for the XY chain. For the symbols in the figure, see the caption of Fig.2a.

$$kT_c/J = 0.908 - 0.2/n^2 , \qquad (11)$$

the critical temperature for the case n=∞ is given by 0.91. On the other hand, the high temperature expansion theory predicts $kT_c/J \approx 0.9$ [10]. Thus, the MC result is at least qualitatively satisfactory. In the case of the antiferromagnetic crystal, the convergence of the simulation is very slow compared with that for the case of ferromagnet. Hence, the MC results for the antiferromagnetic case will be not so reliable as those for the ferromagnetic case. In Fig.4, we also show the sketch of the specific heat data for the antiferromagnetic crystal (broken lines) for n=1, 2, and 3. The deduced critical temperatures are kT_c/J=1.3, 1.05, and 0.9 for the corresponding n values.

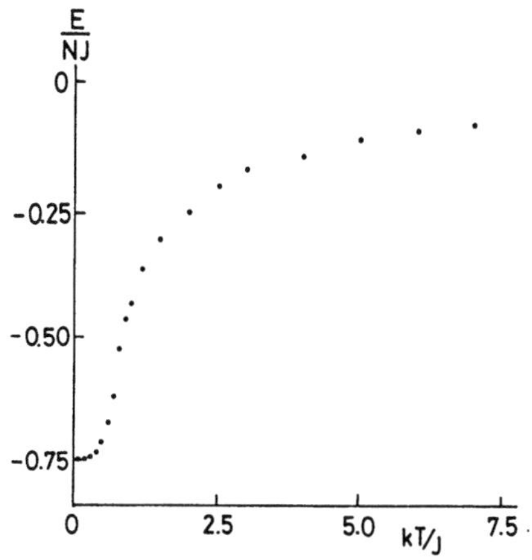

Fig.3a MC data of the energy for the simple cubic ferromagnet (n=2).

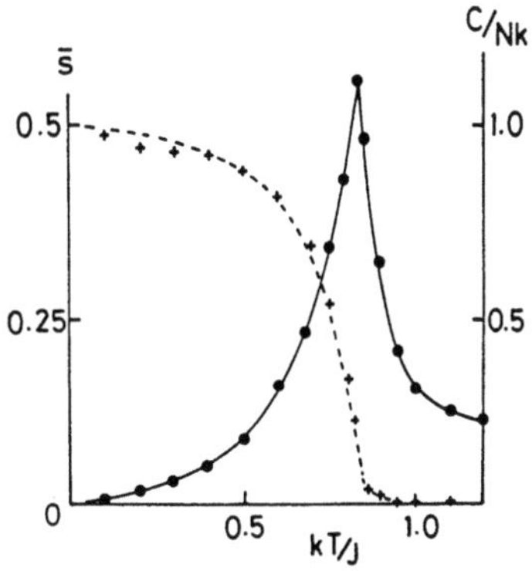

Fig.3b MC data of the spin average and specific heat for the simple cubic ferromagnet (n=2).

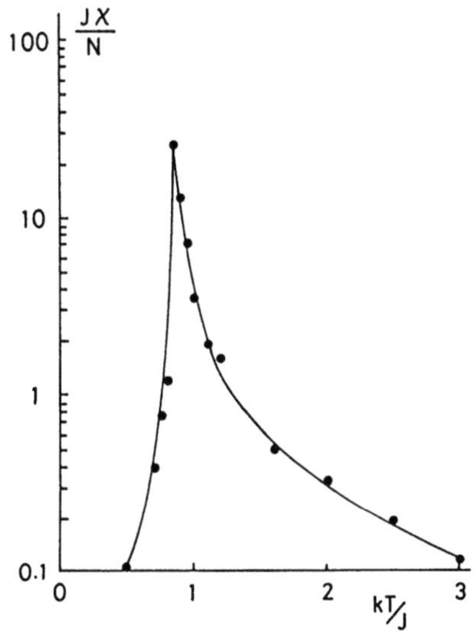

Fig.3c MC data of the susceptibility for the simple cubic ferromagnet (n=2).

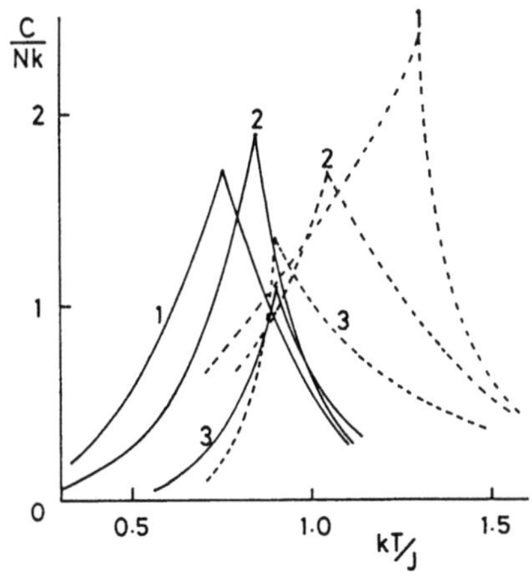

Fig.4 The sketch of the specific heat data for the ferromagnet (solid lines) and antiferromagnet (broken lines). The number on each line denotes the Trotter number.

3. Transverse-Ising Model

The Hamiltonian of the ferromagnetic TI model is written as

$$H = -J\Sigma_{ij}\sigma_i^z\sigma_j^z - \Gamma\Sigma_i\sigma_i^x , \qquad (12)$$

where σ_i^z etc are Pauli spin operators and Γ denotes the transverse field. As SUZUKI [3] has explicitly shown, the d-dimensional TI model is transformed into the (d+1)-dimensional Ising model of which the effective Hamiltonian is written as

$$H_n' = \Sigma_h [\Sigma_{ij} K_n \sigma_{ih}^z \sigma_{jh}^z + \Sigma_i (K_n' \sigma_{ih}^z \sigma_{ih+1}^z + \gamma_n \sigma_{ih}^z)] , \tag{13}$$

where $K_n = J/nkT$, $\gamma_n = \Gamma/nkT$, and $K_n' = (1/2)\ln[\cosh(\Gamma/nkT)]$. Thus, the partition function of the effective Ising system is given by

$$Z = \lim_{n\to\infty} A^{nN} \, Tr[\exp(H_n')], \tag{14}$$

where $A = [(1/2)\sinh(2\Gamma/nkT)]^{1/2}$. Using the standard formulas of thermodynamics, we obtain the physical quantities such as specific heat and susceptibility.

We show, in the following table, the numerical values of K_n/K_n' for the Trotter size n=12. As can be seen from this table, K_n' is considerably larger than K_n except for the extremely high field and low temperature case. Thus, the effective (d+1)-dimensional lattice is strongly anisotropic. That is, this lattice has a linear-chain-like character in the Trotter direction. Hence, the conventional single spin flip method will be inappropriate in the MC simulations for this lattice. In order to take

Table 1 Ratio of the effective coupling (K_n) in the real axis to the effective coupling (K_n') in the Trotter axis for the case n=12.

kT/J	Γ/J=0.0001	0.5	1.0	1.5
0.5	0.0303	0.1340	0.1851	0.2369
1.0	0.0143	0.0524	0.0670	0.0860
1.5	0.0092	0.0310	0.0384	0.0447

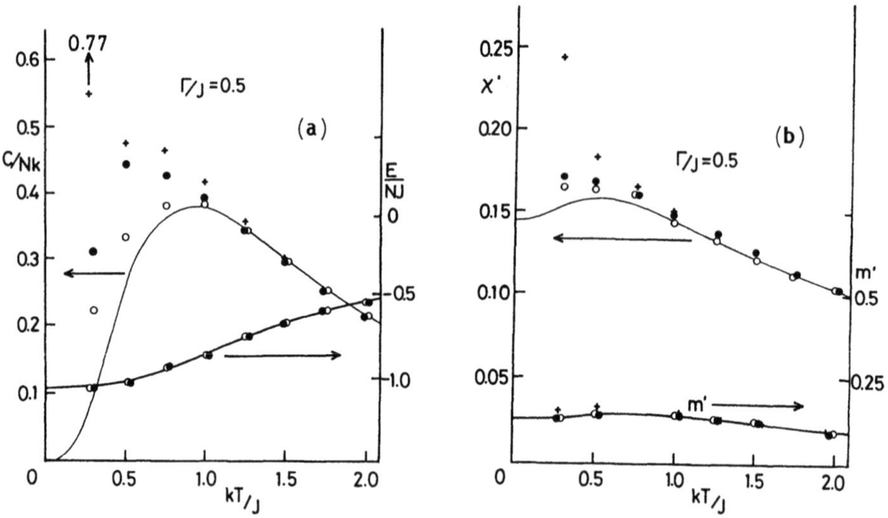

Fig.5a The energy and specific heat for the TI chain. The cross, solid circle, and open circle denote the MC data for n=4, 8, and 12, respectively. The solid lines denote the rigorous results.

Fig.5b The transverse susceptibility, $\chi' = J\chi_{xx}/N$, and the average of σ^x, m', for the TI chain.

into account all possible spin states, we use a spin cluster flip method. That is, we consider a spin cluster composed of n spins on a Trotter axis. There are 2^n energy levels for this spin cluster. Each cluster is surrounded by 2dn spins in the case of hypercubic crystal. If the spin configuration of these 2dn spins is known, the spin state of the cluster can be determined by the MC method based on the heat bath algorithms [22]. Although the heat bath algorithms are commonly used for various lattice gauge models [22,23], they are not used for spin systems except for few examples [24-27]. We confine our calculation to the Trotter size n=4, 8, and 12 in this study. The MC average was taken over 5000 MC steps/cluster after the discard of 5000 MC steps/cluster. We show the MC results for the TI chain in Fig.5a and Fig.5b for the case of $\Gamma/J=0.5$. We think that the MC result reasonably agrees with the rigorous result.

The ferromagnetic TI model for the two- or three-dimensional lattice shows a phase transition at a finite temperature, if the transverse field is smaller than the critical field Γ_c. Our MC calculation for the two- and three-dimensional lattices are performed only for the Trotter size n=12. In Fig.6a, we show the MC data of the specific heat and parallel susceptibility, $\chi=J\chi_{zz}/N$, in the case of the square lattice with N=24×24 spins. The deduced critical temperature is $kT_c/J\simeq2.2$ from the peak position of the above quantities. The average of σ^z, m, and the average of σ^x, m',

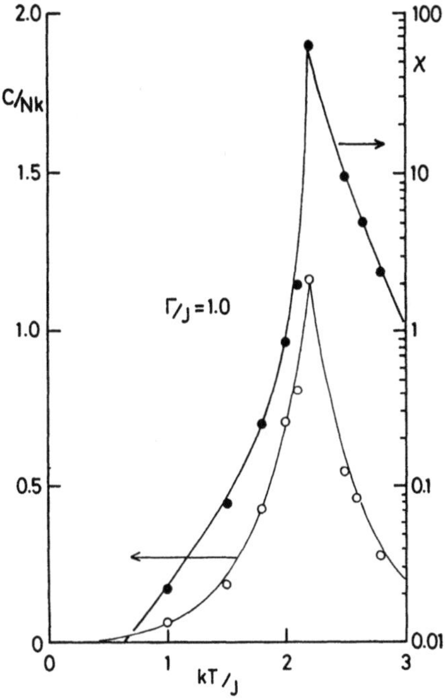

Fig.6a The specific heat and parallel susceptibility, $\chi=J\chi_{zz}/N$, for the two-dimensional ferromagnetic TI model. The MC data of the specific heat are denoted by the open circles and the MC data of the parallel susceptibility are denoted by the solid circles. The solid lines are guides for the eye.

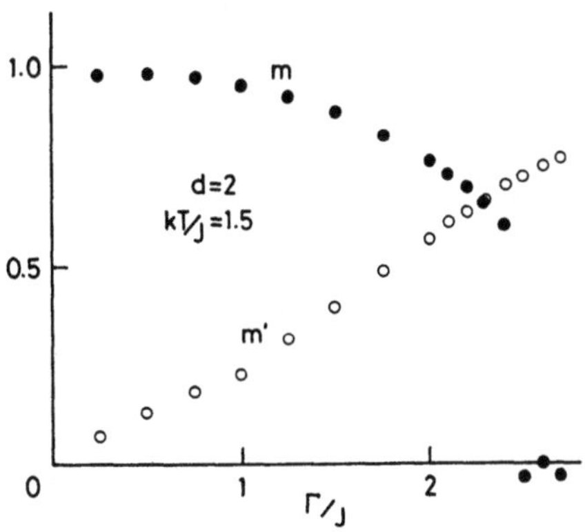

Fig.6b The average of σ^z, m, and the average of σ^x, m', for the two-dimensional ferromagnetic TI model.

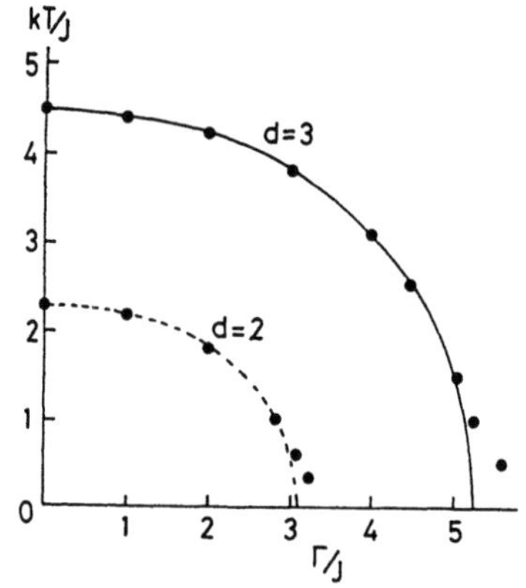

Fig.7 The critical lines for ferromagnetic TI model. The MC data are denoted by the solid circles. The solid and broken lines denote the calculations by ELLIOTT and WOOD [18].

for the square lattice are shown in Fig.6b. The obtained critical line in the T-Γ space agrees, except for low temperature case, with that predicted by PFEUTY and ELLIOTT [19], as seen in Fig.7. The MC result for the three-dimensional lattice is also summarized in Fig.7.

It will be very interesting to study the antiferromagnetic triangular (AFT) TI lattice, because the ground state degeneracy of the AFT Ising lattice is infinite and no spin ordered state exists at finite temperatures if no external field is applied. However, if a transverse field is applied, the ground state degeneracy will be considerably reduced and hence the low temperature properties of this lattice will be strongly affected. We performed the MC simulations by applying a small parallel staggered field whose magnitude is H. Here the symmetry of the staggered field is of the antiferromagnetic honeycomb lattice. The MC result for the specific heat is shown in Fig.8a. A sharp peak is observed at $kT/J \simeq 0.9$ for the case $\Gamma/J=0.75$ and $H/J=0.05$. On the other hand, the peak was not observed for the case $\Gamma/J=0.0001$ and $H/J=0.05$, as seen in Fig.8a. Hence, the peak of the specific heat is due to the transverse field. Similarly, a sharp peak is observed in the MC data of the parallel staggered susceptibility, although

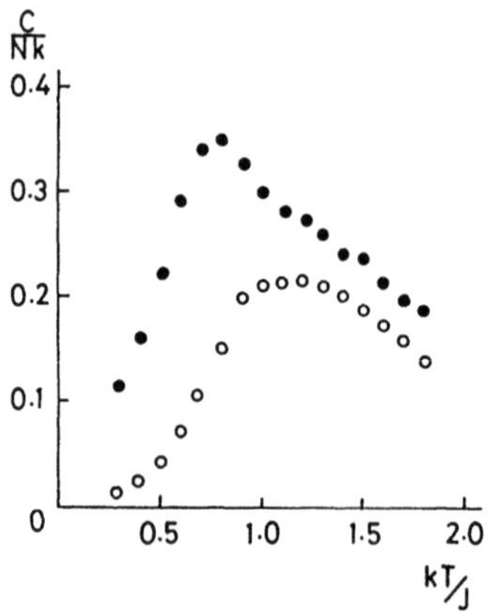

Fig.8a The specific heat for the antiferromagnetic triangular TI model. The MC data for $\Gamma/J=0.75$ are denoted by the solid circles and the MC data for $\Gamma/J=0.0001$ are denoted by the open circles.

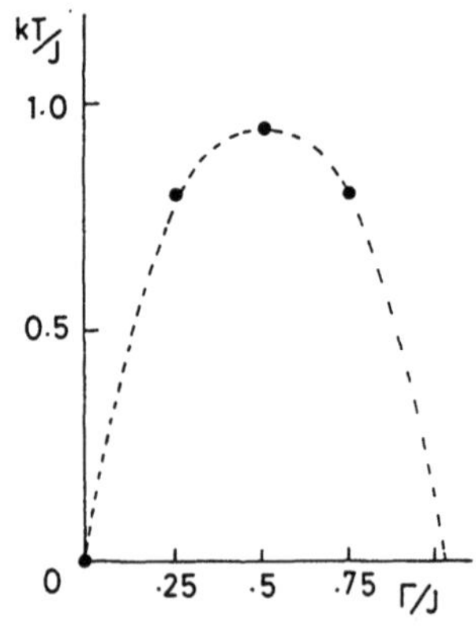

Fig.8b The supposed phase diagram (broken line) of the antiferromagnetic triangular TI model. The MC data are denoted by the solid circles.

the figure of the susceptibility is omitted in this article. These MC data seem to indicate some kind of ordering of the parallel-spin-component, which is induced by transverse field. We assume that the temperature at which the specific heat shows a sharp peak is the critical temperature. The supposed T-Γ phase diagram of the AFT TI model is shown in Fig.8b for the case H=0.

Finally, the MC calculation for the TI model where the spin magnitude is equal to 1 [28] has been performed by using the spin cluster flip method. The computed results will be reported elsewhere.

Acknowledgement

The authors would like to thank Prof. M. Suzuki and Dr. S. Miyashita of University of Tokyo for their valuable discussion. The authors are also much indebted to K. Nishino and J. J. Kim for their help in preparing the manuscript.

References

1. D. C. Mattis: The Theory of Magnetism I and II, Springer Ser. Solid-State Sci., Vol.17 and Vol.55 (Springer, Berlin, Heidelberg 1981 and 1985)
2. K. Binder: In Monte Carlo Methods in Statistical Physics, ed. by K. Binder, (Springer, Berlin, Heidelberg 1979) p. 1
3. M. Suzuki: Prog. Theor. Phys. 56, 1454 (1976)
4. M. Suzuki, S. Miyashita, and A. Kuroda: Prog. Theor. Phys. 58, 701 (1977)
5. H. F. Trotter: Proc. Am. Math. Soc. 10, 545 (1950)
6. M. Suzuki: J. Stat. Phys. 43, 883 (1986)
7. H. De Raedt and A. Lagendijk: Phys. Rep. 127, 233 (1985)
8. J. C. Bonner and M. E. Fisher: Phys. Rev. A135, 640 (1964)
9. J. J. Cullen and D. P. Landau: Phys. Rev. B27, 297 (1983)
10. G. S. Rushbrooke, G. A. Baker, Jr. and P. J. Wood: Phase Transitions and Critical Phenomena: eds. C. Domb and M. S. Green, Vol.3 (Academic Press, New York, 1972) p. 246
11. S. Katsura: Phys. Rev. 127, 1508 (1962)
12. E. Lieb, T. Schultz and D. Mattis: Ann. Phys. 16, 407 (1961)
13. M. E. Fisher: Physica (Utrecht) 26, 613 (1960)
14. P. Pfeuty: Ann. Phys. (NY): 57, 79 (1962)
15. M. Suzuki: Phys. Lett. 34A, 94 (1971)
16. A. Wiesler: Phys. Lett. 89A, 359 (1982)
17. Y. Miyatake, Y. Yamada, K. Nishino, M. Toyonaga, and O. Nagai: J. Mag. Mag. Mat. 54-57, 687 (1986)
18. R. J. Elliott and C. Wood: J. Phys. C: Solid St. Phys. 4, 2359 (1971)
19. P. Pfeuty and R. J. Elliott: J. Phys. C: Solid St. Phys. 4, 2370 (1971)
20. M. Barma and B. S. Shastry: Phys. Rev. B18, 3351 (1978)
21. M. Suzuki: Phys. Lett. A111, 299 (1985)
22. C. Rebbi: In The Applications of Monte Carlo Methods in Statistical Physics, ed. by K. Binder (Springer, Berlin Heidelberg 1984) p. 277
23. C. Kalle and V. Winkelman: J. Stat. Phys. 28, 639 (1982)
24. M. Suzuki, S. Miyashita, and A. Kuroda: Prog. Theor. Phys. 58, 1377 (1977)
25. S. Miyashita: Prog. Theor. Phys. 63, 797 (1980)
26. Y. Miyatake, M. Yamamoto, J. J. Kim, M. Toyonaga, and O. Nagai: J. Phys. C: Solid St. Phys. 19, 2539 (1986)
27. O. Nagai, Y. Miyatake, Y. Yamada, and H. T. Diep: J. Mag. Mag. Mat. 54-57, 681 (1986)
28. J. Oitmaa and G. J. Coomb: J. Phys. C: Solid St. Phys. 14, 143 (1981)

Magnetic Properties of Ising-like Heisenberg S=1/2 Antiferromagnets on the Triangular Lattice

S. Miyashita, M. Takasu, and M. Suzuki

Department of Physics, University of Tokyo,
Hongo, Bunkyo-ku, Tokyo 113, Japan

1. Introduction

It has been pointed out that the effect of competition in interactions (frustration) causes various interesting behaviours in ordering processes [1]. The antiferromagnet on the triangular lattice (AFT) is one of typical models with frustrated interactions. Classical AFT models with various symmetries of spins have been investigated and the natures of phase transitions of them are interpreted through the symmetries of order parameters[2,3].

In the case of quantum spin systems, in particular, in the S=1/2 case, it has been pointed out that the magnitude of the order parameters of corresponding classical models is reduced by the quantum fluctuation owing to the non-commutativity between the hamiltonian and the order parameter[4]. In the case with frustrated interactions, this reduction is much enhanced. P. W. Anderson introduced the idea that the singlet pair state, $(|+-\rangle-|-+\rangle)/\sqrt{2}$, plays an important role in the ground state of the S=1/2 isotropic Heisenberg AFT model [5]. P. Fazekas and P. W. Anderson discussed that the singlet pair plays an important role even in an Ising-like Heisenberg AFT model [6]. The ground states of these models have been also studied by the finite lattice method, and the absence of long-range order is confirmed [7,8], that is, the spin reduction is 100% in these models. It is noticed that the role of the xy interaction, $S_i^xS_j^x+S_i^yS_j^y$, is quite different in classical and quantum systems. Namely, this interaction seems to destroy any static order in quantum systems, while it enhances the sublattice order in classical cases.

In the present article, properties at finite temperatures of the Ising-like Heisenberg AFT model

$$\mathcal{H} = \Sigma_{<ij>}[J_{xy}(\sigma_i^x\sigma_j^x + \sigma_i^y\sigma_j^y) + J_z\sigma_i^z\sigma_j^z] - \Sigma_i H_z\sigma_i^z \qquad (1.1)$$

are studied by a quantum Monte Carlo method with the use of the generalized Trotter decomposition [9]. Details of the present Monte Carlo method are given in the article of Takasu, Miyashita and Suzuki[10]. Temperature- and field-dependences of the magnetization, the specific heat and the sublattice order are investigated. In the present model the magnetization process

$M_z(h,T) = \Sigma_i \sigma_i^z$ is one of the quantities in which the quantum fluctuation has a prominent effect. Hereafter a notation $h = 2H_z/J_{xy}$ is used and J_{xy} is used as a unit of the energy and is often put to be unity. $M_z(h,T=0)$ has been investigated by the finite lattice method, and it has been shown that the quantum fluctuation destroys the ferrimagnetic order [8] which exists in the corresponding classical case[3]

$$\mathcal{H} = \Sigma_{<ij>}[J_{xy}(S_i^x S_j^x + S_i^y S_j^y) + J_z S_i^z S_j^z] - H_z \Sigma_i S_i^z, \quad |S| = 1. \qquad (1.2)$$

Owing to the complicated structure of the interaction, the present model shows a so-called metamagnetic magnetization process with a 1/3-plateau[11]. In the 1/3-plateau, the quantum fluctuation is suppressed, and the system shows the sublattice long-range order. A phase diagram with respect to the temperature and field is estimated from the behaviour of the above quantities.

The negative sign problem, which causes practical difficulty for Monte Carlo simulation,as is mentioned in the article of Takasu, Miyashita and Suzuki, is also relevant to the present system. It is quite interesting to see how the negative sign problem arises and how far we can do with this problem. Although the finite size scaling method can not be used to analyse the data, some non-trivial properties have been still obtained to discuss the physical properties of the present model.

2. Magnetization Processes

In this section, the temperature dependence of magnetization process, $M_z(h,t)$ is studied. The present model has a rather complicated structure of magnetization process, so-called "metamagnetic" nature, which is due to the competition among the interaction of z-component, $J_z \sigma_i^z \sigma_j^z$, the interaction of xy- component, $J_{xy}(\sigma_i^x \sigma_j^x + \sigma_i^y \sigma_j^y)$ and the field, $-H_z \sigma_i^z$. By the investigation of the ground state, it has been pointed out that there are three critical values of field, h_{c1}, h_{c2}, and h_{c3}[3,8]. (1) $h<h_{c1}$: The system does not show any long-range order and the quantum fluctuation is remarkable. In the corresponding classical system, the spin configuration is given in Fig. 1(a) where a ferrimagnetic long-range order is developed. (2) $h_{c1}<h<h_{c2}$: The system is ferrimagnetic and the quantum fluctuation is suppressed. The spin configuration of this region is given in Fig. 1(b) as well as the classical case. (3) $h_{c2}<h<h_{c3}$: The ferrimagnetic order is destroyed. The system is again in a disordered state and quantum fluctuation is again enhanced. Spin configuration in the corresponding classical system is a "spin flop state" as shown in Fig.1(c). And (4) $h_{c3}<h$: the system is in an induced ferrimagnetic state (Fig. 1(d)).

The values of H_{c1}, H_{c2}, and H_{c3} have been obtained for the corresponding classical system as follows[3], $H_{c1} = 3$, $H_{c2} = 3(2A-1+\sqrt{4A^2+4A-7})/2$, and $H_{c3} = 6A + 3$, where $A = J_z/J_{xy}$ and this notation will be used throughout this

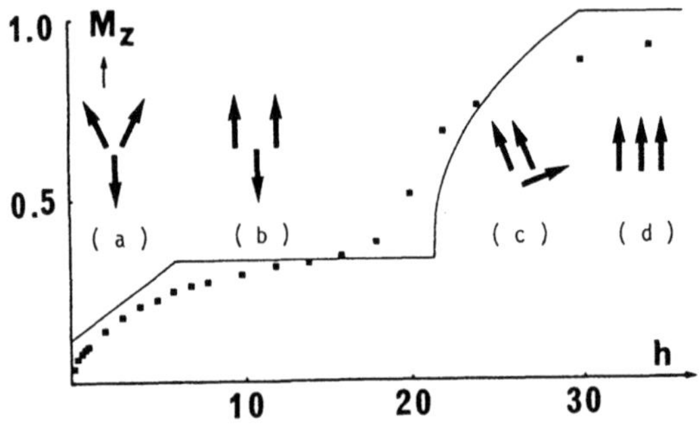

Fig. 1 Magnetization processes for the corresponding classical system: T=0.5(■,N=12×12) and T= 0 (solid line) and spin configurations

article. In the quantum system, only H_{c3} is obtained analytically by a method of Katsura and Suzuki[12] as $H_{c3} = 6A + 3$ and H_{c1} and H_{c2} are estimated from the data of finite lattices [8] as $H_{c1} \cong 3$ and $H_{c2} \cong 5A$.

Reflecting this metamagnetic structure, quantum fluctuation is not monotonic to the field. It is expected that the dependence of data on the Trotter number n is not monotonic to the field. In Fig. 2, n-dependence of magnetization for the case with A = 2.0 and T = 2.0 is given. The data depend strongly on the parameter n for $7 < h < 10$ and $19 < h < 23$. This suggests that quantumness, in other words, the effect of non-commutativity, is very important for the thermodynamic properties in these regions. These data are extrapolated with the formula

$$M_z(n) = M_z(\infty) + a/n^2 + b/n^4 + c/n^6 + \cdots \tag{2.1}$$

as is generally discussed in Suzuki's article[9]. In the cases where less numbers of data have been obtained, e. g., up to n = 2 or 3, shorter forms of (2.1) were used to extrapolate. This extrapolation is rather difficult in the regions with strong n-dependence. In other words, the coefficients in the $1/n^2$-expansion (2.1) become larger for higher order terms, that is, $a < b < c$, so that we are not confident of the convergence. The present data will be elaborated in the future. We can, still, have an idea for $M_z(h)$ with the present data. The magnetization process $\{M_z(n=\infty)\}$ thus obtained is plotted in Fig. 3, where a solid line denotes the magnetization process in the ground state for a lattice of the size 3×6 with the periodic boundary condition. In this figure it is found that this system is not ferrimagnetic but rather paramagnetic for weak fields. The global shape of $M_z(h)$ is consistent with the one in the ground state. This consistency supports that the results of our previous study of the ground state are good in larger lattices and also relieves us from an anxiety that the structure of excitation states might change the shape of $M_z(h)$ largely, even at very low temperatures.

Similarly, the magnetization processes are obtained at other temperatures, which are plotted in Fig. 3. As is suggested by the investigation at T = 0, there is a 1/3-plateau for $T \leqq 2.0$. On the other hand, for $T \geqq 3.0$ the

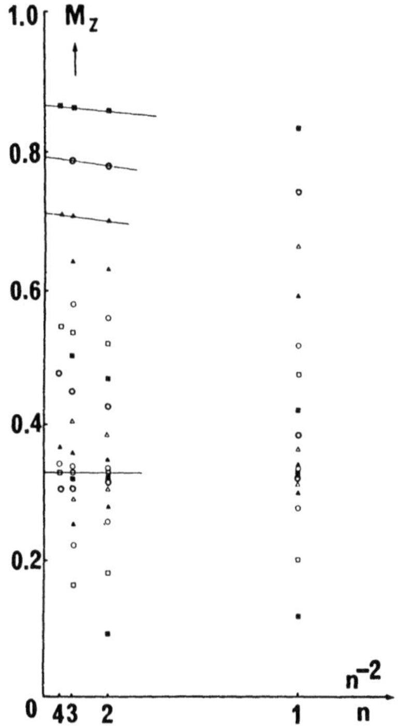

Fig. 2 Field- and n-dependences of M_z. Symbols from bottom to top, ■ , □ , ○ , ▲ , △ , ○ , ··· ■ denote data for h = 3,5,7,8,9,10,11,13,15,17,19,20,21,22,23,25,27,29,31, respectively.

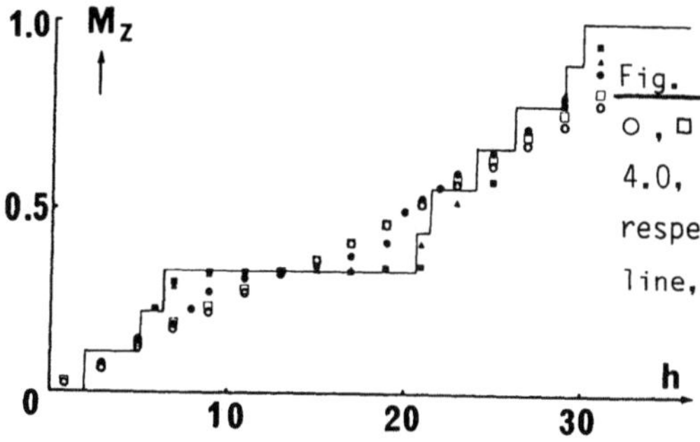

Fig. 3 Magnetization processes: Symbols ○ , □ , ● , ▲ and ■ denote data at T = 4.0, 3.0, 2.0, 1.5 and 1.0, respectively (N=6×6) and T = 0 (solid line, N=3×6)

magnetization processes are quite linear and almost independent of the temperature and they show very weak temperature dependence even at high temperatures. These properties at high temperatures are interesting features of the present model. It would be interesting if this kind of magnetization process were found in materials experimentally.

3. Negative Sign Problem

In order to study the critical phenomena, the size-dependence of data should be investigated, e.g., the finite lattice scaling or Monte Carlo renormalization method as has been demonstrated in other systems such as the ferromagnetic XY model in this symposium. It is, unfortunately, difficult to study larger lattices for the present model because of the so-called negative sign problem. This problem comes from the fact that matrix elements of the Boltzmann factor may be negative, $\langle i | e^{-\beta\mathcal{H}} | j\rangle < 0$. The ratio of the positive part of the partition function, Z_+, and the negative part, Z_-,

$$P = Z_-/(Z_+ + Z_-), \tag{3.1}$$

is one of the indicators of degree of quantum fluctuation. Reflecting the multistage structure of magnetization process mentioned in the section 2, the ratio also shows a non-monotonic behaviour with respect to the field. In Fig. 4, field-dependence of the ratio is plotted for the case $T = 2.0$ and $L = 6$(L is the linear dimension of the size $N = L\times L$). This figure shows that the quantum fluctuation is large for weak fields $h < h_{c1}$ and also for $h_{c2} < h < h_{c3}$. In the present case, the thermal average is obtained by

$$\langle A\rangle = (A_+ - A_-)/(Z_+ - Z_-), \tag{3.2}$$

where A is an observable and $A_+(A_-)$ is an accumulated value for A in the space with plus (minus) Boltzmann factor. In order to obtain a meaningful observation, a large number of samples or Monte Carlo steps, $Z_+ + Z_-$, are necessary to overcome the cancellation in $Z_+ - Z_-$ and $A_+ - A_-$. The details of data sampling are given in the Appendix. As has been generally pointed out, the degree of the cancellation is exponentially serious as the size of lattice increases, that is,

$$\left(\frac{Z_+ - Z_-}{Z_+ + Z_-}\right)_{bL} \cong \left(\frac{Z_+ - Z_-}{Z_+ + Z_-}\right)_L^{b^d} . \tag{3.3}$$

This situation is demonstrated for $L = 6$ and 12 in the case with $T = 2.0$ and $h = 13$;

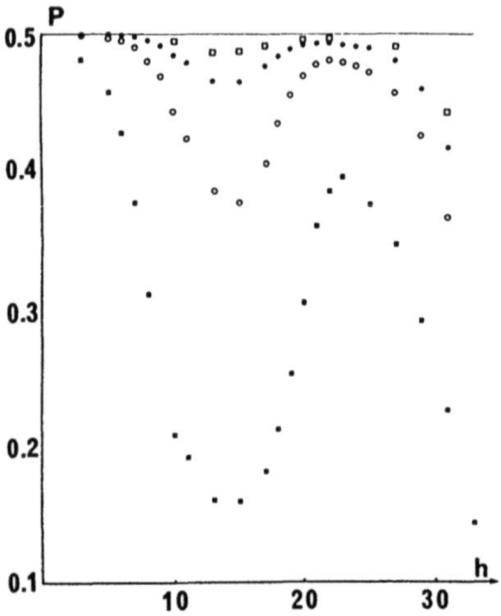

Fig. 4 Field-dependence of the ratio P (3.1). Symbols ■, ○, ●, and □ denote the data for n = 1, 2, 3 and 4, respectively.

L	P	$R=(Z_+-Z_-)/(Z_++Z_-)$
6	0.392 ± 0.001	0.216 ± 0.002
12	0.4985 ± 0.0005	0.003 ± 0.001

Here $0.216^4 = 0.0022$. Thus, it turns out that investigation for L = 12 is almost impossible at the present stage: for example, if R for L = 6 is 0.01 then R for L = 12 is 0.00000001. In this article, we therefore study temperature- and field- dependences of properties on a lattice 6×6 and leave critical phenomena as a future problem.

4. The Specific Heat and Sublattice Order

In order to study changes of characteristics of thermodynamic properties with temperature and field, the specific heat,

$$C = k_B\beta^2\partial^2\ln Z/\partial\beta^2, \tag{4.1}$$

a sublattice order,

$$A_z = 3[(\Sigma_{i\in\alpha}\sigma_i{}^z)^2 + (\Sigma_{i\in\beta}\sigma_i{}^z)^2 + (\Sigma_{i\in\gamma}\sigma_i{}^z)^2]/N, \tag{4.2}$$

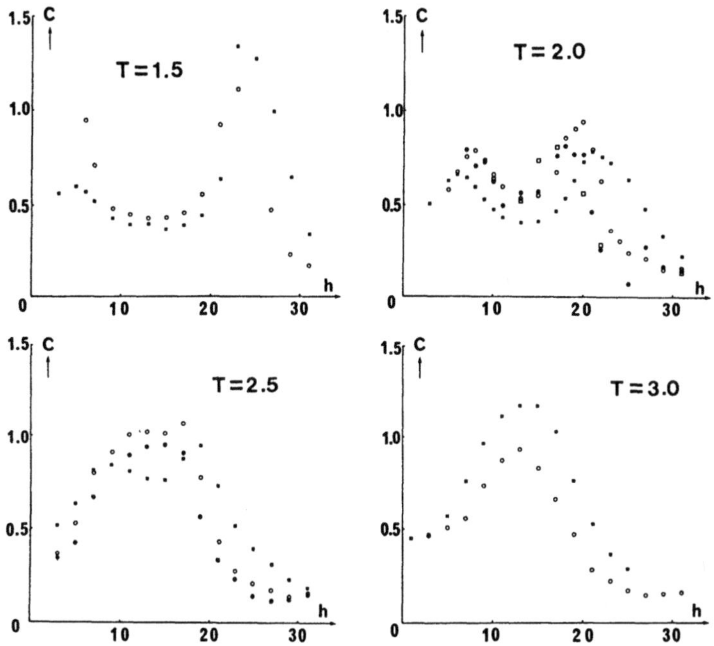

Fig. 5 Field-dependences of the specific heat. Symbols ■, ○, ● and □ denote n = 1, 2, 3 and 4, respectively

where α, β and γ denote sublattices are investigated. In Fig. 5, the data of specific heat at T = 3.0, 2.5, 2.0 and 1.5 are given. It is difficult to extrapolate the data of the specific heat, $n \to \infty$. We can still see a characteristic change with the temperature. At $T \leqq 2.0$ the specific heat has two peaks corresponding to h_{c1} and h_{c2} while there is a peak at higher temperatures. The marginal temperature is about 2.5 where data for n = 1 show two peaks while data for n = 2 do not. These observations give an idea for temperature-field phase diagram. In Fig. 6, the temperature- and field-dependences of A_z are plotted. The minimum point around $h \simeq 25$ corresponds to the spin flop transition in the classical case. With these observations, the temperature-field phase diagram is estimated in Fig. 7. Because analyses with size-dependence are not available, this phase diagram shows only qualitative nature of the system. Positions of peaks of the specific heat are shown by a symbol $\circ$, the positions of 1/3-plateau are shown by solid lines and the minimum points of A_z are shown by a symbol $\triangle$. The shaded region can be called "quantum paramagnetic" where quantum effect seems to have an intrinsic role, and it might have some difference from the usual paramagnetic phase at high temperatures. At least, configurations of the quantum paramagnetic region in Monte Carlo simulation in d+1-dimensional space are very different from those at high temperatures. It would be an interesting problem to find the difference in some physical quantities in the future.

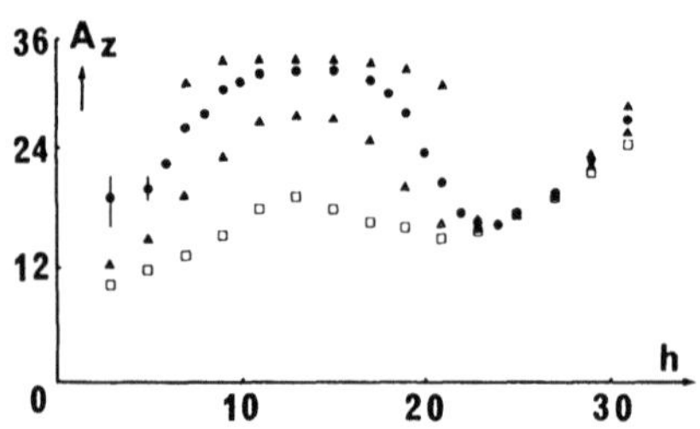

Fig. 6 Temperature and field-dependences of A_z. Symbols $\square$, Δ, $\bullet$ and $\blacktriangle$ denote data at T = 3.0, 2.5, 2.0 and 1.5, respectively.

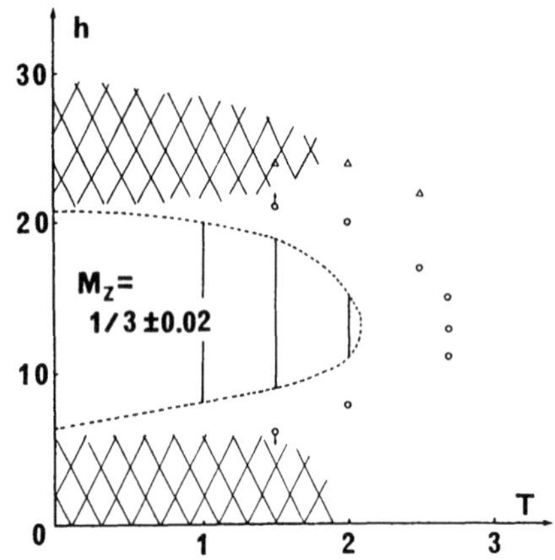

Fig. 7 Phase diagram (T,h). Symbols $\circ$ denote positions of peak of the specific heat. Symbols Δ denote minimum points of A_z. Quantum fluctuation is enhanced in the shaded regions.

5. Summary and Discussion

So far we have studied the antiferromagnetic case, $J_{xy} > 0$ and $J_z > 0$ in (1.1). We hope that the present study gives an insight to analyses of experiments on related materials[13] and also that some properties discussed here, in particular, temperature-dependence of the magnetization process, can be found experimentally.

Further, there is also competition in interactions in the cases with (i) $J_z < 0$ and $J_{xy} > 0$ and with (ii) $J_z > 0$ and $J_{xy} < 0$. These cases have been studied for the classical systems, and various interesting properties have been pointed out[2]. For example, in the case (i) a reentrant phase transition occurs and in the case (ii) there exists a tetracritical point. It would be also interesting to study these cases in quantum spin systems(S=1/2). In the case (i), it can be shown by Katsura and Suzuki's method[12] that the ground state is the ferromagnetic order of the z-component when $J_z < -2J_{xy}$, while another structure [7,14] gives the ground state for $J_z > -2J_{xy}$. This value is the same as the corresponding classical value. A quantum system in the case (ii) should be interpreted as a quantum lattice gas model with a quantum exchange(J_{xy}) and a repulsive interaction(J_z). It has been pointed out that the corresponding classical model shows ordering in both the z-component and xy-components[2], which correspond to diagonal and off-diagonal long-range orders[15] respectively. Thus it would be interesting to study this coexistence in a quantum system which will be reported in the future. It should be also mentioned that there is no negative sign in the case (ii) in the present Monte Carlo method.

Appendix. Negative Sign Problem and Definition of a Monte Carlo Step

As was mentioned in the section 3, large numbers of Monte Carlo steps (MCS = $Z_+ + Z_-$) are necessary to overcome the cancellation in $Z_+ - Z_-$ and $A_+ - A_-$ in (3.2). Usually a Monte Carlo step is defined to be N_0 up-dates of spins, where N_0 is a number of the order of lattice size, because data of physical quantities are too much correlated and the sampling is not economical if we take data with shorter periods. In the present case, as for physical quantities, the situation is the same as the usual cases, but as for the normalizations Z_+ or Z_-, the correlation is expected to be much shorter because a spin flip may change the sign. Thus the usual definition of a Monte Carlo step is crude for the present case. Thus, we define a Monte Carlo step by a shorter set of spin up-dates e.g., (3 global flips, 3 loop flips and 3 local flips) which will be abbreviated by (3,3,3). The structure of the d+1-dimensional lattice and definitions of above flips are given in the article of Takasu, Miyashita and Suzuki. In Fig. 8, an example of time evolution in Monte Carlo simulation for the case T = 2.0 and h = 3 is given, where we find the degree of correlations of the energy and of the sign. We could optimize the definition of a Monte Carlo step with use of information obtained from these considerations, but in the present study we have used mostly a set with (3,3,3) and for the case n=4 we have used a set with (2,4,4).

Finally, a list of Monte Carlo steps performed in the present study is given in Table 1.

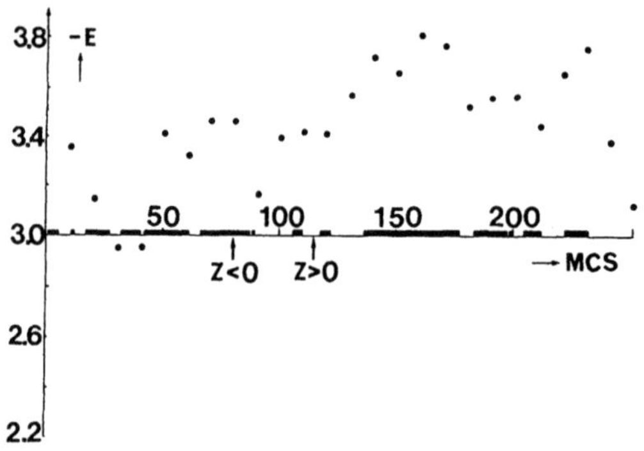

Fig. 8 Temporal evolutions of the sign and the energy (●). Bold lines denote the steps with negative sign.

Table 1 List of Monte Carlo steps, the ratio P given in (3.1) and the specific heat

T	n	h	MCS($\times 10^5$)	P	Specific heat
2.0	3	5.0	48	0.49910±0.0003	1.839±0.4
2.0	3	8.0	116	0.49474±0.0003	0.705±0.1
2.0	3	10.0	136	0.48423±0.0001	0.624±0.02
2.0	3	15.0	80	0.46455±0.0005	0.593±0.01
2.0	3	20.0	146	0.4922 ±0.005	0.765±0.1
2.0	3	24.0	114	0.49079±0.00005	0.285±0.02
2.0	3	27.0	124	0.48212±0.0002	0.157±0.01
2.0	3	31.0	38	0.41680±0.001	0.152±0.01
2.0	2	5.0	90	0.49645±0.0002	0.585±0.1
2.0	2	10.0	36	0.44101±0.003	0.658±0.03
2.0	2	20.0	18	0.46985±0.001	0.946±0.01
2.0	2	27.0	86	0.45838±0.0005	0.197±0.01
2.0	1	27.0	27	0.34782±0.0001	0.479±0.05

Acknowledgements

Numerical calculations were done by HITAC M280H at National Laboratory for High Energy Physics, which is sincerely acknowledged. One of the authors (S. M.) thanks Professor K. Hirakawa, Professor T. Goto, Dr. T. Sakakibara, Dr. M. Arai and Dr. M. Furusaka for their stimulating discussions and encouragements. He also thanks Dr. H. Nishimori for helping to prepare the data at T = 0 shown in Fig. 3. The present work is partially supported by Grant-in-Aid from Ministry of Education, Science and Culture.

References

1) G. Toulouse: Commun. Phys. 2, 115 (1977)
S. Alexander and P. Pincus: J. Phys. A13, 363 (1980)
J. Villain: J. Phys. C10, 1717 (1977)

S. Miyashita: In Topological Disorder in Condensed Matter, ed. F. Yonezawa and T. Ninomiya (Springer-Verlag 1983) p191 and references therein
2) S. Miyashita: Prog. Theor. Phys. Supple. 87, 112 (1986)
3) S. Miyashita: J. Phys. Soc. Jpn. 55, 3605 (1986)
4) P. W. Anderson: Phys. Rev. 86, 694 (1952)
R. Kubo: Phys. Rev. 87, 568 (1952)
T. Oguchi: J. Phys. Chem. Solid 24, 1049 (1963)
R. B. Pearson: Phys. Rev. B16, 1109 (1977)
J. Oitmaa and D. D. Betts: Can. J. Phys. 56, 897 (1978)
M. Suzuki and S. Miyashita: Can. J. Phys. 56, 902 (1978)
S. Fujiki and D. D. Betts: Can. J. Phys. 64, 876 (1986)
M. Suzuki: J. Stat. Phys. 42, 1047 (1986)
5) P. W. Anderson: Mater. Res. Bull. 8, 153 (1973)
6) P. Fazekas and P. W. Anderson: Philos. Mag. 30, 423 (1974)
7) T. Oguchi, H. Nishimori and Y. Taguchi: J. Phys. Soc. Jpn. 55, 323 (1986)
S. Fujiki: submitted to Can. J. Phys.
8) H. Nishimori and S. Miyashita: J. Phys. Soc. Jpn. 55, 4450 (1986)
9) M. Suzuki: the article in this issue and references therein
10) M. Takasu, S. Miyashita and M. Suzuki: the article in this issue
M. Suzuki: J. Stat. Phys. 43, 883 (1986)
11) I. Ono and T. Oguchi: Phys. Lett. 38A, 38 (1972)
T. Sakakibara and M. Date: J. Phys. Soc Jpn. 53, 3599 (1984)
12) S. Katsura and M. Suzuki: J. Phys. Soc. Jpn. 28, 255 (1970)
13) K. Hirakawa, H. Kadowaki and K. Ubukoshi: J. Phys. Soc. Jpn. 54, 3526 (1985)
I. Yamada, K. Ubukoshi and K. Hirakawa: ibid. 54, 3526 (1985)
K. Hirakawa and H. Kadowaki: Neutron Scattering, Proc. Int. Conf. ed. G. H. Lander and R. A. Robinson (North-Holland, 1986)
T. Goto and T. Sakakibara: private communication
14) S. Miyashita: J. Phys. Soc. Jpn. 53, 44 (1984)
15) H. Matsubara and T. Tsuneto: Prog. Theor. Phys. Supple. 46, 411 (1970)

Thermodynamic Properties of the Spin-1/2 Heisenberg Antiferromagnet on the Triangular Lattice

M. Takasu, S. Miyashita, and M. Suzuki

Department of Physics, Faculty of Science, University of Tokyo, Bunkyo-ku, Tokyo 113, Japan

1. Introduction

The Monte Carlo method, which was originally developed for classical systems, has been applied to various quantum systems. In particular, a method based on the generalized Trotter formula has been widely used[1-4]. The Metropolis algorithm for the classical system generates the configurations according to the Boltzmann factor $\exp(-\beta\mathcal{H})$. For a quantum system, it is difficult to calculate the matrix elements of the Boltzmann factor for a large system. It has been shown [1] that the Boltzmann factor can be estimated approximately by using the generalized Trotter formula. One can then map the quantum system into some classical system, in which one can use the Metropolis algorithm. The simulation of this transformed classical system is different from the simulation of usual classical systems in several respects:

(1) The transformed classical system has special interaction, or in other words, conservation laws. We need special care for the spin flips to consider the ergodicity.

(2) There is a possibility for the Boltzmann factor of the transformed classical system to become negative.

The purpose of the present paper is to investigate these aspects of this method in the case of the spin-1/2 quantum Heisenberg model on the triangular lattice;

$$\mathcal{H} = -J \sum_{<ij>} \sigma_i \cdot \sigma_j , \qquad (1.1)$$

where σ is the Pauli matrix and $J>0$ and $J<0$ for ferromagnetic and antiferromagnetic model, respectively. This model has been studied by many authors[5-8,10-16]. Physically, this antiferromagnetic model is interesting, because of the cooperation of the frustration and quantum effect. The ground state of the antiferromagnetic model was predicted by P. W. ANDERSON [5] to be a quantum liquid-like state of singlet pairs, and was studied numerically by FAZEKAS and ANDERSON[6] and by MARLAND and BETTS[7]. One of the authors (S.M.)[8] estimated the ground state energy by a variational method[9]. LEE et al.[10] have also studied the system using Handscomb's method. Recently, numerical diagonalizations were performed by OGUCHI et al.[11] and also by FUJIKI and BETTS [12]. Thermofield dynamics was applied to this system by one of the authors (M.S.) [13], and a similar method using random base was applied by IMADA[14]. Experimentally, $NaTiO_2$, which is a substance corresponding to this model, was found by HIRAKAWA et al.[15].

The method of our simulation is described in Section 2. Application of the method to the ferromagnetic Heisenberg model is shown in Section 3. The nature of negative signs peculiar to frustrated systems is described in Section 4. Thermodynamic quantities for some finite antiferromagnetic systems are presented in Section 5. Finally, summary and discussion are given in Section 6.

2. Quantum Monte Carlo Method

Quantum Monte Carlo method is based on the following generalized Trotter formula[1];

$$\exp(-\beta \mathcal{H}) \simeq [\prod_j \exp(-\beta \mathcal{H}_j/n)]^n, \tag{2.1}$$

where the Hamiltonian $\mathcal{H}$ consists of local interactions $\{\mathcal{H}_j\}$ as follows

$$\mathcal{H} = \sum_j \mathcal{H}_j. \tag{2.2}$$

Using this formula, the d-dimensional quantum system is mapped into a (d+1)-dimensional Ising system with many-body interactions. We obtain results for finite number of n and extrapolate the data for n infinity. The additional direction is sometimes called the "Trotter direction," and the parameter n in (2.1) is called "Trotter number."

It is necessary to choose such local Hamiltonians $\{\mathcal{H}_j\}$ that the matrix elements can be calculated exactly. For the triangular lattice, we choose the following four-spin cluster Hamiltonian $\mathcal{H}_j$, as shown in Fig. 1.

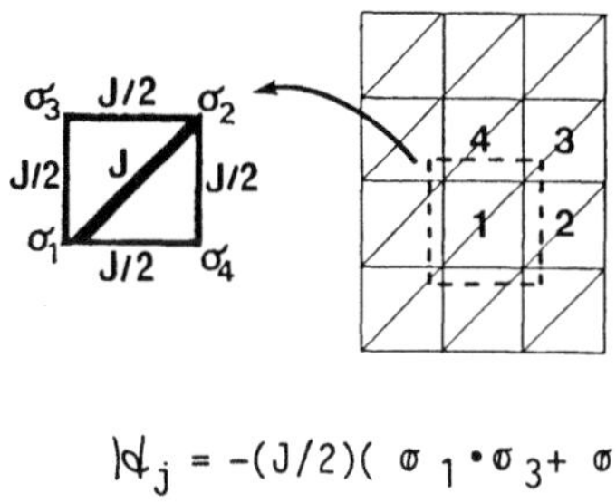

Fig. 1 A cluster of the local interactions taken from the triangular lattice. The numbers 1 to 4 on the lattice denote the order of the local interactions

$$\mathcal{H}_j = -(J/2)(\boldsymbol{\sigma}_1\cdot\boldsymbol{\sigma}_3 + \boldsymbol{\sigma}_1\cdot\boldsymbol{\sigma}_4 + \boldsymbol{\sigma}_2\cdot\boldsymbol{\sigma}_3 + \boldsymbol{\sigma}_2\cdot\boldsymbol{\sigma}_4) - J\,\boldsymbol{\sigma}_1\cdot\boldsymbol{\sigma}_2. \tag{2.3}$$

Since this cluster includes two triangles, we can take into account the frustration more effectively than in the case of one-bond clusters. Taking the base $(\sigma_1, \sigma_2, \sigma_3, \sigma_4)$, the local Boltzmann factor is expressed by the following 16×16 matrix.

$$\rho_j = \exp(-\beta \mathcal{H}_j) \tag{2.4}$$

$$
= \begin{bmatrix}
p & & & & \\
& \begin{matrix} q_1 & q_2 & q_3 & q_3 \\ q_2 & q_1 & q_3 & q_3 \\ q_3 & q_3 & q_4 & q_5 \\ q_3 & q_3 & q_5 & q_4 \end{matrix} & & & \\
& & \begin{matrix} r_1 & r_2 & r_2 & r_2 & r_2 & r_3 \\ r_2 & r_4 & r_5 & r_6 & r_6 & r_2 \\ r_2 & r_5 & r_4 & r_6 & r_6 & r_2 \\ r_2 & r_6 & r_6 & r_4 & r_5 & r_2 \\ r_2 & r_6 & r_6 & r_5 & r_4 & r_2 \\ r_3 & r_2 & r_2 & r_2 & r_2 & r_1 \end{matrix} & & \\
& & & \begin{matrix} q_4 & q_5 & q_3 & q_3 \\ q_5 & q_4 & q_3 & q_3 \\ q_3 & q_3 & q_1 & q_2 \\ q_3 & q_3 & q_2 & q_1 \end{matrix} & \\
& & & & p
\end{bmatrix}
\begin{matrix} 1\;2\;3\;4 \\
\begin{pmatrix}
+ & + & + & + \\ + & + & + & - \\ + & + & - & + \\ + & - & + & + \\ - & + & + & + \\ + & + & - & - \\ + & - & + & - \\ - & + & + & - \\ + & - & - & + \\ - & + & - & + \\ - & - & + & + \\ + & - & - & - \\ - & + & - & - \\ - & - & + & - \\ - & - & - & + \\ - & - & - & -
\end{pmatrix}
\end{matrix}
$$

where $p=e^{-3K}$, $q_1=e^{-3K}/4+e^{-K}/2+e^{K}/4$, $q_2=e^{-3K}/4-e^{-K}/2+e^{K}/4$, $q_3=e^{-3K}/4-e^{K}/4$, $q_4=e^{-3K}/4+e^{K}/4+e^{3K}/2$, $q_5=e^{-3K}/4+e^{K}/4-e^{3K}/2$, $r_1=e^{-3K}/6+e^{K}/2+e^{3K}/3$, $r_2=e^{-3K}/6-e^{3K}/6$, $r_3=e^{-3K}/6-e^{K}/2+e^{3K}/3$, $r_4=e^{-3K}/6+e^{-K}/4+7e^{3K}/12$, $r_5=e^{-3K}/6+e^{-K}/4-5e^{3K}/12$, $r_6=e^{-3K}/6-e^{-K}/4+e^{3K}/12$, and $K=-J/kT$.

The temperature dependence of these transition probabilities is plotted in Fig. 2 for the antiferromagnetic case. At high temperatures, these transition probabilities approach either 1 or 0, while they take large positive or negative values at low temperatures.

The order of local interactions $\{\mathcal{H}_j\}$, which is not unique, is taken as shown in Fig. 1. This representation has the property that the sum of z-components of four spins must be conserved at each interaction of local Hamiltonians. Therefore, if we generate random configurations, the

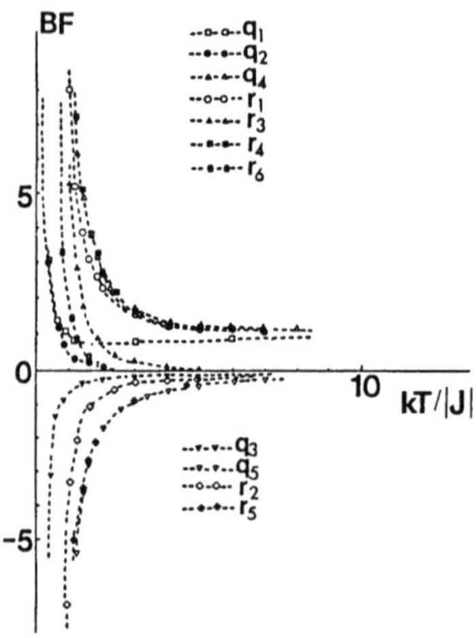

Fig. 2 Temperature dependence of the transition probability in the antiferromagnetic case

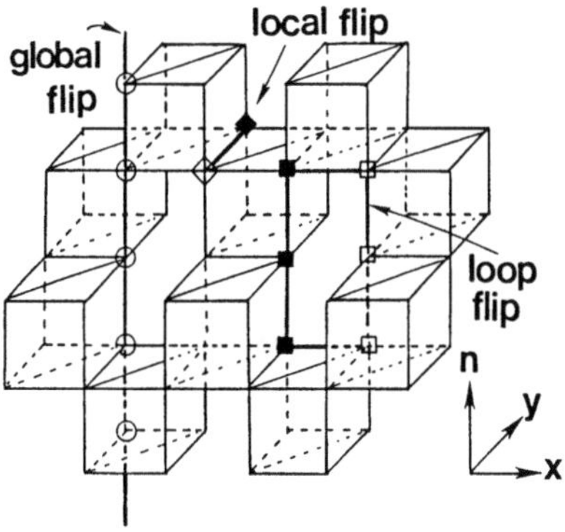

Fig. 3 Structure of equivalent (2+1)-dimensional Ising spin system and the types of flips: ○ denote spins for global flip, and must be of the same sign; ■ and □ denote spins for loop flip, where ■ must have a sign different from □ ; and ◆ and ◇ denote spins for local flip, where ◆ must have a sign different from ◇ .

probability of obtaining an allowed configuration is very small. Thus we start from an allowed configuration and try to generate possible configurations by flipping several spins.

In Fig. 3, we show three kinds of flips which are necessary to move a string of down-spins in all possible ways in the environment of up-spins. A "global flip" involves 4n spins aligned in the Trotter direction, and changes the value of M_z by creating or annihilating a string of down-spins. The deformation and displacement of the strings are taken into account by a "local flip", which flips two adjacent spins, and also by a "loop flip", which flips six spins.

3. Ferromagnetic Heisenberg Model

Our method is first applied to the ferromagnetic Heisenberg model on the triangular lattice, since this model is free from the negative sign problem of Boltzmann factors which will be explained in Section 4. The size of the lattice is 6×6 with periodic boundary condition, and 10^4 to 10^6 Monte Carlo steps (MCS) were performed. One MCS consists of N global flips, 2Nn local flips and 2Nn loop flips, where N is the number of sites and n is the Trotter number.

The energy, the specific heat, and the magnetization $\langle M^2\rangle$ are plotted in Fig. 4 as functions of temperature. In extrapolating the data with different Trotter number n, we have used the following $1/n^2$-theorem proved by one of the present authors (M.S.)[17];

$$C_n=C+a/n^2+b/n^4+\dots \quad , \tag{3.1}$$

where C_n is the value obtained for the finite decomposition of (2.1). Our data in Fig. 5 up to n=3 already show the $1/n^2$ dependence, and it is possible to estimate the value of C rather accurately.

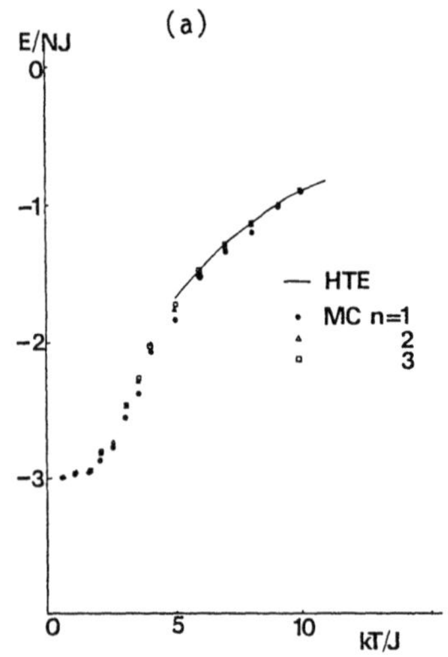

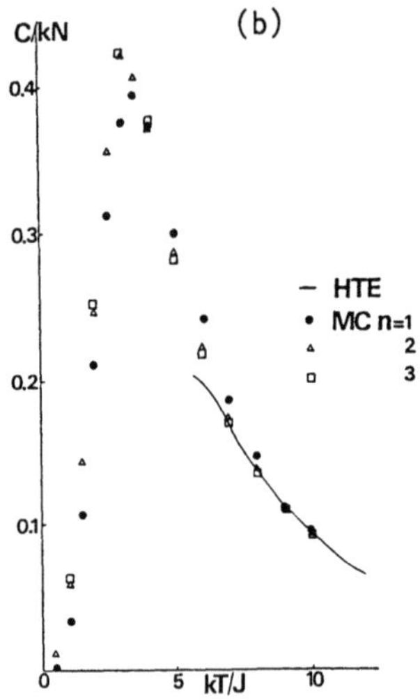

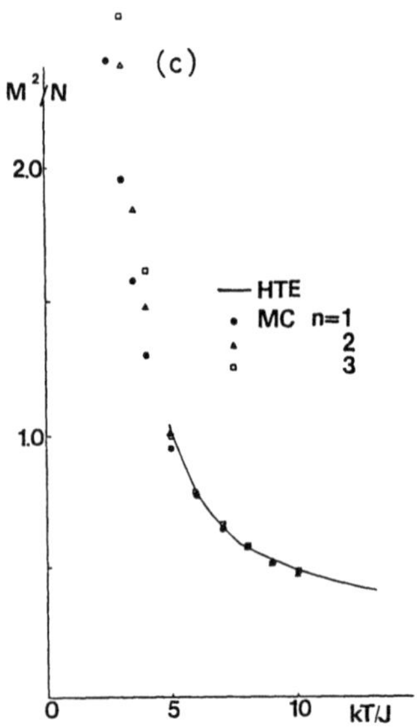

Fig. 4 Temperature dependence of (a) energy, (b) specific heat, (c) magnetization of the ferromagnetic Heisenberg model on the triangular lattice

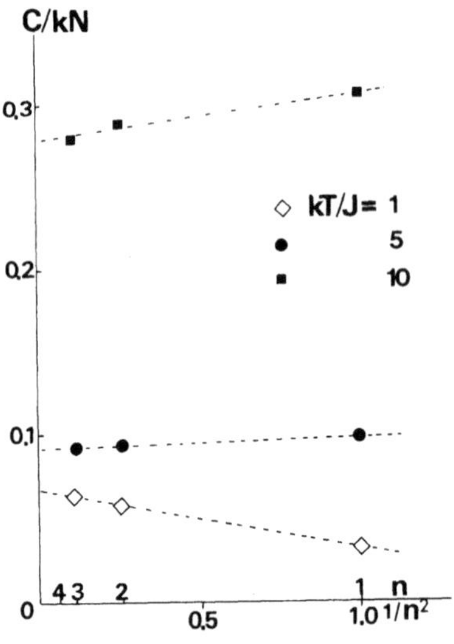

Fig. 5 $1/n^2$-extrapolation of the specific heat of Fig. 4(b)

4. The Nature of the Negative Signs

As we have seen in Section 2, the Boltzmann factor is approximated by a product of local Boltzmann factors. Although the diagonal elements of the local Boltzmann factors are always positive, the off-diagonal matrix elements are not necessarily positive for some interactions. For a non-frustrated lattice, the product of local Boltzmann factors is always positive. For frustrated lattices, the product of them may become negative, and the summation over the intermediate states gives a positive value. Namely, we have

$$\langle j|\exp(-\beta \mathcal{H})|j\rangle = \sum_{k_m} \langle j|\rho_1|k_1\rangle \ldots \langle k_{s-1}|\rho_s|j\rangle = Z_+ - Z_- > 0, \tag{4.1}$$

where ρ_m is the local Boltzmann factor and Z_+ (or Z_-) is the summation of the positive (or negative) products, respectively.

In our representation, for the ferromagnetic Heisenberg model, the matrix elements of the Boltzmann factor of a cluster are always positive. For the antiferromagnetic Heisenberg model, a matrix element of the Boltzmann factor becomes negative, if it is a non-diagonal element with an exchange of an up-spin and a down-spin. For example,

$$\langle -{+}{+}{+}|\exp(-\beta \mathcal{H}_j)|{+}{-}{+}{+}\rangle < 0. \tag{4.2}$$

In Fig. 6, we show a configuration which has a negative Boltzmann factor. The circles denote down-spins in the environment of up-spins, and the string of down-spins goes around a triangle of the original triangular lattice.

Now we explain how the negative Boltzmann factors contribute to the cancellation of the data. In quantum Monte Carlo simulation, we generate each configuration with the probability proportional to the local Boltzmann factor. Since the probability must be positive, we take the absolute value of the Boltzmann factor as the probability, and we register the quantity into the group of positive (or negative) phase. After some number of Monte Carlo flips, we obtain A_+ (or A_-), which is the sum of a physical quantity A in the space of positive (or negative) Boltzmann factors, respectively, and also Z_+ (or Z_-), which is the number of configurations of each space. The Monte Carlo average $\langle A \rangle$ of the quantity A is expressed as follows:

$$\langle A \rangle = (A_+ - A_-)/(Z_+ - Z_-). \tag{4.3}$$

In Fig. 7, the ratio of the negative Boltzmann weights $Z_-/(Z_+ + Z_-)$ is plotted as a function of temperature for different Trotter numbers. As the temperature decreases, this ratio increases and approaches 1/2, which means that the cancellation in (4.3) becomes serious. At the same time, A_+ and A_- become rather close. Thus the calculation of the quotient in (4.3)

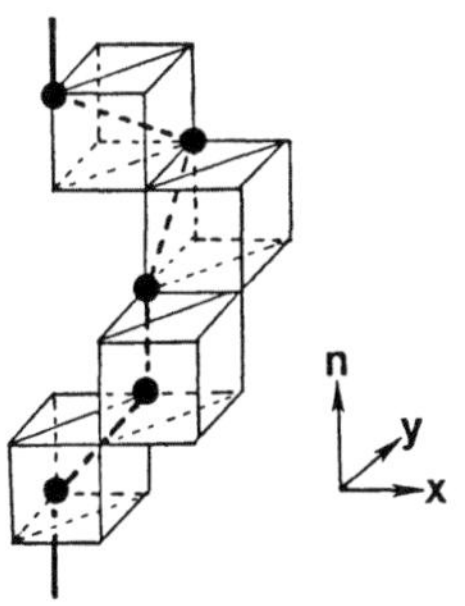

Fig. 6 A configuration in the (2+1)-dimensional Ising system which has a negative Boltzmann weight, where ● denote down-spins in the environment of up-spins

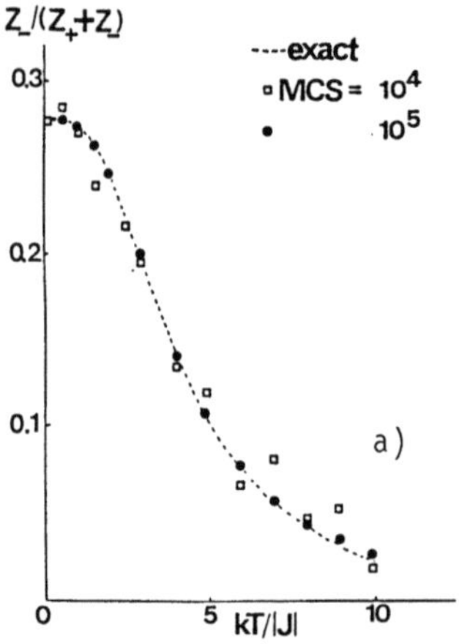

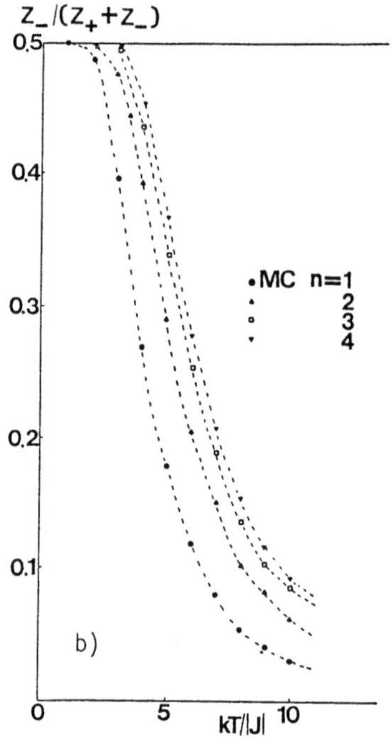

Fig. 7 Temperature dependence of the ratio of the negative weights $Z_-/(Z_+ + Z_-)$ for (a) 2×2 and (b) 6×6 lattice

becomes inaccurate. This tendency becomes more serious as the lattice size becomes large, because $Z_+ - Z_-$ has the following exponential dependence on the total number of spins, N [3];

$$(Z_+ - Z_-)/(Z_+ + Z_-) \simeq e^{-bN} \quad , \quad b>0 \, . \tag{4.4}$$

In the cases where the cancellations are serious, the averages in both groups (positive and negative phase) have almost the same values. One might then argue that the average could be taken independently in the space of positive weights and in the space of negative weights as follows:

$$\langle A\rangle \simeq a_+ \simeq a_- \, , \tag{4.5}$$

where

$$a_+ = A_+/Z_+ , \; a_- = A_-/Z_- . \tag{4.6}$$

However, if we rewrite (4.3) as

$$\langle A\rangle = a_+ + Z_- \, \delta a/\delta Z = a_- + Z_+ \, \delta a/\delta Z \, , \tag{4.7}$$

where $\delta a = a_+ - a_-$ and $\delta Z = Z_+ - Z_-$, then we find that the quantity $Z_- \, \delta a/\delta Z$ or $Z_+ \, \delta a/\delta Z$ may not be negligible even when δa is very small. We observe in Fig. 8 that the average in the space of positive weights a_+ differs from the average in the space of negative weights a_- at high temperatures and that these two averages differ from the average in the whole space $\langle A\rangle$ at low temperatures. Therefore, we cannot take the average in the space of each definite sign.

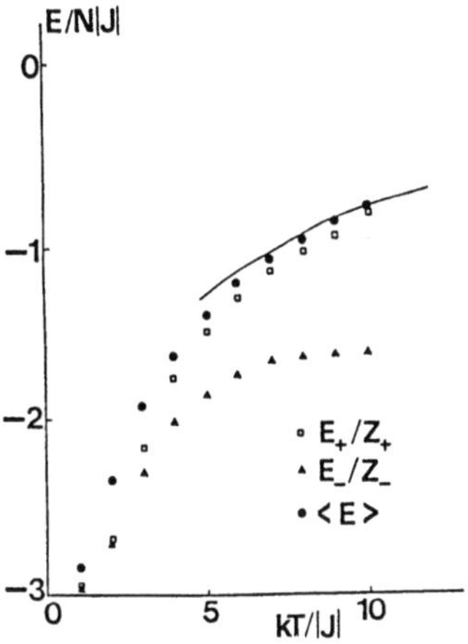

Fig. 8 Temperature dependence of the energy averaged in the space of the positive (or negative) Boltzmann weights E_+/Z_+ (or E_-/Z_-), denoted by □ (or ▲). ● denote the average in the whole space

Because of this kind of inaccuracy caused by negative weights, we must perform a large number of Monte Carlo runs at low temperature. The results of the simulation for the frustrated system are shown in the next section.

5. Antiferromagnetic Heisenberg Model

Using the method described in Section 2 and keeping in mind the peculiar features of the method as explained in Section 4, we apply the Monte Carlo method to the quantum antiferromagnetic Heisenberg model on the triangular lattice. The lattice size is again 6×6. Monte Carlo steps range from 10^6 to 10^7 according to the temperatures and Trotter numbers.

As explained in Section 2, we used three types of flips such as global flips, local flips and loop flips. An example of the acceptance ratios of these flips is plotted as a function of temperature for antiferromagnetic Heisenberg model in Fig. 9. We find that the acceptance ratio of global flips becomes large at high temperature, and also that the acceptance ratios of the local flips and loop flips become large at low temperatures. For a

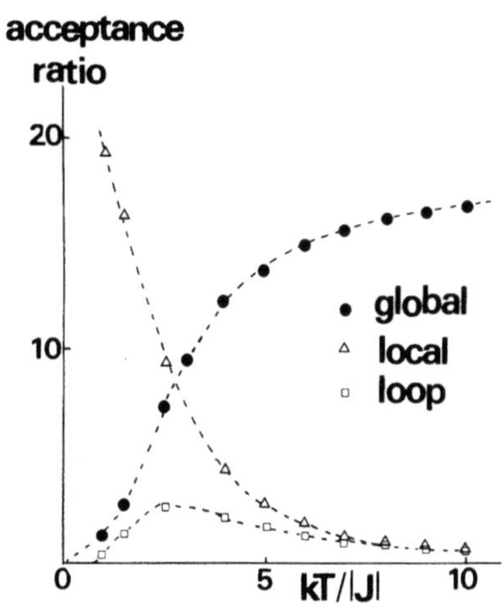

Fig. 9 Temperature dependence of the acceptance ratio of three kinds of spin flips for the 6×6 and n=1 antiferromagnetic spins

finite lattice, the acceptance ratio of the loop flip has a peak at intermediate temperatures. As the lattice becomes larger, this ratio is expected to behave similarly to the ratio of the local flips.

The energy is plotted in Fig. 10(a) as a function of temperature. At high temperatures, the result is in agreement with the high-temperature expansion[18]. As we increase the Trotter number n, the energy approaches larger values. The ground state energy is fixed to the energy of the local cluster. However, when we extrapolate the data of the energy of finite n using the $1/n^2$-rule as seen in Section 3, the ground state energy approaches the true value, which is obtained by other methods[11-14].

The total magnetization $\langle M^2\rangle = 3\langle M_z{}^2\rangle$ is plotted in Fig. 10(b). As the temperature decreases, the total magnetization decreases and falls into a state in which the singlet pairs play an important role.

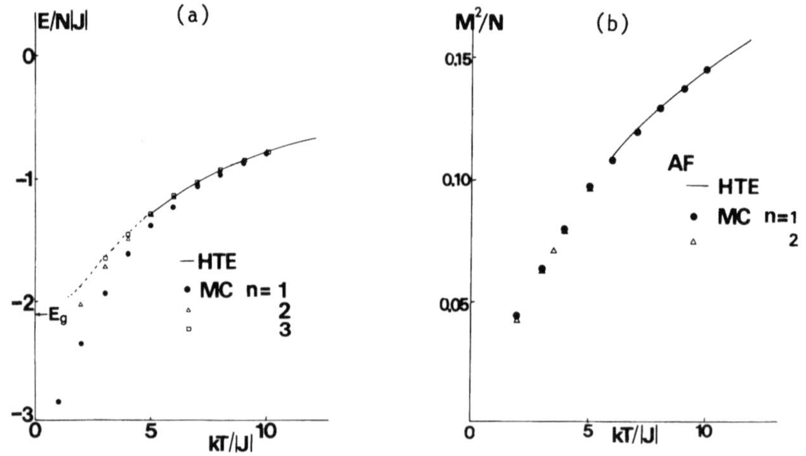

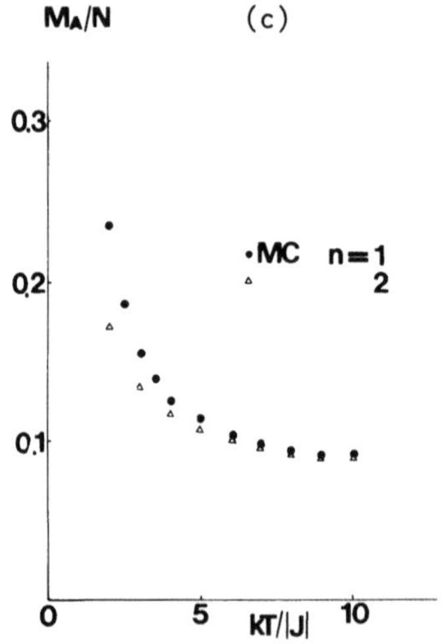

Fig. 10 Temperature dependence of (a) energy, (b) square magnetization, (c) sublattice magnetization of the antiferromagnetic Heisenberg model on the triangular lattice

On the other hand, the sublattice magnetization $\langle M_A^2 \rangle$, where A is one of the three sublattices, increases as the temperature decreases, as shown in Fig. 10(c). However, as we increase the Trotter number, that is, as the system becomes closer to the real quantum system, this increase is suppressed. This shows that the sublattice magnetization is not a good order parameter in the quantum model, contrary to the classical case[19].

6. Summary and Discussion

We have studied the quantum Heisenberg models on the triangular lattice by means of quantum Monte Carlo simulation. We have used the 4-spin cluster in order to take into account the frustration effect of the antiferromagnetic model. Several kinds of spin flips were used to insure ergodicity in the space of zero winding number. The same algorithm can be used for both the ferromagnetic and antiferromagnetic models. The nature of the negative signs of Boltzmann factors, peculiar to the frustrated lattices, was discussed. Thermodynamic properties such as energy, magnetization and sublattice magnetization were calculated for the antiferromagnetic model and it was seen that this quantum model shows sublattice ordering different from that of the corresponding classical model.

As described in Section 4, the quantum Monte Carlo simulation needs a large computation time at low temperatures for frustrated models owing to the negative sign problem. Although there is a similar situation in the case of fermion systems[3,20], the negative signs of fermion systems come from the global structure of the system, while the negative signs of the frustrated spin system come from the local structure of the lattice. For the n=1 case of the quantum Villain model, ONOGI et al.[21] found a special transformation of spin variables, by which the negative signs of Boltzmann factors can be avoided. However, for the present model on the triangular lattice, such a transformation is not yet found.

The quantum Monte Carlo method is considered to be more powerful when applied to non-frustrated models, and has an advantage over the numerical diagonalization method, because it can be applied to large systems. The study of the XY ferromagnetic model on the triangular lattice using similar algorithm is now in progress.

Acknowledgements

The authors would like to thank T. Onogi for helpful discussions. The numerical calculations have been performed on the FACOM M190 of the Laboratory of the International Collaboration on Elementary Particle Physics, University of Tokyo. The present study is partially supported by Grant-in-Aid from the Ministry of Education, Science and Culture.

References

1. M. Suzuki: Commun. Math. Phys. 51, 183 (1976); 57, 193 (1977); Prog. Theor. Phys. 56, 1454 (1976); J. Math. Phys. 26, 601 (1985); Phys. Rev. B31, 2985 (1985); J. Stat. Phys.(1986) [review article]

2. M. Suzuki, S. Miyashita, A. Kuroda: Prog. Theor. Phys. 58, 1377 (1977)
3. J. E. Hirsch, D. J. Scalapino, R. L.Sugar, R. Blankenbecler: Phys. Rev. Lett. 47, 1628 (1981); Phys. Rev. B26, 5033 (1982)
4. H. De. Raedt and A. Lagendijk: Phys. Rep. 127, 233 (1985) and references therein
5. P. W. Anderson: Mater. Res. Bull. 8,153 (1973)
6. P. Fazekas, P. W. Anderson: Philos. Mag. 30, 423 (1974)
7. L. G. Marland, D. D. Betts: Phys. Rev. Lett. 43, 1618 (1979)
8. S. Miyashita: J. Phys. Soc. Jpn. 53, 44 (1984)
9. M. Suzuki, S. Miyashita: Can. J. Phys. 56, 902 (1978)
10. D. H. Lee, J. D. Joannopoulos, J. W. Negele: Phys. Rev. B30, 1599 (1984)
11. T. Oguchi, H. Nishimori and Y. Taguchi: J. Phys. Soc. Jpn. 55, 323 (1986); J. Mag. Mag. 54, 1353 (1986)
12. S. Fujiki, D. D. Betts: Can. J. Phys.
13. M. Suzuki: J. Stat. Phys. 43, 883 (1986); and in Progress in Quantum Field Theory ed. by H. Ezawa and S. Kamefuchi (North Holland, 1986). See also J. Phys. Soc. Jpn. 54, 4483 (1985); Phys. Rev. B35 (1987)
14. M. Imada: J. Phys. Soc. Jpn., (1986).
15. K.Hirakawa, H. Kadowaki, K. Ubukoshi: J. Phys. Soc. Jpn. 54, 3526 (1985) I. Yamada, K. Ubukoshi, K. Hirakawa: ibid, 54, 3571 (1985)
16. M. Takasu, S. Miyashita, M. Suzuki: Prog. Theor. Phys. 75, 1254 (1986)
17. M. Suzuki, Phys. Lett. 113A, 299 (1985)
18. G. S. Rushbrooke, G. A. Baker Jr., P. J. Wood: in Phase Transitions and Critical Phenomena Vol.3, ed. by C. Domb and M. S. Green (Academic Press, London, 1974)
19. H.Kawamura, S. Miyashita: J. Phys. Soc. Jpn. 53, 4138 (1984)
20. M. Imada: J. Phys. Soc. Jpn. 53, 2861 (1984) V. Elser: Phys. Rev. A34, 2293 (1986)
21. T. Onogi, S. Miyashita, M. Suzuki: J. Stat. Phys. 45, 183 (1986)

Randomness and Frustration in Triangular Antiferromagnets

M. Imada

Institute of Physics, College of Liberal Arts, University of Saitama, Shimo-Okubo, Urawa, Saitama 338, Japan and
Institute for Solid State Physics, University of Tokyo, Roppongi, Tokyo 106, Japan

Thermodynamic properties of the spin-1/2 triangular antiferromagnet with the isotropic and Ising-like exchange interactions are investigated and the behavior as a quantum-spin-liquid is discussed. The interplay between the full frustration and the effect of exchange randomness is also discussed in relation to spin glass order. The Ising triangular antiferromagnet seems to show spin glass order at least at T=0 even with a small exchange randomness. The spin-1/2 triangular antiferromagnet shows anomalous thermodynamic behavior with a striking enhancement of the uniform susceptibility at low temperatures.

1. Introduction

Recent studies on random Ising spin models in three dimensions showed strong evidence for symmetry breaking at finite temperatures due to spin glass order [1]. The recognition several years ago of the importance of frustration in random spin systems stimulated extensive studies on regular but frustrated spin systems. The two-dimensional Heisenberg triangular antiferromagnet offered a good and typical example for the study of the effect of frustration on thermodynamic properties and phase transitions. In fact it revealed some peculiar and unexpected properties in classical systems [2-6], some of which seem to reflect general features of the full frustration. Except for the Ising triangular antiferromagnet, however, the ground state of classical Heisenberg spin systems with various anisotropies is rather trivial and has antiferromagnetic-like long-range order, because the frustration is partly removed by the continuous degrees of freedom of vector spins. The isotropic Heisenberg vector model (CHAFT), for example, has the ground state characterized by "120°-structure" with three sublattices, which may be regarded as a variation of the antiferromagnetic structure.

Most classical spin systems which show various antiferromagnetic, ferromagnetic or helical long-range order have quantum counterparts that also preserve the same symmetry breaking. An exception is the one-dimensional antiferromagnetic system, where both spin-1/2 and spin-1 systems have the singlet ground state without symmetry breaking [7,8] due to the one-dimensional fluctuation. Some years ago Anderson et al. [9,10] pointed out that the spin-1/2 Heisenberg antiferromagnet on the triangular lattice (QHAFT) might be another example where the classical long-range order is not preserved. He also discussed the possibility of symmetry breaking of a quite different type with off-diagonal long-range order line Bose condensation. Since the long-range order in CHAFT is based on reconciliation with the frustration, the spin-1/2 system might provide another valuable example where the classical order is destroyed by strong quantum fluctuation. As we will see below, QHAFT, in fact, shows quite different behavior from CHAFT. It seems to stay as a quantum spin liquid state at low temperatures with the singlet ground state and without symmetry-breaking. Because such a quantum spin liquid state has not been studied in detail, except for the spin-1/2 antiferromagnetic chain, the triangular lattice system serves to clarify an aspect of the quantum spin systems.

The numerical procedure of the calculation useful for this problem is explained in Sect.2. Thermodynamic properties of the triangular antiferromagnet are discussed in Sect.3. They show features characteristic of the quantum spin liquid both in the isotropic and Ising-like Heisenberg systems. The detailed analysis of the content of Sect.3 is published elsewhere [11]. We show in Sect.4 that one of the characteristic features of the quantum spin liquid state is seen in the sensitivity to randomness, which seems to arise from the "localization" of spin singlet pairs. The Ising triangular antiferromagnet seems to show spin glass order at least at T=0 even with small exchange randomness. This means that the randomness stabilizes the appearance of a symmetry-broken state like the spin glass order in fully frustrated systems, because we know that the pure Ising antiferromagnet does not show any kind of symmetry breaking. In a quantum system like QHAFT, striking enhancement of the uniform spin susceptibility is induced by the randomness.

2. Numerical Procedure

Although frustrated quantum spin systems could show unexpected and interesting thermodynamic properties, our understanding seems to be still at a primitive stage. Because neither an intuitive picture has been established nor powerful analytic methods such as perturbation or renormalization group analyses apply, numerical methods seem to be the only reliable way to test various possibilities and physical concepts at low temperatures in frustrated quantum spin systems. Numerically exact calculation of relatively small systems and the quantum Monte Carlo simulation of spin systems developed by Suzuki [12] could be powerful techniques. However, it turns out that application of the quantum Monte Carlo method is not a straightforward extension of simulations in non-frustrated spin systems [13]. The difficulty shares essentially the same origin as that in Monte Carlo methods of the path integral in other problems. In the path integral representation, the problem is that the partition function is often expressed by a large number of positive and negative terms [14]. Although the total sum of the path integral should be positive definite, the Monte Carlo sampling of the sum of positive and negative terms within practically possible computation time results in a meaningless value with a large fluctuation. This, in fact, seems to reflect some essential feature of the frustration, at least in available representations of the base. If we knew the eigenstates and used the base of eigenvectors, such a difficulty would not exist. However, such a statement is meaningless when the solution of the problem is not known. This circumstance at present causes trouble in performing the simulation in the fully quantum region of large systems with frustration.

We will discuss below a way to improve finite size calculation avoiding this difficulty. The summation to get the thermodynamic average of A may be expressed as

$$\langle A\rangle = \sum_s \langle s|AW|s\rangle / Z, \tag{1}$$

$$Z = \sum_s \langle s|W|s\rangle \tag{2}$$

and by using the Hamiltonian H and the inverse temperature β, W is given by

$$W = \exp(-\beta H). \tag{3}$$

In the path integral representation, W is written as

$$W = \prod_{i=1}^{M-1} (w_M q_i) w_M, \tag{4}$$

$$w_M = \exp(-\beta H/M) \tag{5}$$

and q_i is unity given by the summation over a complete set as

$$q_i = \sum_{\ell_i} |\ell_i\rangle\langle\ell_i|. \tag{6}$$

Because the difficulty occurs when one replaces the summation over $|\ell_1\rangle, |\ell_2\rangle, \ldots, |\ell_{M-1}\rangle$ by the Monte Carlo sampling, the difficulty disappears if we take the Monte Carlo sampling only for the trace summation over $|s\rangle$ and operate with W explicitly. The operation of W on $|s\rangle$ is performed by employing the approximant of the type

$$W = w_M{}^M \tag{7}$$

and

$$w_M \simeq \sum_{j=0}^{L} \frac{1}{j!} (-\beta H/M)^j, \tag{8}$$

which provides sufficiently accurate expressions by keeping M and/or L large. The operation of W on $|s\rangle$ is reduced to successive operations and summations of the Hamiltonian matrix H. Because a state generated from $|s\rangle$ operated on by H's should be represented by a linear combination of a complete basis set, we have to keep N_t coefficients in memory at each step, where N_t is the total number of states. In the case of a spin-half system composed of N sites, $N_t=2^N$. The computation time to get $W|s\rangle$ is essentially proportional to N_t.

At first sight, it seems to be useless to do such a calculation, because we cannot treat large systems because the number N_t grows exponentially with increasing N, and for small systems the exact diagonalization is available. It turns out [15], however, that this procedure makes it possible to treat spin-1/2 systems twice as large as those tractable by exact diagonalization. This makes the extrapolation to the thermodynamic limit substantially easier in the case of the triangular antiferromagnet. For the exact diagonalization of the Hamiltonian matrix, it is known that the memory size needed is proportional to $N_t{}^2$. In the case of the method explained above, which we call the quantum transfer Monte Carlo method (QTMC) [15], the memory size is proportional to N_t. Thus, we can treat twice as large systems as by exact diagonalization. The computation time by QTMC is proportional to $N_s N_t$, where N_s is the number of Monte Carlo samples for the trace summation. It turns out that N_s may be reduced to of the order of 100 to reach sufficiently small statistical errors that the extrapolation becomes possible in typical examples. QTMC provides, in fact, a better way to obtain thermodynamic properties. To control and reduce statistical errors arising from the Monte Carlo sampling of the trace summation, each state $|s\rangle$ in the Monte Carlo sampling is given not by a pure state of a given base $\{|\phi_i\rangle\}$ but given by a mixed state that is a random linear combination of N_t states:

$$|s\rangle = \sum_{i=1}^{N_t} C_{si}|\phi_i\rangle, \quad (9)$$

where C_{si} is a random number provided that any two samples $|s_1\rangle$ and $|s_2\rangle$ are orthogonal to each other. It should be noted that QTMC itself becomes essentially the same procedure at T=0 as an exact method for the ground state such as the Lanczos method. At finite temperatures, this method shares the same spirit as the thermofield method proposed by Suzuki [16,17].

3. Regular Triangular Antiferromagnet

The Hamiltonian of the spin-1/2 QHAFT is given by

$$H = \frac{1}{2}\sum_{\langle i,j\rangle} J[\lambda(\sigma_i^x\sigma_j^x + \sigma_i^y\sigma_j^y) + \sigma_i^z\sigma_j^z], \quad (10)$$

where $\vec{\sigma}_i=(\sigma_i^x,\sigma_i^y,\sigma_i^z)$ is the Pauli spin interacting with the nearest neighbor and placed on the triangular lattice with J=1.0. The ground state properties of the isotropic Heisenberg case (λ=1.0) and the Ising-like case ($0<\lambda<1$) were speculated to be resonating valence bond systems by Anderson [9] and Fazekas and Anderson [10]. Some slightly different types of variational approach [16,18], and the exact diagonalization of some finite systems [19,20] were also used for the same problem. The most systematic calculation of the ground state available so far was performed by Oguchi et al. [20], where the exact calculation up to 21 spins was combined with extrapolation to infinite size. In their result, they suggest that the ground state of the isotropic Heisenberg system does not have the long-range order of the sublattice ordering. An apparent symmetry breaking with the sublattice ordering is therefore unlikely. However, we need thermodynamic properties to see whether other types of symmetry breaking, which are usually accompanied by a singular thermodynamic behavior, occur or not.

The QTMC method explained in Sect.2 was applied to obtain thermodynamic properties [11]. Finite size systems up to 18 spins with various shapes of the lattice were treated and they were extrapolated by using the railroad trestle extrapolation method proposed by Anderson [9]. We first discuss results of [11] for the isotropic Heisenberg case (λ=1.0). The specific heat and susceptibility seem to be smooth and have a broad maximum, as seen in Figs.1 and 2, which have a qualitative similarity to the one-dimensional antiferromagnetic chain. The railroad trestle extrapolation works rather well and any apparently singular behavior of these properties is ruled out. The squared sublattice magnetization

$$\mu = \{\langle(M_z^A)^2\rangle + \langle(M_z^B)^2\rangle + \langle(M_z^C)^2\rangle\}/3N, \quad (11)$$

where M_z^A, M_z^B and M_z^C are the z-component of magnetizations of each sublattice A, B and C, seems probably to have a finite value at any temperature, as plotted in Fig.3. This behavior contrasts with the classical vector model [5], where μ shows well-developed short-range order at low temperatures and diverges as $T\to 0$. Figs.1-3 show typical behavior of a quantum spin liquid.

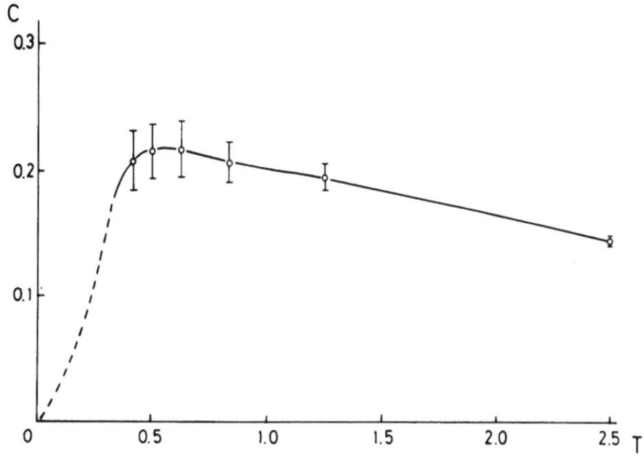

Fig.1 Temperature dependence of the specific heat of the spin-1/2 triangular antiferromagnet (QHAFT) with λ=1.0 (isotropic case). Error bars come from the ambiguity in the extrapolation procedure. Dashed curve is an estimate from the energy data not shown here.

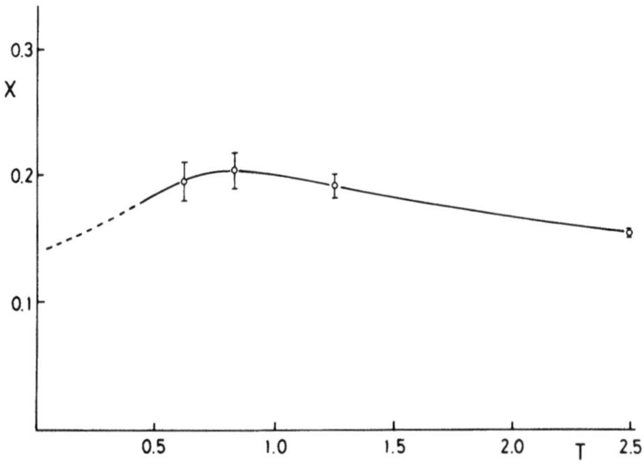

Fig.2 The uniform susceptibility against temperature for the isotropic QHAFT.

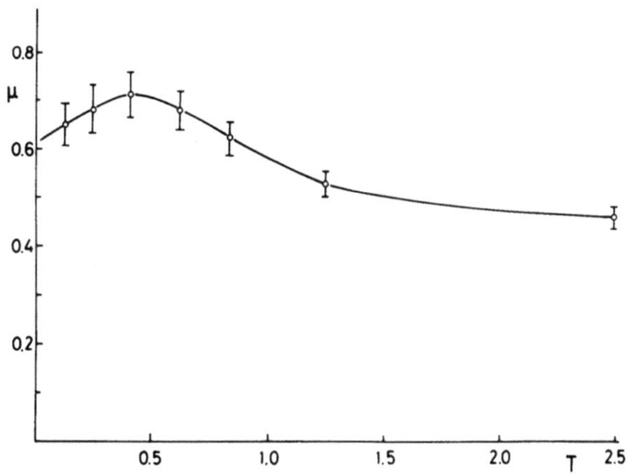

Fig.3 The squared sublattice magnetization μ vs temperature for the isotropic QHAFT.

When the Ising-like anisotropy exists and increases, the short-range order of the sublattice magnetization gradually increases. It is not surprising because the correlation of the sublattice magnetization decays algebraically as $\langle\sigma_i\sigma_{i+j}\rangle \sim |j|^{-1/2}$ and μ diverges in the ground state of the Ising limit (λ=0) [3]. However, μ is strikingly suppressed as compared to the classical counterpart. For the anisotropy λ=0.25. The energy is plotted in Fig.4 as a function of temperature. It clearly indicates the double peak structure of the specific heat. In the classical vector model with Ising-like anisotropy, it is known [6] that μ diverges below the temperature of the upper specific heat peak, which is denoted by T_2. In the spin-1/2 system with λ=0.25, μ is given in Fig.5, where μ is still rather small well below $T_2 \sim 1.0$. Although the extrapolation to infinite size from the finite size data becomes difficult at low temperatures, as seen in Fig.5, the divergence of μ at finite temperatures seems to be unlikely, at least for $T \gtrsim 0.2$. Except for the double peak structure in the specific heat and the quantitative increase of μ, the spin-1/2 Ising-like Heisenberg system seems to show qualitatively similar behavior to the isotropic case (λ=1.0).

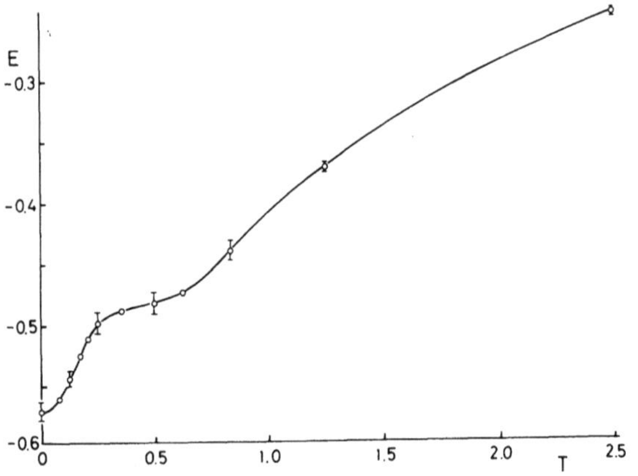

Fig.4 Temperature dependence of the energy for the spin-1/2 triangular antiferromagnet with Ising-like anisotropy (λ=0.25)

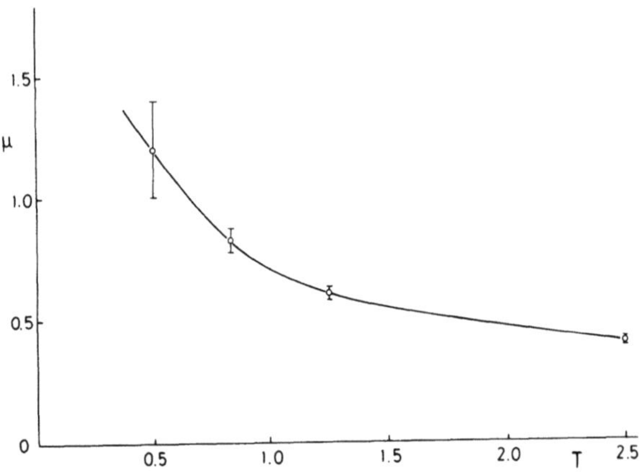

Fig.5 Temperature dependence of the squared sublattice magnetization μ for spin-1/2 triangular antiferromagnet with λ=0.25.

4. Random Triangular Antiferromagnet

A remarkable property of a quantum spin-liquid state may be instability to various kinds of perturbations, such as the spin-Peierls-like distortion. Regular distortions of the triangular lattice may induce either antiferromagnetic order or the spin-Peierls state, depending on the type of distortion. This problem will be discussed elsewhere. Here we investigate the sensitivity of thermodynamic properties to small randomness of the antiferromagnetic exchange interaction. The Hamiltonian is given by

$$H = \frac{1}{2} \sum_{<ij>} J_{ij}[\lambda(\sigma_i^x\sigma_j^x + \sigma_i^y\sigma_j^y) + \sigma_i^z\sigma_j^z], \quad (12)$$

where the difference from (10) is that J is random and bond dependent with the mean value $\bar{J}=1.0$. We introduce the randomness by distribution of the exchange coupling $P(J_{ij})$ given by the square function with width R illustrated in Fig.6. We consider here random triangular systems under the condition that $R/2<\bar{J}$, where the exchange coupling is always antiferromagnetic.

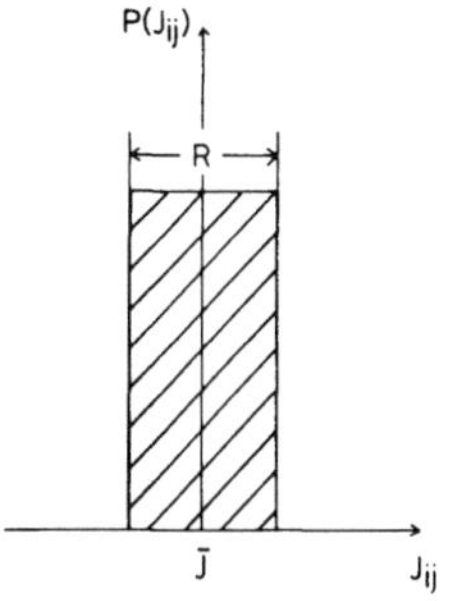

Fig.6 The probability distribution function P of the exchange coupling. The exchange couplings J_{ij} have uncorrelated random values with this square distribution.

We first discuss the Ising limit (λ=0). Two-dimensional random Ising systems such as the ±J model and the Gaussian model on a square lattice are thought to show the spin glass phase transition at T=0 [21-24]. For R=0, the spin correlation function $<\sigma_i\sigma_{i+j}>$ of the triangular system behaves asymptotically as $j^{-1/2}$ at T=0 [3]. Therefore the correlation function of the spin glass order defined by

$$C_{SG}(j) \equiv [|<\sigma_0\sigma_j>|^2]_J$$

is apparently proportional to j^{-1} in this pure case, where $[\ ]_J$ denotes the average over the distribution of the randomness. Because we expect that the randomness removes the enormous degree of degeneracy known in the pure Ising case, the spin glass order may be more stabilized in the random system than in the pure case. For small R, this system seems to be quite different from the ±J model or the Gaussian model on the bipartite lattice. Because an apparent symmetry breaking such as the antiferromagnetic order is completely suppressed by the frustration, the spin glass order could be realized more easily by small randomness than in a simple bipartite lattice. The interplay of the frustration and the randomness may be seen more directly in the Ising triangular antiferromagnet with small randomness.

The spin glass correlation function $C_{SG}(j)$, as well as other thermodynamic properties, is investigated by a recursive transfer matrix method, which gives numerically exact thermodynamic quantities, and was used some years ago in the two-dimensional ±J model and the Gaussian model on a square lattice by Morgenstern and Binder [21]. Finite size L×L lattices with L≤16 are treated with the shape illustrated in Fig.7. The periodic boundary condition is imposed along the right and the left boundaries and the free boundary condition for the top and the bottom boundaries. We took 20-100 independent distributions of randomness to take the sample average. The spin glass correlation function $C_{SG}(j)$ is plotted for several choices of temperature in Fig.8. It clearly shows that $C_{SG}(j)$ is strongly enhanced by the randomness as compared to the ground state of the pure system shown by a dashed curve. By assuming the exponential decay of C_{SG}, which seems to be reasonable in Fig.8, we plotted the inverse correlation length for R=1.0 and R=0.5 in Fig.9. Although it is not clear whether the spin glass transition occurs at finite temperatures, it is consistent with the spin glass long-range order at T=0 even for small randomness. It is confirmed that the randomness induces the symmetry breaking in this case and the spin glass long-range order appears. Recently the critical exponent of the spin glass transition at T=0 in a two-dimensional Ising square lattice with various distribution functions of randomness was reported [22-24]. The spin glass correlation length ξ is reported to behave as $\xi \sim T^{-\nu}$ with the exponent $\nu \sim 4$ or larger. Because the triangular antiferromagnet with small randomness may show a different behavior as is suggested above, it would be interesting to investigate in detail the critical exponent of this system in future.

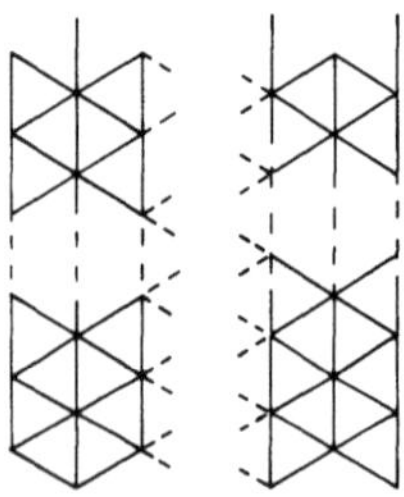

Fig.7 The lattice shape used in the transfer matrix calculation of the Ising triangular antiferromagnet

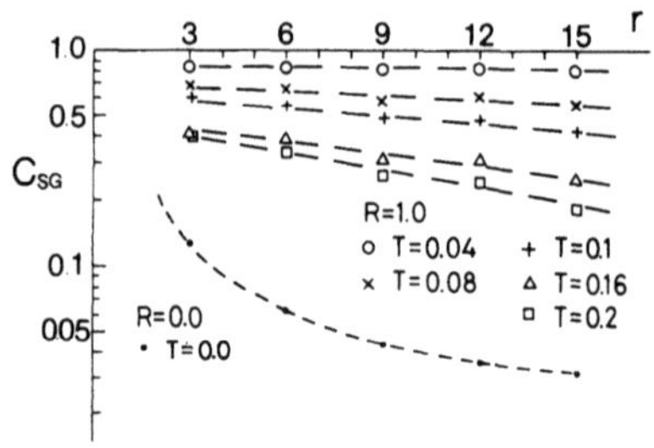

Fig.8 The spin glass correlation function $C_{SG}(r)$ vs r for several temperatures of random Ising triangular antiferromagnet of the 16×16 lattice. The dashed line shows the exact results of C_{SG} for r=3n spacings at R=0 and T=0 [3].

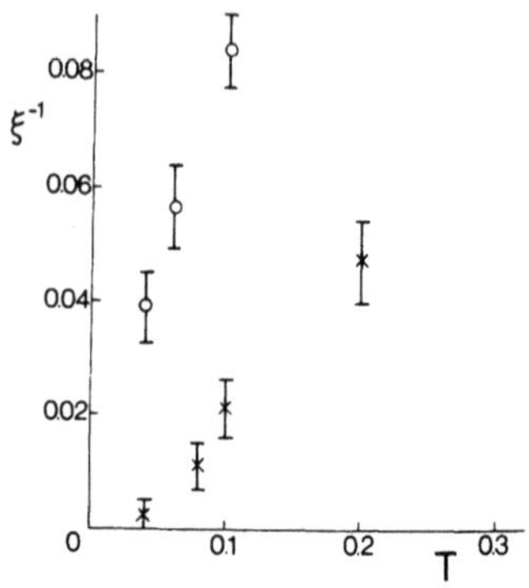

Fig.9 Temperature dependence of the inverse correlation length ξ^{-1} for R=1.0 (cross points) and R=0.5 (open circles).

We next discuss the spin-1/2 quantum systems with randomness in the isotropic ($\lambda=1$) and the Ising-like case ($0<\lambda<1$) by applying the QTMC method considered in Sects.2 and 3 [25]. The z-component of the uniform susceptibility χ_Z was calculated for several m×n finite size systems of the shape shown in Fig.10 with the free boundary condition. For the isotropic Heisenberg antiferromagnet ($\lambda=1$, $\bar{J}=1$), χ_Z illustrated in Fig.11 shows a striking enhancement, due to the randomness, at low temperatures and seems to diverge as $T\to0$ when R becomes large. In the Ising-like antiferromagnet, the divergence of the susceptibility is more enhanced than the isotropic case.

The system size dependence of the enhanced susceptibility is seen in Fig.12, where we have plotted the threshold value R_0, beyond which χ_Z shows divergent-like behavior as $T\to0$. Although the system sizes investigated here are not large enough, it is likely that the divergence occurs with very small (possibly infinitesimal) randomness in the infinite system. The scaling power ν of the susceptibility in the thermodynamic limit, where ν is defined by $\chi_Z \propto T^{-\nu}$, remains for future study. A similar behavior is found in the spin-1/2 one-dimensional random antiferromagnet [26-28]. Dasgupta and Ma [27] argued that the susceptibility diverges as $T\to0$ even for small randomness. Anomalous behavior of the thermodynamic properties is due to

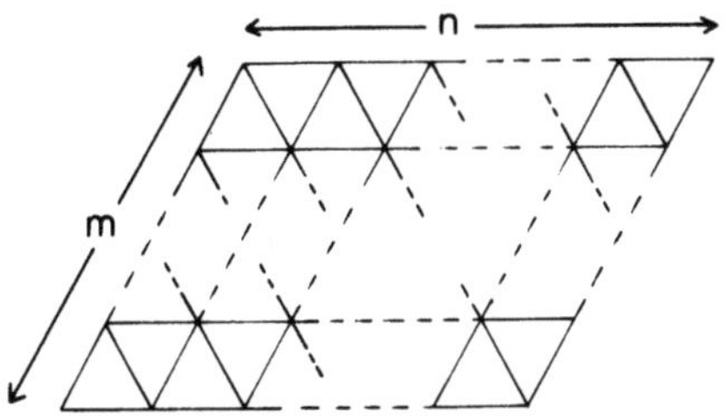

Fig.10 The lattice shape used in the QTMC calculation of spin-1/2 random systems.

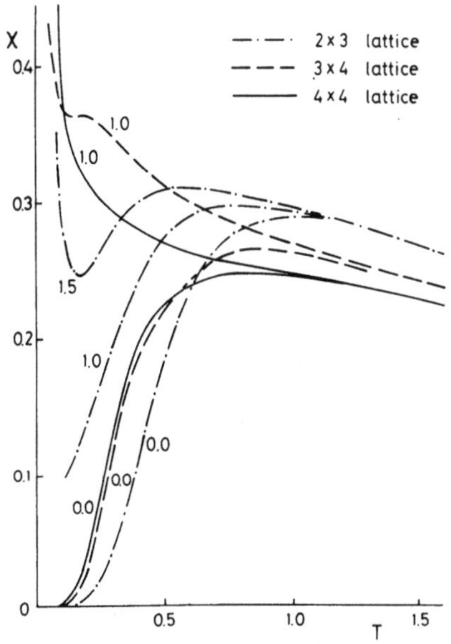

Fig.11 The z-component uniform susceptibility vs temperature in the isotropic spin-1/2 system for several choices of the randomness R, which is shown beside the curves, and for a few choices of the lattice sizes.

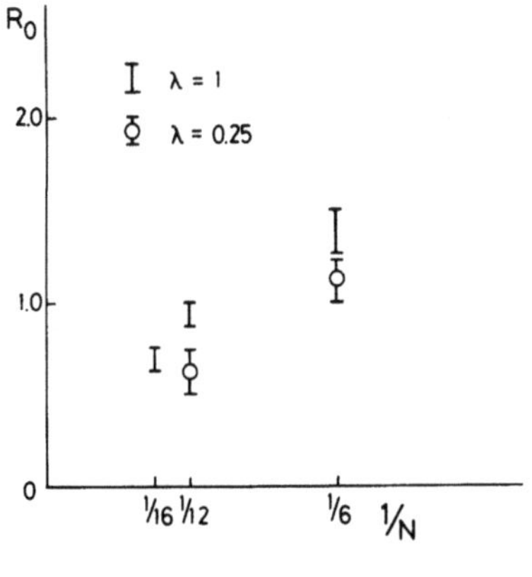

Fig.12 The size dependence of the randomness threshold R_0 for the divergent-like behavior of the uniform susceptibility. The results for N=3×2, 4×3 and 4×4 lattices are illustrated. See the text.

the fact that an enormous number of non-singlet states are almost degenerate with the ground state.

The sensitivity to the randomness strongly suggests that the system tends to be filled with resonating valence bonds in the uniform system, while singlet pairs are frozen with some ratio of relatively free spins in the random system. When relatively stronger bounds form singlet pairs, they leave other spins coupled with weaker bonds as isolated spins at low temperatures. Such isolated spins also interact with each other through the fluctuation of singlet pairs. Therefore the spin-1/2 triangular and the one-dimensional system may show quite different behavior. It would be an interesting problem to examine the quantum effect on the spin glass order seen in the Ising limit of the triangular antiferromagnet with small random exchange. So far the spin glass correlation function in this quantum regime has not been examined because of the rather small system size limit in the numerical calculations. It should be clarified whether the divergence of the uniform susceptibility as $T \to 0$ in QHAFT, in particular, with Ising-like anisotropy, is accompanied by the spin glass order at $T=0$ or not. With regard to the sensitivity to the randomness, an experimentally observed divergence of the susceptibility for the isotropic Heisenberg system [29] at low temperatures is suggestive, although other extrinsic possibilities such as magnetic impurities should also be carefully examined.

5. Concluding Remarks

Low-temperature properties of the spin-1/2 triangular antiferromagnet seem now to be characterized by the "quantum spin liquid" without symmetry breaking. Its quantum-liquid-like behavior and the sensitivity to various perturbations such as randomness are clarified in some aspects. The analysis discussed here is confined to the results extrapolated from the finite size calculation. An important task is to develop techniques to treat quite large systems. The triangular lattice systems seem to be suited to investigate the interplay of the frustration and the randomness as well as quantum fluctuation effects on the spin glass order. Full understanding of the problems, such as possibilities of symmetry breaking in random and quantum systems, or in systems coupled to lattice distortions, should be reached in future with the analysis of large systems.

References

1. R.N. Bhatt, A.P. Young: Phys. Rev. Lett. 54, 924 (1985)
   A.T. Ogielski, I. Morgenstern: Phys. Rev. Lett. 54, 928 (1985)
2. G. Wannier: Phys. Rev. 79, 357 (1950); B7, 5017 (1973)
3. J. Stephenson: J. Math. Phys. 5, 1009 (1964)
4. S. Miyashita, H. Shiba: J. Phys. Soc. Jpn. 53, 1145 (1984)
5. H. Kawamura, S. Miyashita: J. Phys. Soc. Jpn. 53, 4113 (1984)
6. S. Miyashita, H. Kawamura: J. Phys. Soc. Jpn. 54, 3385 (1985)
7. H.A. Bethe: Z. Phys. 71, 205 (1931)
   L. Hulthen: Arkiv Math. Astron, Fys. 26A, 938 (1938)
8. F.D.M. Haldane: Phys. Rev. Lett. 50, 1153 (1983)
   M.P. Nightingale, H.W.J. Blöte: Phys. Rev. B33, 659 (1986) and references therein.
9. P.W. Anderson: Mater. Res. Bull. 8, 153 (1973)
10. P. Fazekas, P.W. Anderson: Phil. Mag. 30, 423 (1974)
11. M. Imada: J. Phys. Soc. Jpn. 56, No.1 (1987)
12. M. Suzuki: Prog. Theor. Phys. 56, 1454 (1976); see also this proceedings.
13. M. Takasu, S. Miyashita, M. Suzuki: Prog. Theor. Phys. 75, 1254 (1986)

14. See for example, R.P. Feynman and A.R. Hibbs: Quantum Mechanics and Path Integrals (McGraw-Hill, New York, 1965) p.292.
15. M. Imada, M. Takahashi: J. Phys. Soc. Jpn. 55, 3354 (1986)
16. M. Suzuki: J. Stat. Phys. 42, 1047 (1986)
17. M. Suzuki, H. Betsuyaku: Phys. Rev. B34, 1829 (1986)
18. S. Miyashita: J. Phys. Soc. Jpn. 53, 44 (1984)
19. L.G. Marland, D.D. Betts: Phys. Rev. Lett. 43, 1618 (1979)
20. T. Oguchi, H. Nishimori, Y. Taguchi: J. Phys. Soc. Jpn. 55, 323 (1986)
21. I. Morgenstern, K. Binder: Phys. Rev. B22, 288 (1980)
22. A.J. Bray, M.A. Moore: J. Phys. C 17, L463 (1984)
23. D.A. Huse, I. Morgenstern: Phys. Rev. B32, 3032 (1985)
24. R.R.P. Singh, S. Chakravarty: Phys. Rev. Lett. 57, 245 (1986)
25. M. Imada: Submitted to J. Phys. Soc. Jpn.
26. L.N. Bulaevskii, A.V. Zvarykina, Yu. S. Karimov, R.B. Lyubovskii, I.F. Shchegolev: Sov. Phys. -JETP 35, 384 (1972)
G. Clark, J. Hammann, J. Sanny, L. Tippe: Quasi One-Dimensional Conductors II Proceedings, Dubrovnik 1978, edited by S. Barisic et al. (Springer, 1979)
27. C. Dasgupta, S. Ma: Phys. Rev. B22, 1305 (1980)
28. J.E. Hirsch, J.V. José: Phys. Rev. B22, 5339 (1980)
J.E. Hirsch, R. Kariotis: Phys. Rev. B32, 7320 (1985)
29. K. Hirakawa, H. Kadowaki, K. Ubukoshi: J. Phys. Soc. Jpn. 54, 3526 (1985)
I. Yamada, K. Ubukoshi, K. Hirakawa: J. Phys. Soc. Jpn. 54, 3571 (1985)

Two New Methods for Computing Finite Quantum Spin Systems

T. Oguchi and H. Kitatani

Department of Sciences and Mathematics, The Technological University of Nagaoka, Kami-Tomioka, Nagaoka 940-21, Japan

1. Introduction

One of the methods for computing the thermodynamic quantities of the infinite quantum spin systems is to compute rigorously the thermodynamic quantities for a finite system which consists of N spins and then extrapolate them to the limit $N\to\infty$. The linear chain lattice is the most suitable for such an extrapolation, because we may utilize all the data for N=2,3,4,·· ···. On the other hand, for two- or three-dimensional lattices, N increases rapidly as the linear size of the system is increased, so that the computations are limited by the memory space of the computer. We want to stress that if we could obtain the numerical values of physical quantities up to $N\lesssim 30$, we would be able to get them for the infinite size system even if we could not carry out the computations for N>30, because we could probably extrapolate the data for $N\lesssim 30$ with considerable accuracy to the infinite-size system.

In the present paper, we propose two new methods, both of which make computations for larger systems possible.

2. The First Methods

In the first method, as is shown in Fig. 1, the system consisting of N spins is divided into two subsystems A and B, each of which consists of N/2 spins, respectively. They are computed rigorously, and the interactions between them are treated by perturbational methods.

The Hamiltonian H_{tot} of the total system can be written as

$$H_{tot} = H_A + H_B + H' , \tag{2.1}$$

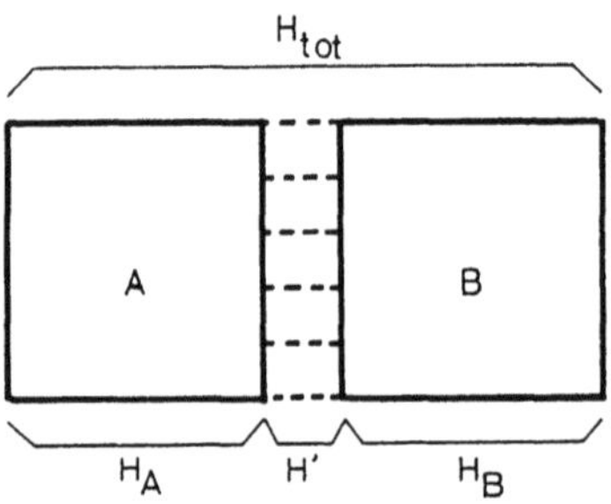

Fig. 1 The total system which is divided into two subsystems

where H_A and H_B represent the Hamiltonians of the subsystems and H' is the Hamiltonian of the interactions between A and B. Since $[H_A, H_B]=0$, we may compute the partition functions of the two subsystems separately. Therefore, our requirement of computer storage can be reduced to $2^{N/2}\times 2^{N/2}$ instead of $2^N\times 2^N$ in the case of S=1/2 (the product is needed to solve the eigenvalue equation of the Hamiltonian).

We expand the density operator into a power of the perturbation Hamiltonian, H'. We will write the unperturbed Hamiltonian just as H instead of H_A+H_B, for convenience. We use the representation which makes H_2 diagonal. The partition function Z may be written up to $O(H'^2)$ as

$$Z = \sum_m \exp(-KH_{mm}) \times \begin{cases} 1 + K^2/2 \sum_{n\neq m} |H'_{nm}|^2 & \text{when } H_{nn}=H_{mm} \\ 1 + K \sum_{n\neq m} |H'_{nm}|^2/(H_{nn}-H_{mm}) & \text{when } H_{nn}\neq H_{mm} \end{cases} \quad (2.2)$$

($K=J/k_BT$, where the Hamiltonian is in the form $H=2J\sum S_i\cdot S_j$). Unfortunately we have not completed the calculation of Z, so that here we show only the energy of the ground state E_g of the anti-ferromagnetic Heisenberg linear chain system with S=1/2. The result reveals that this method gives a very accurate approximation.

We obtain E_g taking the limit of $T\rightarrow 0$ as follows:

$$E_g = \lim_{T\to 0} [-k_BT \log(Z)] = E_0 + \Delta E_2,$$

$$E_0 = H_{00}, \quad \Delta E_2 = \sum_{n\neq 0} |H'_{0n}|^2/(H_{00}-H_{nn}) \; . \quad (2.3)$$

The numerical results are given in Table 1.

Table 1 The energy of the ground state E_g given by the present method and the rigorous value $E_{g.r}$ for N spins. The error means $(E_{g,r}-E_g)/E_{g.r}$. see also Fig. 2

N	E_0	ΔE_2	E_g	$E_{g,r}$	error	$E_{g.r}/N$	$(E_g+\Delta E_2)/N$
4	-3.0000	-0.1875	-3.1875	-3.2321	0.0138	-0.8080	-0.8438
8	-6.4642	-0.2212	-6.6854	-6.7499	0.0096	-0.8437	-0.8633
12	-9.9744	-0.2343	-10.2087	-10.2840	0.0073	-0.8570	-0.8702
16	-13.4988	-0.2409	-13.7407	-13.8235	0.0060	-0.8640	-0.8738
20	-17.0322	-0.2448	-17.2770	-17.3640	0.0050	-0.8682	-0.8761

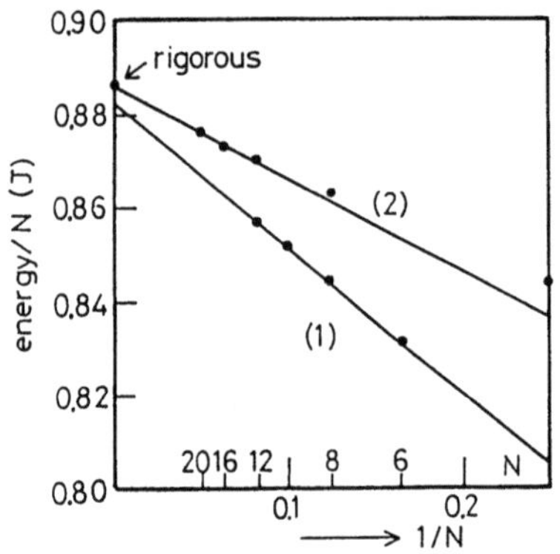

Fig. 2 The energy of the ground state versus 1/N: (1) $E_{g,r}/N$, (2) $(E_g+\Delta E_2)/N$

As is clearly seen in Table 1, our result E_g is a very accurate approximation to the rigorous $E_{g,r}$, i.e. the relative error for N=4 is less than 1.4%, and it decreases as N is increased, because the expected value of the unperturbed Hamiltonian H_A+H_B increases so that the perturbation H' becomes less important as N is increased.

Figure 2 shows the plots of $E_{g,r}/N$ and $(E_g+\Delta E_2)/N$ versus 1/N. Both plots are a little convex against the 1/N axis. The extrapolations by straight lines to $N=\infty$ show that the latter is better than the former, as is expected.

In the case of linear chains, it is simpler to solve them under the periodic boundary condition[1] than the free boundary condition. However, in some cases in which frustration exists such as in the antiferromagnetic Heisenberg model on the triangular lattice, it is not correct to solve them under the periodic boundary condition and use the simple extrapolations as MARLAND and BETTS[2] did. Although OGUCHI et al.[3] obtained them under the free boundary condition and used the railroad trestle extrapolation[4], the present method is expected to be more accurate.

In order to obtain the partition function, we have to compute the shifts of all energy levels. We have been continuing with the computations.

3. The Second Method

Next, we present the second method by which we can also calculate the thermodynamic quantities of a finite quantum spin system with fairly small computer storage requirements compared with calculating them by direct diagonalization. To put it more concretely, we divide a system into several subsystems. We apply the Trotter-formula to them[5], and calculate thermodynamic quantities rigorously for some Trotter-number n. For example, to diagonalize a quantum system with 18 spins(2X9) explicitly, we must prepare about 9.5×10^3MB(mega-byte). By our method, only about 17MB is necessary. Our method is also powerful for calculating the system which contains frustration, because we calculate the Trotter-formula rigorously, namely, we are free from minus-sign problems which appear in the usual quantum Monte Carlo methods[6].

Now, we explain our method by using a system with 10 spins (Fig. 3). We define A-, B- and C-clusters (which consist of 4, 4 and 2 spins respectively) as is shown in Fig. 3(a). A-cluster can take $2^4=16$ states (16 spin configurations). We denote the state of the A-cluster by a state variable p, namely, p takes the values 1,2,⋯,16 each of which corresponds to one of 16 states. Similarly, we denote the states of B- and C- clusters by q and r, respectively. We divide the system into two subsystems (Fig. 3(b)). The C-cluster is contained in both the subsystems. Then the Hamiltonian H_{tot} can be written as

$$H_{tot} = H_1(p,r) + H_2(q,r) , \tag{3.1}$$

where $H_1(H_2)$ depends only on state variables p and r(q and r) because $H_1(H_2)$ consists of only A- and C-(B- andC-)clusters. Now the partition function Z can be written as

$$\begin{aligned} Z &= \mathrm{Tr}[\exp(-KH_{tot})] \\ &= \lim_{n\to\infty} \mathrm{Tr}[\exp(-KH_1(p,r)/n)\cdot\exp(-KH_2(q,r)/n)]^n , \end{aligned} \tag{3.2}$$

where we have applied the Trotter-formula to our system.

From now on, we explain a method to calculate the partition function Z_n for finite Trotter number n rigorously;

$$Z_n = \mathrm{Tr}[\exp(-KH_1(p,r)/n)\cdot\exp(-KH_2(q,r)/n)]^n. \tag{3.3}$$

When we calculate Z_n directly, $\exp(-KH_1(p,r)/n)$ and $\exp(-KH_2(q,r)/n)$ must be represented by $2^{10}\times2^{10}$ matrices. Therefore, we first write out $\exp(-KH_1(p,r)/n)$ and $\exp(-KH_2(q,r)/n)$ in terms of only r explicitly. Then, they are written as the following 4×4 matrices:

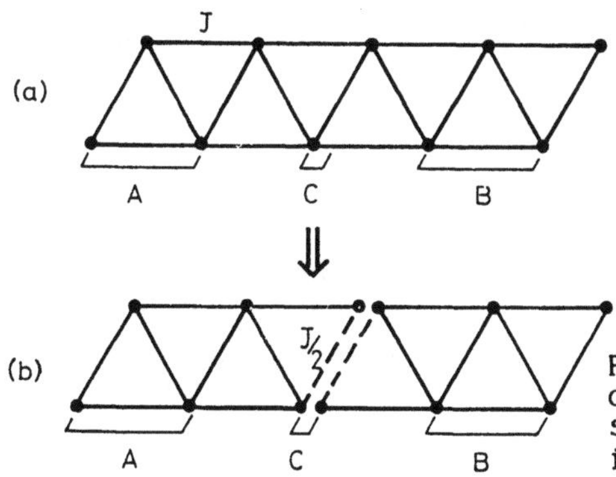

Fig. 3 A spin system which consists of 10 spins: (a)total system,(b) the system divided into two subsystems.

$$\exp(-KH_1(p,r)/n) = \begin{bmatrix} U_{1,1}(p) & U_{1,2}(p) & U_{1,3}(p) & U_{1,4}(p) \\ U_{2,1}(p) & U_{2,2}(p) & U_{2,3}(p) & U_{2,4}(p) \\ U_{3,1}(p) & U_{3,2}(p) & U_{3,3}(p) & U_{3,4}(p) \\ U_{4,1}(p) & U_{4,2}(p) & U_{4,3}(p) & U_{4,4}(p) \end{bmatrix}$$

$$\exp(-KH_2(q,r)/n) = \begin{bmatrix} V_{1,1}(q) & V_{1,2}(q) & V_{1,3}(q) & V_{1,4}(q) \\ V_{2,1}(q) & V_{2,2}(q) & V_{2,3}(q) & V_{2,4}(q) \\ V_{3,1}(q) & V_{3,2}(q) & V_{3,3}(q) & V_{3,4}(q) \\ V_{4,1}(q) & V_{4,2}(q) & V_{4,3}(q) & V_{4,4}(q) \end{bmatrix}, \quad (3.4)$$

where the columns and rows are arranged according to the order ++, +-, -+ and -- for the state variable r. Here both the state variables p and q take 2^4 states. Therefore, $\{U_{i,j}(p)\}$ and $\{V_{i,j}(q)\}$ are actually $2^8 \times 2^8$ matrices in the matrix representation. It should be noticed, however, that $\{U_{i,j}(p)\}$ and $\{V_{i,j}(q)\}$ only depend on the state variables p and q, respectively, so that they commute with each other. Putting (3.4) into (3.3), we obtain

$$Z_n = \mathrm{Tr}\Big[\sum_{n1,n2,\cdots,ny=1}^{4} \{U_{n1,n2}(p)\cdot V_{n2,n3}(q)\cdot U_{n3,n4}(p)\cdot V_{n4,n5}(q) \cdots\cdots U_{nx,ny}(p)\cdot V_{ny,n1}(q)\}\Big]$$

$$= \sum_{n1,n2,\cdots,ny=1}^{4} \mathrm{Tr}[U_{n1,n2}(p)\cdot V_{n2,n3}(q)\cdot U_{n3,n4}(p)\cdot V_{n4,n5}(q) \cdots\cdot U_{nx,ny}(p)\cdot V_{ny,n1}(q)], \quad (3.5)$$

where the summation comes from the product of $\exp(-KH_1/n)=\{U_{i,j}\}$ and $\exp(-KH_2/n)=\{V_{i,j}\}$ in (3.3), and we have taken the trace over the state variable r (namely the first and the last suffixes must be the same, i.e. n1), so that the trace in (3.5) must be taken over the state variables p and q, i.e. $2^8 \times 2^8$ matrices. Now we use the above important property that $\{U_{i,j}(p)\}$ and $\{V_{i,j}(q)\}$ commute with each other. Then, we find

$$Z_n = \sum_{n1,n2,\cdots,ny=1}^{4} \mathrm{Tr}[U_{n1,n2}(p)\cdot U_{n3,n4}(p)\cdots\cdots U_{nx,ny}(p) \cdot V_{n2,n3}(q)\ V_{n4,n5}(q)\cdots\cdots V_{ny,n1}(q)]. \quad (3.6)$$

As is written before, $\{U_{i,j}(p)\}$ only depends on the state variable p, and $\{V_{i,j}(q)\}$ only depends on the state variable q. Therefore, every $U_{i,j}(p)$ and $V_{i,j}(q)$ may be written as the direct products of two spaces, each of which is in terms of the state variables p and q, respectively (hereafter we call them p-space and q-space respectively);

$$U_{i,j}(p) = u_{i,j}(p) \otimes 1^q,$$

$$V_{i,j}(q) = 1^p \otimes v_{i,j}(q) \;, \qquad (3.7)$$

where 1^p and 1^q are the $2^4 \times 2^4$ unit matrices in p- and q-spaces respectively, and $u_{i,j}(p)(v_{i,j}(q))$ is a $2^4 \times 2^4$ matrix in p-(q-) space. Using (3.7), (3.6) can be written as

$$Z_n = \sum_{n1,n2,\cdots,ny=1}^{4} \mathrm{Tr}[(u_{n1,n2}(p)\cdot u_{n3,n4}(p)\cdots\cdots u_{nx,ny}(p)) \otimes (v_{n2,n3}(q)\cdot v_{n4,n5}(q)\cdots\cdots v_{ny,n1}(q))]$$

$$= \sum_{n1,n2,\cdots,ny=1}^{4} \mathrm{Tr}^p[u_{n1,n2}(p)\cdot u_{n3,n4}(p)\cdots\cdots u_{nx,ny}(p)] \cdot \mathrm{Tr}^q[v_{n2,n3}(q)\cdot v_{n4,n5}(q)\cdots\cdots v_{ny,n1}(q)], \qquad (3.8)$$

where Tr^p and Tr^q denote the trace performed over p- and q-space, respectively.

Equation (3.8) is the main result of this method. Namely, the number of memories we must prepare is not that of the corresponding system but that of divided subsystems. By this method, we can greatly decrease the number of memories, and calculate Z_n for larger systems rigorously. Here we have divided the system into two subsystems. To divide the system into three or more subsystems is also possible.

As a simple example, we calculate Z_n of the anti-ferromagnetic Heisenberg system (S=1/2) shown in Fig. 3(a) for Trotter number n=1,2,3 and 4 at T/J=0.5 (Fig. 4). In this case, the system consists of only 10 spins. Therefore, the exact partition function Z can be obtained, and $Z=2.206\times10^8$. On the other hand, the extrapolated value from Z_n (n=2,3,4) is 2.231×10^8. This value is about 1.1% larger than the true value.

We have been carrying out computations for the anti-ferromagnetic Heisenberg(XY) model; the results will be reported in the near future.

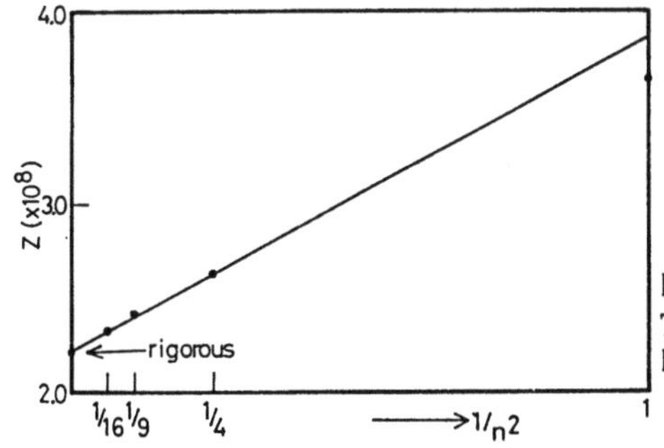

Fig. 4 Z and Z_n versus $1/n^2$ at T/J=0.5 for the system shown in Fig.3

Appendix

In this Appendix, we give a simple example to explain the actual calculation of the second method. We calculate Z_2 (namely, the Trotter number n=2,) for the system shown in Fig. A.1.

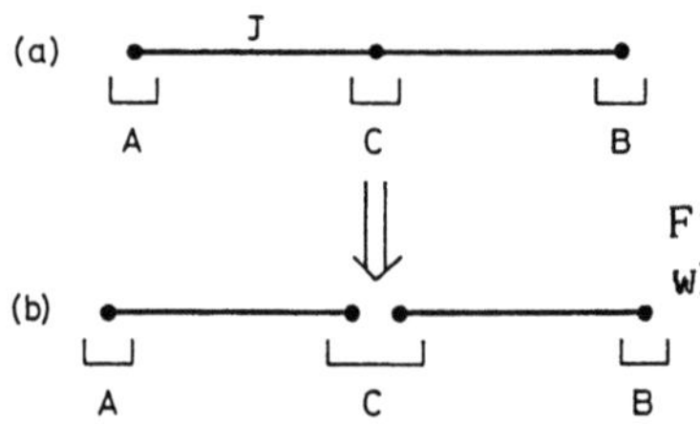

Fig. A.1 A system(a) with three spins, which is divided into two subsystems(b)

In this case, the C-cluster consists of only one spin, so that r takes only two values. From (3.5), we must calculate

$$Z_2 = \sum_{n1,n2,n3,n4=1}^{2} \mathrm{Tr}[U_{n1,n2}(p)\cdot V_{n2,n3}(q)\cdot U_{n3,n4}(p)\cdot V_{n4,n1}]. \tag{A.1}$$

Now, we calculate the trace in the summation of (A.1) for the case n1=1, n2=2, n3=2 and n4=1. Since $\{U_{i,j}\}$ and $\{V_{i,j}\}$ commute with each other, we obtain

$$\begin{aligned} &\mathrm{Tr}[U_{1,2}(p)\cdot V_{2,2}(q)\cdot U_{2,1}(p)\cdot V_{1,1}(q)] \\ &= \mathrm{Tr}[U_{1,2}(p)\cdot U_{2,1}(p)\cdot V_{2,2}(q)\cdot V_{1,1}(q)]. \end{aligned} \tag{A.2}$$

Here all $\{U_{i,j}\}$ and $\{V_{i,j}\}$ are $2^2 \times 2^2$ matrices. For example, $U_{1,2}(p)$ is written as

$$U_{1,2}(p) = \begin{bmatrix} 0 & 0 & 0 & 0 \\ 0 & 0 & 0 & 0 \\ x & 0 & 0 & 0 \\ 0 & x & 0 & 0 \end{bmatrix}. \qquad \begin{aligned} x &= -\exp(K/2)\sinh(K) \\ y &= \exp(K/2)\cosh(K) \\ z &= \exp(-K/2) \end{aligned} \tag{A.3}$$

Now we use (3.7). Then $U_{1,2}(p)$ may be written as

$$U_{1,2}(p) = \begin{bmatrix} 0 & 0 \\ x & 0 \end{bmatrix} \otimes \begin{bmatrix} 1 & 0 \\ 0 & 1 \end{bmatrix} = u_{1,2}(p) \otimes 1^q. \tag{A.4}$$

Similarly, $U_{2,1}(p)$, $V_{2,2}(q)$ and $V_{1,1}(q)$ can be also written as

$$U_{2,1}(p) = \begin{bmatrix} 0 & x \\ 0 & 0 \end{bmatrix} \otimes \begin{bmatrix} 1 & 0 \\ 0 & 1 \end{bmatrix} = u_{2,1}(p) \otimes 1^q,$$

$$V_{2,2}(q) = \begin{bmatrix} 1 & 0 \\ 0 & 1 \end{bmatrix} \otimes \begin{bmatrix} y & 0 \\ 0 & z \end{bmatrix} = 1^p \otimes v_{2,2}(q),$$

$$V_{1,1}(q) = \begin{bmatrix} 1 & 0 \\ 0 & 1 \end{bmatrix} \otimes \begin{bmatrix} z & 0 \\ 0 & y \end{bmatrix} = 1^p \otimes v_{1,1}(q). \qquad (A.5)$$

Using the identity

$$(D \otimes E) \cdot (F \otimes G) = (D \cdot F) \otimes (E \cdot G),$$

equation (A.2) can be written as

$$\begin{aligned} &\mathrm{Tr}[U_{1,2}(p) \cdot U_{2,1}(p) \cdot V_{2,2}(q) \cdot V_{1,1}(q)] \\ &= \mathrm{Tr}[\{u_{1,2}(p) \cdot u_{2,1}(p)\} \otimes \{v_{2,2}(q) \cdot v_{1,1}(q)\}] \\ &= \mathrm{Tr}^p[u_{1,2}(p) \cdot u_{2,1}(p)]\ \mathrm{Tr}^q[v_{2,2}(q) \cdot v_{1,1}(q)]. \qquad (A.6) \end{aligned}$$

Since

$$u_{1,2}(p) \cdot u_{2,1}(p) = \begin{bmatrix} 0 & 0 \\ 0 & x^2 \end{bmatrix},$$

$$v_{2,2}(q) \cdot v_{1,1}(q) = \begin{bmatrix} yz & 0 \\ 0 & yz \end{bmatrix},$$

we have

$$\begin{aligned} &\mathrm{Tr}^p[u_{1,2}(p) \cdot u_{2,1}(p)] = x^2, \\ &\mathrm{Tr}^q[v_{2,2}(q) \cdot v_{1,1}(q)] = 2yz. \qquad (A.7) \end{aligned}$$

Therefore, we obtain

$$\mathrm{Tr}[U_{1,2}(p) \cdot V_{2,2}(q) \cdot U_{2,1}(p) \cdot V_{1,1}(q)] = e^K \sinh(K) \sinh(2K). \quad (A.8)$$

By similar calculations, we can obtain Z_2.

References

1. J.C. Bonner and M.E. Fisher: Phys. Rev. 135, A640 (1964)
2. L.G. Marland and D.D. Betts: Phys. Rev. Lett. 43, 1618 (1979)
3. T. Oguchi, H. Nishimori and Y. Taguchi: J. Phys. Soc. Jpn. 55, 323 (1986)
4. P.W. Anderson: Mater. Res. Bull. 8, 153 (1973)
5. M. Suzuki: J. Stat. Phys. 43, 883 (1986)
6. For example, see M. Takasu, S. Miyashita and M. Suzuki: Prog. Theor. Phys. 75, 1254 (1986)

Monte Carlo Method for Quantum Spin Systems with Use of the Bloch Coherent State Representation

H. Takano

Department of Physics, Faculty of Science and Technology, Keio University, Yokohama 223, Japan

1. Introduction

Since the pioneering work of SUZUKI, MIYASHITA and KURODA [1], there have been many attempts to apply the Monte Carlo method to quantum spin systems [2], which utilize the Suzuki-Trotter formula [3]:

$$\exp\left(\sum_{j=1}^{p} A_j\right) = \lim_{n\to\infty}\left(e^{A_1/n} e^{A_2/n} \cdots e^{A_p/n}\right)^n. \qquad (1)$$

Here, A_j's are bounded operators. Most of the applications dealt with cases where the magnitude of the spin S is small. Their method is efficient for small S.

In this paper, another Monte Carlo method for quantum spin systems proposed by the author [4] is explained. This method utilizes a path integral representation of the partition function of a system [5] based on the Bloch coherent state representation [6, 7]. In this method, S appears only as a parameter. Therefore, the method may be useful in studying how the behavior of a system changes as the magnitude of the spin S is changed.

In the present Monte Carlo method, however, the convergence of thermal averages becomes slow for large systems because of the fluctuation of the phase of the path weight, if the number of time slices n in the path integral is larger than 2. Therefore, it is practically difficult to make an extrapolation to the limit $n \to \infty$ from the calculations with various values of n. In this paper, we propose a new method to estimate corrections to an approximation with a fixed value of n. This estimation method is useful not only for the present Monte Carlo method but also for the other Monte Carlo and transfer matrix methods which utilize the path integral representation of the partition function of a system.

It is believed that the quantum effects do not change the universality class of the usual critical phenomena at finite temperatures. Concerning the two-dimensional spin-S X-Y model, however, it is still a open question whether the spin-1/2 and the spin infinity models belong to the same universality class or not. In the limit $S \to \infty$, which is the classical limit, this model is expected to show the Kosterlitz-Thouless type transition [8], where the susceptibility diverges exponentially in approaching the transition temperature from above. The results of the high temperature series expansion support this exponential divergence of the susceptibility [9]. On the contrary, in the case of S = 1/2, which is the quantum limit, the high temperature series suggests that the susceptibility diverges with the usual power law [10]. If this suggestion is correct, the spin-1/2 model does not belong to the same universality class as the spin infinity model and the nature of the transition changes as the magnitude of the spin S decreases from infinity to 1/2.

Under these circumstances, it is a very interesting problem how the properties of the transition of the two-dimensional X-Y model are changed as the magnitude of the spin S is changed. The present Monte Carlo method may be useful to see this change induced by the quantum effects in the large S region.

In §2, a brief review of the Bloch coherent state representation is given. The present Monte Carlo method is explained in §3. A new method to estimate the corrections to an approximate path integral with a finite value of n is discussed in §4. Preliminary results of an application of the present Monte Carlo method to the two-dimensional X-Y model are presented in §5. Summary and discussion are given in §6.

2. The Bloch Coherent State Representation

In the following, a brief summary of the Bloch coherent state representation [5, 6] is given.

Let $\mathbf{S} = (S_x, S_y, S_z)$ be a quantum spin operator of magnitude S and $|S\rangle$ be a normalized eigenstate of S_z with the largest eigenvalue S: $\mathbf{S}^2|S\rangle = S(S+1)|S\rangle$ and $S_z|S\rangle = S|S\rangle$. Then, the Bloch coherent state $|\Omega\rangle$, which is specified by a set of two angles $\Omega = (\theta, \phi)$ with $0 \leqq \theta \leqq \pi$ and $0 \leqq \phi < 2\pi$, is defined by rotating $|S\rangle$ as

$$|\Omega\rangle \equiv \exp[i\theta(S_x\sin\phi - S_y\cos\phi)]|S\rangle. \tag{2}$$

The Bloch states are normalized and complete:

$$\langle\Omega|\Omega\rangle = 1 \tag{3}$$

and

$$\int d\Omega|\Omega\rangle\langle\Omega| = \mathbf{1}, \tag{4}$$

where

$$\int d\Omega \equiv \frac{2S+1}{4\pi}\int d\phi d\cos\theta. \tag{5}$$

They are overcomplete and not orthogonal:

$$\langle\Omega'|\Omega\rangle = \{\cos\frac{\theta}{2}\cos\frac{\theta'}{2} + e^{i(\phi - \phi')}\sin\frac{\theta}{2}\sin\frac{\theta'}{2}\}^{2S}, \tag{6}$$

where $\Omega = (\theta, \phi)$ and $\Omega' = (\theta', \phi')$.

For any operator A in the spin space, there exists a c-number function $G(\Omega)$ which satisfies

$$A = \int d\Omega G(\Omega)|\Omega\rangle\langle\Omega|. \tag{7}$$

The trace of the operator A is given by an integral of the function G:

$$\mathrm{Tr}\, A = \int d\Omega G(\Omega). \tag{8}$$

In order to find a c-number function corresponding to an operator, the product theorem [6] is useful. Let A and B be operators in the spin space and $G_B(\Omega)$ be the c-number function corresponding to the operator B. Then the c-number function $G_{AB}(\Omega)$ corresponding to the operator AB can be constructed from $G_B(\Omega)$ by

$$G_{AB}(\Omega) = A(\mathbf{S} \rightarrow \mathbf{S}^{(A)})G_B(\Omega). \tag{9}$$

Here, $A(\mathbf{S} \rightarrow \mathbf{S}^{(A)})$ is an operator obtained by replacing the spin operator $\mathbf{S}$ in the operator A by an operator $\mathbf{S}^{(A)}$ acting on the space of Ω, which is defined by

$$\mathbf{S}^{(A)} = (S + 1)\mathbf{m} + \frac{1}{2}\mathbf{L} + \frac{i}{2}(\mathbf{m} \times \mathbf{L}) \tag{10}$$

with

$$\mathbf{m} = (\sin\theta\cos\phi,\ \sin\theta\sin\phi,\ \cos\theta) \tag{11}$$

and

$$\mathbf{L} = i(\sin\phi\frac{\partial}{\partial\theta} + \cot\theta\cos\phi\frac{\partial}{\partial\phi},\ -\cos\phi\frac{\partial}{\partial\theta} + \cot\theta\sin\phi\frac{\partial}{\partial\phi},\ -\frac{\partial}{\partial\phi}). \tag{12}$$

We can calculate $G(\Omega)$ for A from (9) by setting $B = \mathbf{1}$ and $G_B(\Omega) = 1$.

3. Monte Carlo Method

By using the Bloch coherent state representation, we can express the partition function of a quantum spin system as a path integral over classical spin variables [5]. The partition function Z of a system is given by

$$Z = \mathrm{Tr}\ \exp(H) = \mathrm{Tr}\ [\exp(H/n)]^n, \tag{13}$$

Here, H is the Hamiltonian of the system including the inverse temperature factor $-1/(k_BT)$. Let the classical representation (7) for the operator $\exp(H/n)$ be given by $\exp(H_n(\Omega))$:

$$\exp(H/n) = \int d\Omega \exp(H_n(\Omega))|\Omega\rangle\langle\Omega|. \tag{14}$$

Then, the partition function is expressed as an integral over classical spin variables:

$$Z = \mathrm{Tr} \int\{d\Omega\}[\exp(H_n(\Omega_1))|\Omega_1\rangle\langle\Omega_1|\cdots\exp(H_n(\Omega_n))|\Omega_n\rangle\langle\Omega_n|]$$

$$= \int\{d\Omega\} \prod_{i=1}^{n} [\exp(H_n(\Omega_i))\langle\Omega_i|\Omega_{i+1}\rangle], \tag{15}$$

with $\Omega_{n+1} = \Omega_1$ and $\{d\Omega\} = \Pi_{i=1}^n d\Omega_i$. Equation (15) defines a "weight" $W(\{\Omega\})$ for a configuration $\{\Omega\}$ of classical spin variables:

$$W(\{\Omega\}) = \prod_{i=1}^{n} [\exp(H_n(\Omega_i))\langle\Omega_i|\Omega_{i+1}\rangle]. \tag{16}$$

The thermal average of an operator A is defined by

$$\langle A\rangle = \mathrm{Tr}\ A\exp(H)\ /\ \mathrm{Tr}\ \exp(H). \tag{17}$$

By introducing a c-number function $\tilde{A}(\Omega)$ corresponding to the operator A, which is defined by

$$A\exp(H/n) = \int d\Omega \tilde{A}(\Omega)\exp(H_n(\Omega))|\Omega\rangle\langle\Omega|, \tag{18}$$

we can express Tr Aexp(H) as

$$\begin{aligned} \mathrm{Tr}\, A\exp(H) &= \mathrm{Tr}\, A\exp(H/n)[\exp(H/n)]^{n-1}, \\ &= \mathrm{Tr} \int\{d\Omega\}[\tilde{A}(\Omega_1)\exp(H_n(\Omega_1))|\Omega_1\rangle\langle\Omega_1|\exp(H_n(\Omega_2))|\Omega_2\rangle\langle\Omega_2| \\ &\qquad \cdots \exp(H_n(\Omega_n))|\Omega_n\rangle\langle\Omega_n|] \\ &= \int\{d\Omega\}\tilde{A}(\Omega_1)\prod_{i=1}^{n}[\exp(H_n(\Omega_i))\langle\Omega_i|\Omega_{i+1}\rangle]. \end{aligned} \tag{19}$$

Note that the "weight" (16) appears in (19).

Our Monte Carlo method for quantum spin systems is based on the classical spin representations of the partition function (15) and the thermal average (19). In the Monte Carlo simulation, the probability $P(\{\Omega\})$ for a classical spin configuration $\{\Omega\}$ to be sampled should reflect the importance of the configuration with respect to the weight $W(\{\Omega\})$. Because $W(\{\Omega\})$ is complex in general, $P(\{\Omega\})$ cannot be chosen to be proportional to $W(\{\Omega\})$ itself. Instead, we may choose $P(\{\Omega\})$ to be proportional to the absolute value of the real part of $W(\{\Omega\})$:

$$P(\{\Omega\}) \propto |\mathrm{Re}[W(\{\Omega\})]|. \tag{20}$$

The change of the variables $\Omega_i \to \Omega_{n+2-i}$ in (15) and (19) gives

$$\begin{aligned} Z &= \int\{d\Omega\}\prod_{i=1}^{n}[\exp(H_n(\Omega_i))\langle\Omega_i|\Omega_{i+1}\rangle] \\ &= \int\{d\Omega\}\prod_{i=1}^{n}[\exp(H_n(\Omega_i))\langle\Omega_{i+1}|\Omega_i\rangle] \\ &= \int\{d\Omega\}\prod_{i=1}^{n}\mathrm{Re}[\exp(H_n(\Omega_i))\langle\Omega_i|\Omega_{i+1}\rangle] \end{aligned} \tag{21}$$

and

$$\begin{aligned} \mathrm{Tr}\, A\exp(H) &= \int\{d\Omega\}\tilde{A}(\Omega_1)\prod_{i=1}^{n}[\exp(H_n(\Omega_i))\langle\Omega_i|\Omega_{i+1}\rangle] \\ &= \int\{d\Omega\}\tilde{A}(\Omega_1)\prod_{i=1}^{n}[\exp(H_n(\Omega_i))\langle\Omega_{i+1}|\Omega_i\rangle] \\ &= \int\{d\Omega\}\tilde{A}(\Omega_1)\prod_{i=1}^{n}\mathrm{Re}[\exp(H_n(\Omega_i))\langle\Omega_i|\Omega_{i+1}\rangle], \end{aligned} \tag{22}$$

because $\langle\Omega_{i+1}|\Omega_i\rangle = \langle\Omega_i|\Omega_{i+1}\rangle^*$. Thus, the thermal average $\langle A\rangle$ can be calculated from the Monte Carlo simulation with the sampling probability (20) through

$$\langle A\rangle = \langle\tilde{A}(\Omega_1)\rangle_{MC}. \tag{23}$$

Here, the average $\langle\ \rangle_{MC}$ for an arbitrary function f of $\{\Omega\}$ is defined by

$$\langle f(\{\Omega\})\rangle_{MC} = \frac{\langle f\mathrm{Re}(W)/|\mathrm{Re}(W)|\rangle_P}{\langle \mathrm{Re}(W)/|\mathrm{Re}(W)|\rangle_P}, \tag{24}$$

where $\langle\ \rangle_P$ denotes the average with respect to the sampling probability (20).

For a system of many spins, it is very difficult to find the exact expression of $H_n(\Omega)$, which is defined by (14). Therefore, we should make some approximation to $H_n(\Omega)$. A cumulant expansion [11] can be used as a systematic approximation [4]. Let $H^m(\Omega)$ be the c-number function corresponding to the operator H^m:

$$H^m = \int d\Omega H^m(\Omega)|\Omega\rangle\langle\Omega|. \tag{25}$$

Then, the operator exp(H/n) is expanded in 1/n as

$$\exp(H/n) = 1 + (H/n) + (1/2)(H/n)^2 + (1/6)(H/n)^3 + \cdots. \tag{26}$$

$$= \int d\Omega[1 + n^{-1}H(\Omega) + (1/2)n^{-2}H^2(\Omega) + (1/6)n^{-3}H^3(\Omega) + \cdots]|\Omega\rangle\langle\Omega|$$

$$= \int d\Omega \exp(H_n(\Omega))|\Omega\rangle\langle\Omega|, \tag{27}$$

which gives the cumulant expansion of $H_n(\Omega)$:

$$H_n(\Omega) = \ln[1 + n^{-1}H(\Omega) + (1/2)n^{-2}H^2(\Omega) + (1/6)n^{-3}H^3(\Omega) + \cdots]$$

$$= n^{-1}H(\Omega) + (1/2)n^{-2}[H^2(\Omega) - \{H(\Omega)\}^2]$$

$$+ (1/6)n^{-3}[H^3(\Omega) - 3H(\Omega)H^2(\Omega) + 2\{H(\Omega)\}^3] + \cdots. \tag{28}$$

By truncating (28) at the m-th order, we obtain the m-th order approximation $H_n^{(m)}(\Omega)$ to $H_n(\Omega)$. The c-number functions $H^m(\Omega)$ can be calculated by means of the product theorem (9) as

$$H^m(\Omega) = [H(\mathbf{S} \to \mathbf{S}^{(A)})]^m \cdot 1. \tag{29}$$

The c-number function $\tilde{A}(\Omega)$, which is defined by (18), is given by

$$\tilde{A}(\Omega) = \exp(-H_n(\Omega))A(\mathbf{S} \to \mathbf{S}^{(A)})\exp(H_n(\Omega)). \tag{30}$$

In the m-th order approximation, $H_n(\Omega)$ in (30) is replaced with its approximation $H_n^{(m)}(\Omega)$.

Let K be the dimensionless coupling constant of the Hamiltonian H. Then the order of the correction to the m-th order approximation is given by

$$H_n(\Omega) \sim H_n^{(m)}(\Omega) + O((K/n)^{m+1}). \tag{31}$$

Because $\exp(H_n(\Omega))$ appears n times as a product in the partition function (15), the correction to the approximate partition function is considered to be $O(K^{m+1}/n^m)$ [3, 4].

The present Monte Carlo method seems to work well for small systems [4]. In large systems, however, the fluctuation of the phase of the weight $W(\{\Omega\})$ becomes large and therefore the convergence of $\langle f\mathrm{Re}(W)/|\mathrm{Re}(W)|\rangle_P$ and $\langle \mathrm{Re}(W)/|\mathrm{Re}(W)|\rangle_P$ in (24) becomes slow. Thus, the convergence of thermal averages becomes very slow in large systems. This makes the method not practical for large systems.

In the case of n = 2, the weight $W(\{\Omega\})$ is always real and not negative if $\exp(H_n(\Omega))$ is real and positive, because $\langle\Omega_1|\Omega_2\rangle\langle\Omega_2|\Omega_1\rangle = |\langle\Omega_1|\Omega_2\rangle|^2$. Thus, the present method is practical for n = 2, because there is no phase fluctuation of the weight $W(\{\Omega\})$. Even though n is fixed, we can improve the approximation used in the present method systematically by making the order m of the approximation higher. If n is fixed to 2, however, we cannot make the extrapolation to the limit $n \to \infty$, where the approximation becomes exact. Instead, in the next section, we propose a new method to estimate the corrections to the approximation with a fixed value of n.

4. An Extrapolation Method

In this section, a method to estimate corrections to an approximate path integral with a fixed value of n is explained. Though the explanation is made for n = 2, the method is also applicable to the case of larger values of n.

We divide the Hamiltonian H into n slices with ratios $\{r_i\}$ as

$$\exp(H) = \prod_{i=1}^{n} \exp(r_i H) \tag{32}$$

with $\Sigma_{i=1}^{n} r_i = 1$. Equation (13) corresponds to the case $r_i = 1/n$. In the case of n = 2, (32) becomes

$$\exp(H) = \exp(rH)\exp((1 - r)H). \tag{33}$$

By replacing each slice with its m-th order approximant, we obtain

$$\exp(H) \sim [\exp(rH)]^{(m)}[\exp((1 - r)H)]^{(m)}, \tag{34}$$

where $[\ ]^{(m)}$ represents the m-th order approximation. Because the corrections to the m-th order approximation starts at the (m + 1)-th order, a thermal average $\langle A\rangle^{(m)}$ calculated from the m-th order approximation (34) is expected to have the following form of corrections:

$$\begin{aligned}\langle A\rangle^{(m)} = \langle A\rangle &\\ &+ C_{m+1}[r^{m+1} + (1 - r)^{m+1}]\\ &+ C_{m+2}[r^{m+2} + (1 - r)^{m+2}]\\ &+ \cdots.\end{aligned} \tag{35}$$

We can estimate the corrections as follows: First, we perform the simulations for various values of r's. Then, by fitting the results of simulations into the form (35), we can estimate the coefficients C_i's. Finally, we can estimate $\langle A\rangle$ by removing the r-dependent terms from $\langle A\rangle^{(m)}$.

Next, we show some simple examples in order to see how this method works. The first example is a single spin **S** of magnitude S, on which a magnetic field h of z-direction is applied: $H = hS_z$. We have calculated $\langle S_z\rangle$ numerically for S = 1/2 and h = 3 with the third order approximation (m = 3). The c-number functions $H_n^{(m)}$ and $\tilde{S}_z$ are given in [4]. Because $\langle S_z\rangle$ is an odd function of h, there exist only odd order corrections. The result for r = 0.5 is $\langle S_z\rangle/S = 0.9095$. The estimation of C_5 from the results for r = 0.5 and 0.4 gives $\langle S_z\rangle/S = 0.9071$. The estimation of C_5

and C_7 from the results for r = 0.5, 0.4 and 0.3 gives the value 0.9057. The estimation of C_5, C_7 and C_9 from the results for r = 0.5, 0.4, 0.3 and 0.2 gives the value 0.9051, which agrees well with the exact value 0.9051. Thus, the present method seems to give a good estimation of the corrections.

The second example is the one-dimensional isotropic Heisenberg model:

$$H = \frac{K}{S(S+1)} \sum_{i=1}^{N} \mathbf{S}_i \cdot \mathbf{S}_{i+1} \tag{36}$$

with $\mathbf{S}_{N+1} = \mathbf{S}_1$, where S is the magnitude of the spin. We have calculated $\langle \mathbf{M}^2 \rangle/(3N)$ for S = 1 and N = 4 from Monte Carlo simulations. Here, the reduced total magnetization is defined by

$$\mathbf{M} = \frac{1}{S(S+1)} \sum_{i=1}^{N} \mathbf{S}_i . \tag{37}$$

We have used the second order approximation (m = 2). The c-number functions $H_n^{(2)}$, $\hat{\mathbf{M}}$ and $(\hat{\mathbf{M}}^2)$ are given in [4]. The results of the Monte Carlo simulations are shown in Fig. 1. The data are obtained from the simulations of 800,000 MCS (Monte Carlo steps per spins) after initial 200,000 MCS. Each extrapolated value is calculated by estimating C_3 from the results for various r's. For K = 0.5, 1.0 and 1.5, the results for r = 0.5, 0.4 and 0.3 are used. The results for r = 0.5, 0.45, 0.42, 0.4 are used for K = 2.0. The extrapolated values seem to agree well with the exact results.

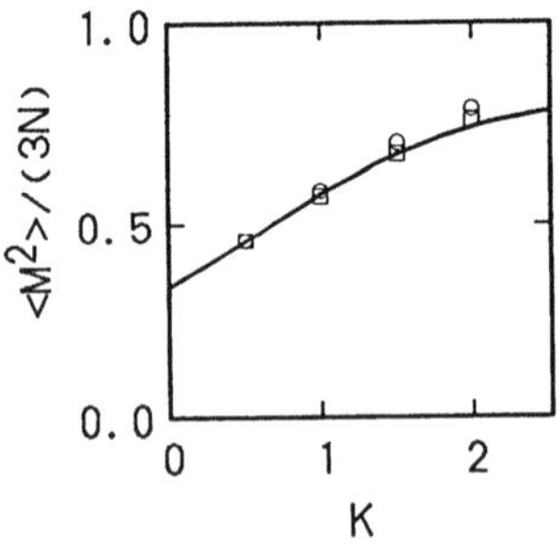

Fig. 1. Coupling constant dependence of $\langle \mathbf{M}^2 \rangle/(3N)$. Circles denote the results for r = 0.5 and squares denote the extrapolated values. Solid curve represents the exact result

5. The X-Y Model in Two Dimensions

In this section, preliminary results concerning the two-dimensional X-Y model are reported. The Hamiltonian of the spin-S X-Y model is given by

$$H = \frac{K}{S(S+1)} \sum_{\langle ij \rangle} (S_{x,i}S_{x,j} + S_{y,i}S_{y,j}), \tag{38}$$

where $\Sigma_{\langle ij \rangle}$ denotes the summation over all nearest neighbor pairs. As mentioned before, what we are interested in is to see how the properties of the transition of the two-dimensional X-Y model are changed as the magnitude of the spin S is changed.

In the following, we consider the X-Y model on the L × L square lattice with periodic boundary conditions. We have performed the Monte Carlo simulations for the case of S = 2. We have calculated

$$\hat{\chi} \equiv \langle M_x{}^2 + M_y{}^2\rangle/(2L^2) = \langle(\Sigma_i S_{x,i})^2 + (\Sigma_i S_{y,i})^2\rangle/[2L^2 S(S + 1)], \tag{39}$$

which is expected to show the same singularity as the susceptibility, for L = 4, 6 and 8 with r = 0.5 and 0.3. The lengths of the simulations are 5,000 ∿ 15,000 MCS. It should be noted that no significant r-dependence was found in the results of the simulations. The results for r = 0.5 are shown in Fig. 2. We can see the strong size-dependence of $\hat{\chi}$ for K ≧ 1.5.

According to the finite-size scaling, a quantity

$$y(K, L, L') \equiv \ln[\hat{\chi}(K, L) / \hat{\chi}(K, L')] / \ln[L / L'] \tag{40}$$

becomes independent of L and L' at the critical point for sufficiently large L and L'. Its value at the critical point is related to the magnetic exponent y_H through

$$y = 2y_H - d, \tag{41}$$

where d is the dimension of the system. If the system shows the Kosterlitz-Thouless type transition, then the low temperature phase is always critical with the temperature-dependent magnetic exponent and the value of the magnetic exponent y_H at the transition point is 15/8, which corresponds to y = 7/4 [8].

Figure 3 shows the K-dependence of y for (L, L') = (4, 6) and (6, 8). The behavior of y suggests that the system is always critical in the low temperature region, which is consistent with the Kosterlitz-Thouless type transition. The transition point seems to be between K = 1 and 1.5.

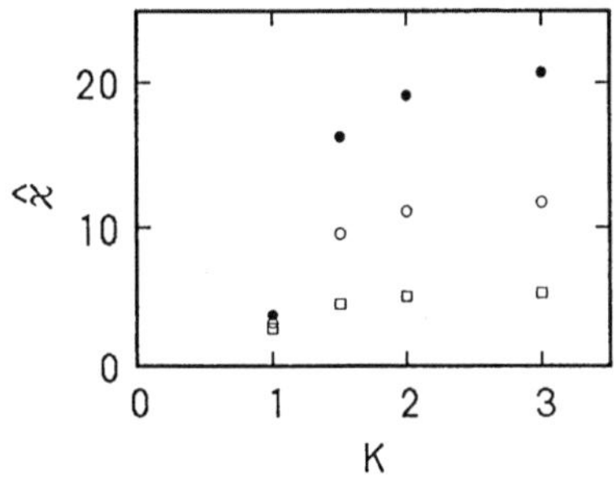

Fig. 2. Coupling constant dependence of $\hat{\chi}$. Squares, open and solid circles correspond to L = 4, 6 and 8, respectively

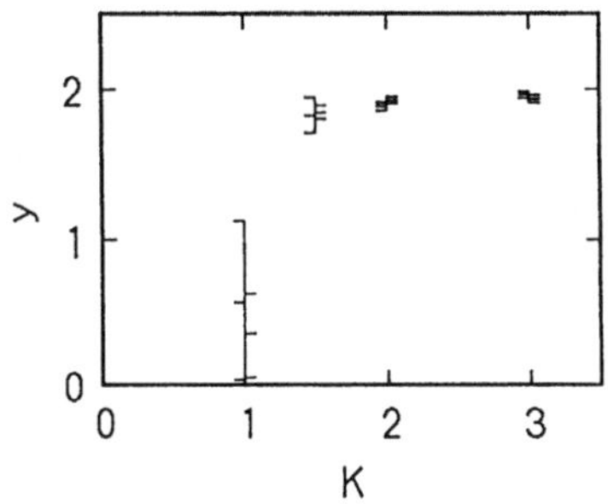

Fig. 3. Coupling constant dependence of y. Right and left sides of error bars correspond to (L, L') = (4, 6) and (6, 8), respectively

6. Summary and Discussion

We have explained a Monte Carlo method based on the Bloch coherent state representation. In this method the magnitude of the spin S appears only as a parameter. Approximation used in this method can be systematically improved by a kind of cumulant expansion. Because of the phase fluctuation in large systems, this method is only practical for the case that the number of the slices n = 2.

A method to estimate corrections to the approximation is proposed. In this method, the number of slices n is fixed and the ratios of decomposition of the original Hamiltonian into n slices are changed. In this paper, this method has been used with the present Monte Carlo method in the case of $n = 2$. It can also be used with other Monte Carlo and transfer matrix methods which utilize the Suzuki-Trotter formula with $n > 2$. Especially, the combination of this method with the transfer matrix method seems appropriate, because there is no statistical error in results of the transfer matrix method.

Finally, the X-Y model on the square lattice is studied by the present method. The preliminary results for $S = 2$ suggest that the low temperature phase is always critical, which is consistent with the Kosterlitz-Thouless type transition. The transition point seems to lie between $K = 1.0$ and 1.5. This value of K is smaller than that for the spin-1/2 X-Y model and larger than that for the plane-rotator model [8, 9, 10], which is expected from the consideration that the quantum fluctuation reduces the transition temperature.

References

1. M. Suzuki, S. Miyashita and A. Kuroda, Prog. Theor. Phys. **58**(1977)1377.
   M. Suzuki, Prog. Theor. Phys. **56**(1976)1454.
2. H. De Raedt and Ad Lagendijk, Phys. Rep. **127**(1985)233.
3. M. Suzuki, Commun. Math. Phys. **51**(1976)183.
4. H. Takano, Prog. Theor. Phys. **73**(1985)332.
5. E.H. Lieb, Commun. Math. Phys. **31**(1973)327.
6. Y. Takahashi and F. Shibata, J. Stat. Phys. **14**(1976)49.
7. F.T. Arrechi, E. Courtens, R. Gilmore and H. Thomas, Phys. Rev. **A6**(1972)2211.
   J.M. Radcliffe, J. Phys. **A4**(1971)313.
8. J.M. Kosterlitz and D.J. Thouless, J. Phys. **C6**(1973)1181.
   J.M. Kosterlitz, J. Phys. **C7**(1974)1046.
9. W.J. Camp and J.P. Van Dyke, J. Phys. **C8**(1975)336.
   A.J. Guttmann, J. Phys. **A11**(1978)545.
10. J. Rogiers, E.W. Grundke and D.D. Betts, Can. J. Phys. **57**(1979)1719.
11. R. Kubo, J. Phys. Soc. Jpn. **17**(1962)1100.

Decoupled Cell Monte Carlo Method for Quantum Spin Systems

S. Homma[1], *K. Sano*[2], *H. Matsuda*[3], *and N. Ogita*[4]

[1]Department of Applied Physics, Nagoya University, Nagoya 464, Japan
[2]Department of Physics, Nagoya University, Nagoya 464, Japan
[3]Department of Biology, Kyushu University, Fukuoka 812, Japan
[4]The Institute of Physical and Chemical Research, Wako 351-01, Japan

A decoupled cell method is proposed as a new method of Monte Carlo simulation for quantum spin systems. Thermodynamic quantities such as internal energy, in-plane and out-of-plane susceptibilities and spin pair correlation functions are calculated by this method in the 1D XY model. The results obtained agree well with the exact ones except for very low temperatures, indicating the validity of this method. Further application of this method strongly suggests (1) the existence of a phase transition in the quantum XY model on the square lattice similar to the Kosterlitz-Thouless transition in the classical XY model, and (2) the existence of a sublattice structure composed of three lattices which is short-ranged both spatially and temporally in the antiferromagnetic Heisenberg model on the triangular lattice.

1. Introduction

Recently, a number of Monte Carlo methods have been proposed and tested for quantum mechanical spin systems. There have been two major categories of such attempts. The first category is to evaluate the thermodynamic quantities as a high-temperature series expansion [1], where the Monte Carlo calculation is used to estimate expansion coefficients. The second category is to use the generalized Trotter formula to transform a d-dimensional quantum system into a mathematically equivalent (d+1)-dimensional classical system [2,3], in which the time-honored Metropolis Monte Carlo procedure can be applicable. Many studies have followed the method of one of these two categories [4-18]. However, these approaches are technically rather complicated, so that it is not easy to treat quantum systems in two or three dimensions.

In view of these circumstances we introduced a simpler method applicable to such higher dimensional quantum systems without complexity. This method , which we call 'Decoupled Cell Method' (DCM), is a direct extension of the classical Metropolis method [19-21].

In this report we give a brief account of DCM and results of our numerical studies of the 1D and 2D XY model and the antiferromagnetic Heisenberg model on the triangular lattice (AHTL).

In the next section we give a basic idea of DCM. In section 3 the results of the 1D XY model are shown and compared with the known rigorous results. In section 4 the results are given of the XY model on the square lattice and the antiferromagnetic Heisenberg model on the triangular lattice. The last section is devoted to discussion.

2. Decoupled Cell Method

To explain the basic idea of DCM, let us consider the anisotropic Heisenberg model on one-dimensional torus of N sites with Hamiltonian

$$H = -\tfrac{1}{2}\sum(J_x\hat{\sigma}_i^x\hat{\sigma}_{i+1}^x + J_y\hat{\sigma}_i^y\hat{\sigma}_{i+1}^y + J_z\hat{\sigma}_i^z\hat{\sigma}_{i+1}^z) - H\sum\hat{\sigma}_i^z , \tag{1}$$

where $\hat{\sigma}_i^x$, $\hat{\sigma}_i^y$, $\hat{\sigma}_i^z$ are the components of the Pauli spin operator of the i-th site. J_x, J_y and J_z are coupling constants, and H is the external magnetic field in z-direction. Generally, we take a representation in which the following component of the spin operator is diagonal:

$$\hat{\sigma}_i = l\hat{\sigma}_i^x + m\,\hat{\sigma}_i^y + n\,\hat{\sigma}_i^z , \tag{2}$$

where l,m and n are direction cosines. Therefore the value of $\hat{\sigma}_i$ is either 1 or -1.

Now, let $L_i^{(\nu)}$ denote a set of sites within ν-th neighbors of the i-th site: $L_i^{(\nu)} = [i-\nu, i-\nu+1,\dots,i,\dots,i+\nu]$, ($\nu=1,2,\dots$). Let $\bar{L}_i^{(\nu)}$ be the set of all the sites not belonging to $L_i^{(\nu)}$. Let S_i^{σ} denote a spin configuration ($\sigma_{i-\nu}$, $\sigma_{i-\nu+1},\dots,\sigma_i,\dots,\sigma_{i+\nu}$) of $L_i^{(\nu)}$ where $\sigma_i = \sigma \in \{\pm 1\}$. Let $\bar{S}_i$ denote a spin configuration of $\bar{L}_i^{(\nu)}$: $\bar{S}_i = (\sigma_1, \sigma_2,\dots,\sigma_{i-\nu-1},\dots \sigma_N)$. Thus the spin configuration of the total system is given by (S_i^{σ}, $\bar{S}_i$). In the canonical distribution at temperature T, the probability of a spin configuration (S_i^{σ},$\bar{S}_i$) is given by

$$P(S_i^{\sigma},\bar{S}_i) = \langle S_i^{\sigma},\bar{S}_i|\exp(-\beta H)|S_i^{\sigma},\bar{S}_i\rangle/\mathrm{Tr}[\exp(-\beta H)], \tag{3}$$

where $\beta = 1/k_BT$; k_B is the Boltzmann constant. We denote the relative probability of the spin configuration (S_i^{σ},$\bar{S}_i$) to ($S_i^{-\sigma}$,$\bar{S}_i$) by

$$q(S_i^{\sigma},\bar{S}_i) = P(S_i^{\sigma},\bar{S}_i)/P(S_i^{-\sigma},\bar{S}_i). \tag{4}$$

This is the ratio of the probability of $\sigma_i = \sigma$ to $\sigma_i = -\sigma$ with all the other spins fixed.

The Monte Carlo method for classical case where $J_x = J_y = 0$, $l = m = 0$, $n = 1$, $\hat{\sigma}_i = \hat{\sigma}_i^z$, that is, for the Ising model, is based on the following: (I) The canonical ensemble can be generated by a time sequence of a Markov chain whose transition probability $W(S_i^{\sigma},\bar{S}_i)$ from (S_i^{σ},$\bar{S}_i$) to ($S_i^{-\sigma}$,$\bar{S}_i$), (i=1, 2,...,N) satisfies the condition of detailed balance at equilibrium such that

$$W(S_i^{\sigma},\bar{S}_i) = \begin{cases} q(S_i^{-\sigma},\bar{S}_i) & (q(S_i^{-\sigma},\bar{S}_i) < 1) \\ 1 & (q(S_i^{-\sigma},\bar{S}_i) \geq 1) \end{cases} \tag{5}$$

and (II) the above transition probability depends only on the spin configuration of $L_i^{(1)}$ such that

$$q(S_i^{\sigma},\bar{S}_i) = \exp[\beta(J_z\sigma(\sigma_{i-1} + \sigma_{i+1}) + 2H\sigma)], \tag{6}$$

so that the whole set of transition probabilities can easily be calculated.

Even in the quantum case the property (I) persists to hold, but the relative probability $q(S_i^\sigma, \bar{S}_i)$ generally depends on $\bar{S}_i$ as well as S_i^σ. However, its dependence on $\bar{S}_i$ will become less and less important as ν increase. Therefore, defining the decoupled cell Hamiltonian of the i-th site by

$$H_i^{(\nu)} = -\frac{1}{2}\sum_{j=i-\nu}^{i+\nu-1}(J_x\hat{\sigma}_j^x\hat{\sigma}_{j+1}^x + J_y\hat{\sigma}_j^y\hat{\sigma}_{j+1}^y + J_z\hat{\sigma}_j^z\hat{\sigma}_{j+1}^z) - H\sum_{j=i-\nu}^{i+\nu}\hat{\sigma}_j^z \quad (7)$$

we may approximate $q(S_i^\sigma, S_i)$ by

$$q(S_i^\sigma) = \langle S_i^\sigma|\exp(-\beta H_i^{(\nu)})|S_i^\sigma\rangle / \langle S_i^{-\sigma}|\exp(-\beta H_i^{(\nu)})|S_i^{-\sigma}\rangle. \quad (8)$$

The right-hand side of this equation can readily be obtained by solving the eigenvalue problem of $H_i^{(\nu)}$ by computer. Then, using $q(S_i^\sigma)$ for $q(S_i^\sigma, \bar{S}_i)$ in (5), we can follow the classical method of Monte Carlo simulation. Namely, a sample of the Markov chain is generated in a computer by repeating the process of random picking up of a site i and flipping of the i-th spin according to (5). The time average over this Markov chain approximates the canonical average of the corresponding physical quantity.

We apply this method (DCM) to the 1D XY model ($S=\frac{1}{2}$) for which the rigorous solutions are available.

3. One-dimensional XY Model ($S=\frac{1}{2}$)

The Hamiltonian of the 1D XY model here to be studied is given by (1) with $J_x = J_y = J$ and $J_z = 0$:

$$H = -\frac{1}{2}J\sum_i(\hat{\sigma}_i^x\hat{\sigma}_{i+1}^x + \hat{\sigma}_i^y\hat{\sigma}_{i+1}^y) - H\sum_i\hat{\sigma}_i^z, \quad (9)$$

where we assume $\hat{\sigma}^\alpha_{N+1} = \hat{\sigma}^\alpha_1$, ($\alpha = x,y,z$), and H is the external magnetic field.

Corresponding to the quantity we need to observe we choose a direction of diagonal representation for spin $\hat{\sigma}$.

(I) Taking the representation in which $\{\hat{\sigma}_i^x\}$ is diagonal, we can calculate the following physical quantities:

Internal energy: $\epsilon = -(1/N)\sum_{i=1}^{N} J\langle\hat{\sigma}_i^x\hat{\sigma}_{i+1}^x\rangle$, (10)

In-plane susceptibility: $\chi_\parallel = (1/N)\lim_{H\to 0}\langle\sum_i\hat{\sigma}_i^x\rangle/H$, (11)

Mean square magnetization: $m_2 = (1/Nk_BT)\langle(\sum_i\hat{\sigma}_i^x)^2\rangle$, (12)

Spin pair correlation function: $C^x(r) = (1/N)\sum_i\langle\hat{\sigma}_i^x\hat{\sigma}_{i+r}^x\rangle$, (13)

where $\langle A\rangle$ denotes the canonical average of A. $\hat{\sigma}_i^x$ and $\hat{\sigma}_i^y$ are given by

$$\hat{\sigma}_i^x = \begin{pmatrix}1 & 0\\ 0 & -1\end{pmatrix}, \quad \hat{\sigma}_i^y = \begin{pmatrix}0 & 1\\ 1 & 0\end{pmatrix}, \quad (14)$$

with $l = 1$, $m = n = 0$ in (2).

(II) By taking the representation in which $\{\hat{\sigma}_i^z\}$ is diagonal we can calculate zero-field susceptibility and spin pair correlation function $C^z(r)$ defined by

$$\chi_\perp = (1/Nk_BT)\left\langle (\sum_i \hat{\sigma}_i^z)^2 \right\rangle , \tag{15}$$

$$C^z(r) = (1/N)\sum_i \langle \hat{\sigma}_i^z \hat{\sigma}_{i+r}^z \rangle , \tag{16}$$

where $\hat{\sigma}_i^x$, $\hat{\sigma}_i^y$, and $\hat{\sigma}_i^z$ are given by

$$\hat{\sigma}_i^x = \begin{pmatrix} 1 & 0 \\ 0 & -1 \end{pmatrix} \quad \hat{\sigma}_i^y = \begin{pmatrix} 0 & 1 \\ 1 & 0 \end{pmatrix} \quad \hat{\sigma}_i^z = \begin{pmatrix} 0 & i \\ -i & 0 \end{pmatrix} . \tag{17}$$

The number of Monte Carlo steps is usually 25000 for each set of (ν , N) and temperature T. The observed quantities ϵ , $C^x(r)$, $\chi_\perp$ and $C^z(r)$, for which exact solutions are available [22,23] suggest that with increasing a size of decoupled cell the results by DCM approach the exact ones except at very low temperatures. We show in Fig.1 the result of $C^z(r=1)$ in conjunction with the rigorous one by McCoy [23].

The results for ϵ , $\chi_\perp$, and $C^x(r)$ were published elsewhere [20,21]. We find that for those quantities there exists a certain demarcation temperature $T_d(\nu)$ depending on the cell size $2\nu + 1$, such that above $T_d(\nu)$, the result of this method is excellent with the very small error whereas below $T_d(\nu)$, the result is rather unreliable. Especially the obtained values of ϵ give a negative specific heat below $T_d(\nu)$. $T_d(\nu)$ decreases with increasing ν . A calculation for $\chi_\parallel$ is performed by observing the ratio of $\langle \hat{\sigma}_i^x \rangle$ to small values of external field H. In Fig.2 we show $\log(\chi_\parallel)$ vs. $\log(1/T)$ for the case of $(\nu =4, N=128)$, where and hereafter temperature T is the reduced temperature and is defined to be k_BT/J.

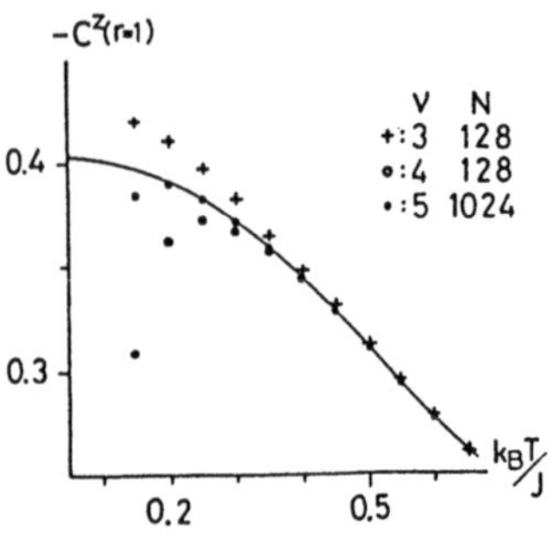

Fig. 1. Out-of-plane spin pair correlation function $C^z(r=1)$ for 1D XY model vs. k_BT/J. The solid line is the exact result by McCoy [23]

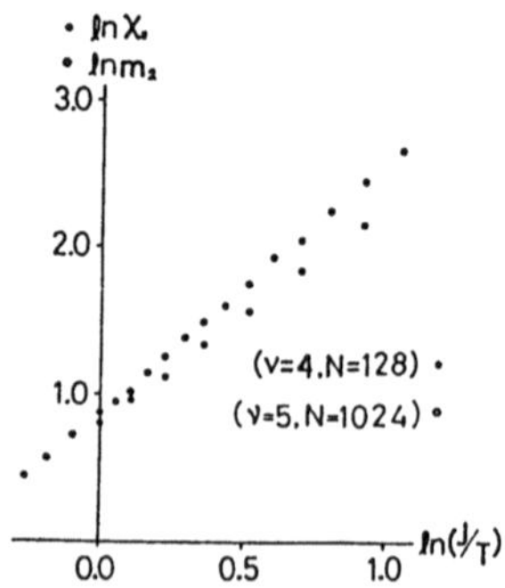

Fig.2. Susceptibility $\chi_\parallel$ (represented by • and mean square magnetization m_2 (represented by o) vs. T for the 1D XY model

From this we find that $\chi_{\parallel}$ is well expressed as

$$\chi_{\parallel} = \chi_{\parallel}^{\circ}\, T^{-\gamma}, \tag{18}$$

where the values of $\chi_{\parallel}^{\circ}$ and γ are 2.235±0.010 and 1.49±0.01, respectively. Here we used 50000 Monte Carlo steps for each H with given T. In Fig.2 we also give the results for the relation between m_2 and T. It is found that m_2 is expressed as $m_2^{\circ}\, T^{-\tilde{\gamma}}$ with $m_2^{\circ} = 2.375$ and $\tilde{\gamma} = 1.73$, respectively. The difference between γ and $\tilde{\gamma}$ means that the quantum effect is of importance in the 1D XY model.

The results thus obtained show that DCM Monte Carlo simulation gives very good estimations of the thermodynamic quantities of the 1D XY model above a certain demarcation temperature which depends on the decoupled cell size. It is therefore worthwhile to use this method for higher dimensional quantum systems for which the exact properties are still obscure. We do it in the next section.

4. Two-Dimensional Quantum Spin Systems

In this section we apply DCM to (1) the 2D quantum mechanical XY model ($S=\frac{1}{2}$) to examine the possibility of a phase transition and (2) AHTL to look for a spin ordering and the possible types of sublattice structure in it, because AHTL is a simple and typical model with frustration among quantum spin systems.

4.1. XY Model on the Square Lattice

In the classical limit ($S=\infty$) Kosterlitz and Thouless (referred to hereafter as KT) showed that the low-temperature phase is characterized by a power law decay in the spin pair correlation function modified by the presence of pairs of tightly bound vortices of opposite sign and at T_c the pairs unbind causing correlations to fall off exponentially in the high temperature phase [24]. This transition (KT transition) is supported by the numerical experiments [25-29]. However, for the extreme quantum limit ($S=\frac{1}{2}$) the presence or absence has remained obscure [12,21,22]. In addition, previous Monte Carlo calculations were mainly restricted to the macroscopic thermodymamic quantities, such as internal energy, specific heat, mean square magnetization and so on. It is therefore desirable to observe directly microscopic thermodynamic quantities such as spatial and temporal spin correlation functions at low and high temperature phases.

The Hamiltonian of the 2D XY model is given by

$$H = -\tfrac{1}{2}J \sum_{\langle i,j\rangle} (\hat{\sigma}_i^x \hat{\sigma}_j^x + \hat{\sigma}_i^y \hat{\sigma}_j^y), \tag{19}$$

where $\langle i,j \rangle$ stands for a nearest neighbor pair. As discussed in the previous section we can observe the thermodynamic quantities defined through Eqs.(10) to (16) by suitably choosing a direction of diagonal representation for spin operators. In Fig.3. we show the size of a decoupled cell and directions along which spin pair correlation is observed by DCM.

The Monte Carlo calculations were performed on the sequence of 8x8, 16x16, 32x32 and 64x64 square lattices with a periodic boundary condition. For each lattice we used 1000 steps to equilibriate and another 30000 to 50000

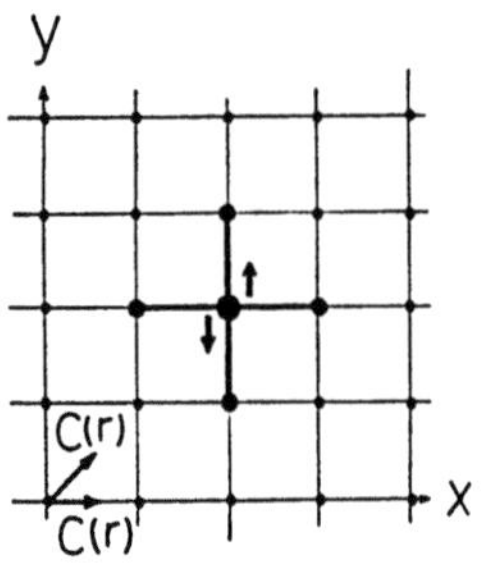

Fig.3. The size of a decoupled cell and the spin pair correlation function $C^x(r)$ for 2D XY model

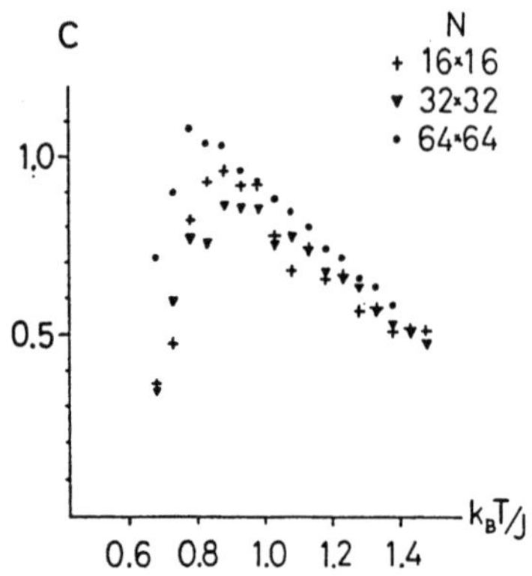

Fig.4. Specific heat per spin vs. temperature for 16x16, 32x32, and 64x64 spin systems for XY model on a square lattice

steps to calculate the averages. The specific heat c as a function of T is shown in Fig. 4, in which c is calculated from the temperature derivative of the computed internal energy ϵ.

The results suggest that the height of the specific heat peak depends weakly on the size of the system.

In Figs.5 and 6 we show $C^x(r)$ for T=0.7 and 1.4 of 64x64 lattice. From these figures we see that at T=0.7, $C^x(r)$ is characterized by a power-law decay, whereas at T=1.4 it decays exponentially. This qualitative difference of $C^x(r)$ at low and high temperature phases suggests the existence of a phase transition in the 2D XY model on the square lattice. At high temperature phase we analyzed our results of $C^x(r)$ assuming $C^x(r) = C^\circ(T) \exp(-r/\xi(T))$, which defines a correlation length $\xi(T)$, for relatively large value of r.

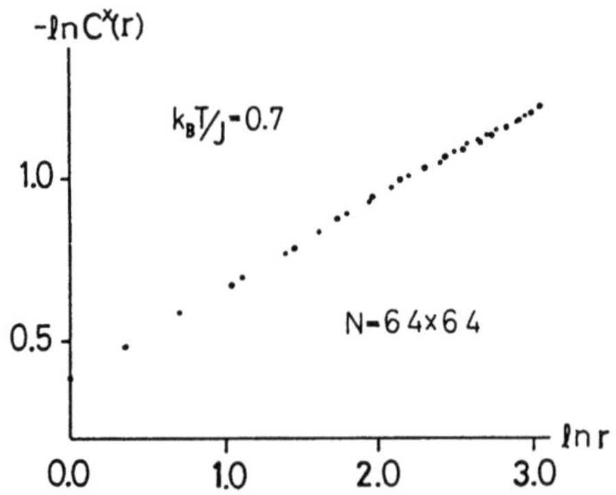

Fig.5. In-plane spin pair correlation function $C^x(r)$ vs. distance r at T = 0.7 for 2D XY model of 64x64 lattice

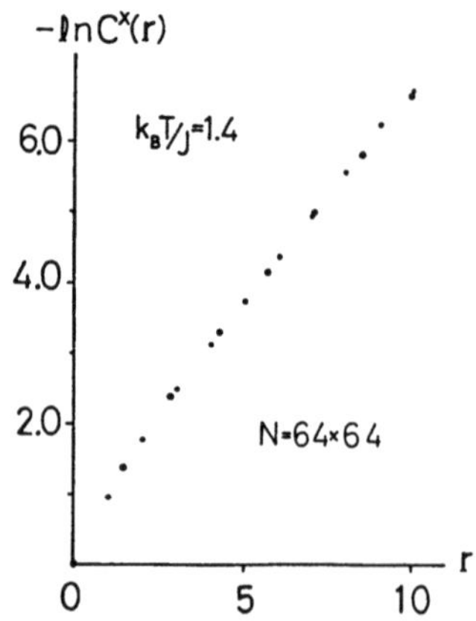

Fig.6. In-plane spin pair correlation function $C^x(r)$ vs. distance r at T = 1.4 for 2D XY model of 64x64 lattice

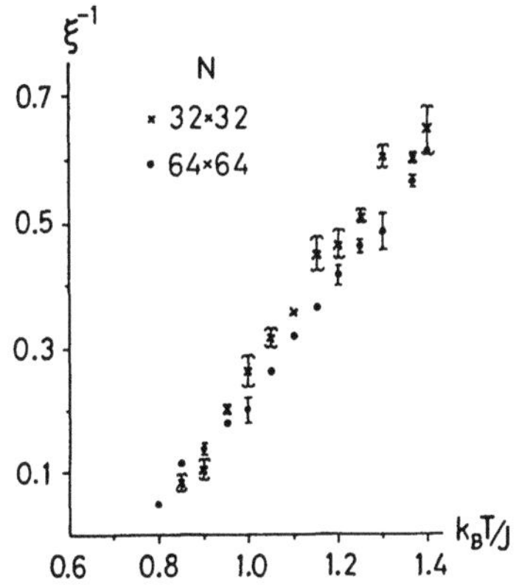

Fig.7. A plot of an inverse of correlation length $(\xi(T))^{-1}$ vs. temperature T for 2D XY model of 32x32 and 64x64 lattices

In Fig.7. we show a plot of inverse of correlation length $\xi(T)$ vs. T for 32x32 and 64x64 lattices. We see from this that with an increase of the size of the system $(\xi(T))^{-1}$ become smooth and the statistical error of the observation diminishes.

From this figure the transition temperature T_c' is estimated as $T_c = 0.75 \pm 0.03$ at which a correlation length $\xi(T_c)$ diverges, although the size of the system in our DCM Monte Carlo simulation is limited to 64x64 lattice points.

4.2. Antiferromagnetics on the Triangular Lattice

The triangular antiferromagnetic model (ATL) has attracted much interest because of the general question: is there a relation between frustration and the possibility of a sublattice structure for ATL that shows an ordering of a different nature from the well-known Néel type ? The problem for ATL Ising model (AITL) was studied by Wannier [30]. The ground state of ATL Heisenberg model (AHTL) was investigated numerically and theoretically [31-36]. The thermodynamic properties of AHTL were investigated by Monte Carlo calculation based on the generalized Trotter formula [18]. Recent experiment of $NaNiO_2$ is considered as a realization of this model [37].

Here we apply DCM Monte Carlo simulation to AHTL to study the thermodynamic properties and the presence or absence of a sublattice structure.

The Hamiltonian is given by

$$H = \frac{1}{2}J \sum_{\langle i,j \rangle} (\hat{\sigma}_i \cdot \hat{\sigma}_j), \tag{20}$$

in (20) we take the representation in which $\{\hat{\sigma}_i^z\}$ is diagonal. We calculate internal energy, specific heat and spin pair correlation function $C^z(r)$ for 33x33 lattice with the decoupled cell depicted in Fig.8. There, we also show the definition of $C_l^z(r)$, l=1,2.

The observed results of internal energy and specific heat will be published elesewhere. We show in Figs.9 and 10 the results of $C_1^z(r)$ and $C_2^z(r)$ of AHTL together with those of AITL for comparison.

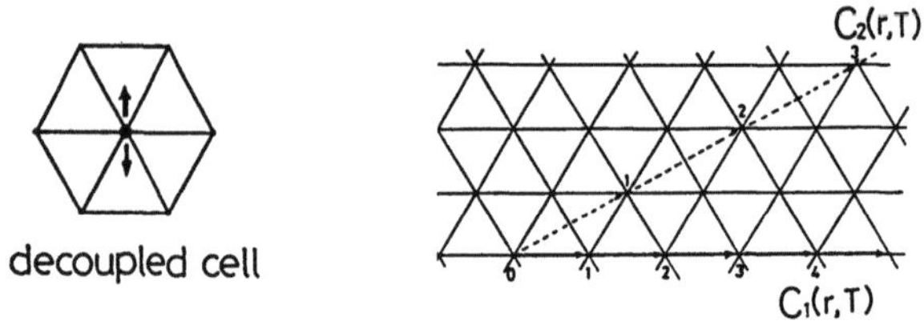

Fig.8. The size of a decoupled cell and the spin pair correlation function $C_l^z(r)$, $l=1,2$ on ATL

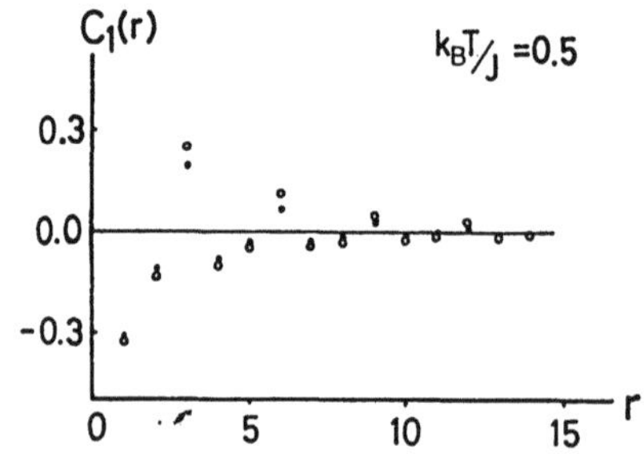

Fig.9 The spin pair correlation function $C_1^z(r)$ of AHTL and AITL vs distance r at T=0.5. Distance r is in units of a nearest neighbour distance. The symbols o and • denote $C_1^z(r)$ of AHTL and AITL, respectively.

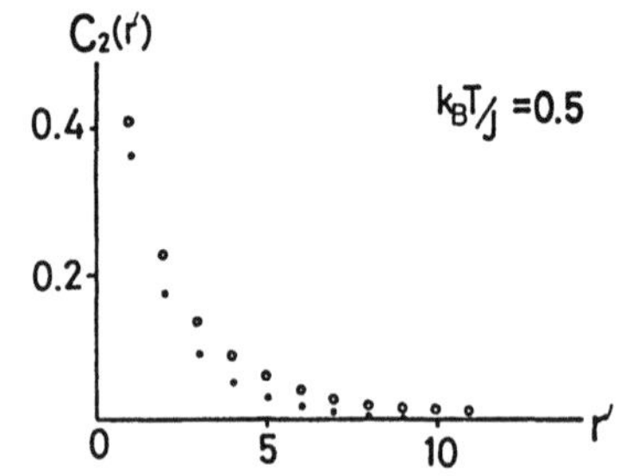

Fig. 10. The spin pair correlation function $C_2^z(r')$ of AHTL and AITL vs distance r' at T=0.5. Distance r' is in units of nearest neighbour distance times $\sqrt{3}$. The meaning of symbols is the same as of Fig.9.

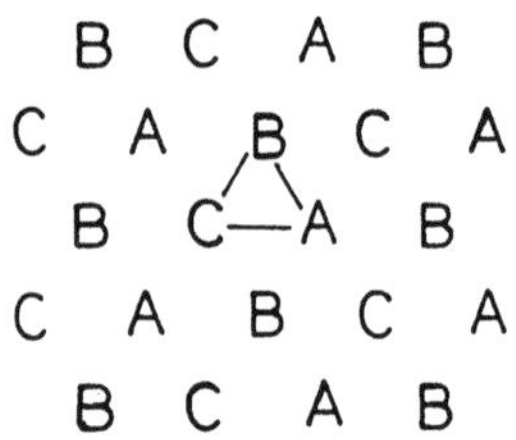

Fig.11. The sublattice structure of ATL

It is interesting to note that the coupling in the perpendicular components (x and y components) in AHTL tends to increase the spin correlation in the parallel component (z component). From these figures we also find that in AHTL and AITL there exists a sublattice structure in the spin ordering, as depicted in Fig.11.

The spin pair correlation within a sublattice is positive, whereas it is negative between different sublattices. This structure is of the same type as the one suggested by Wannier [30]. The amplitudes of C(r) decrease exponentially with increasing r. We also measured the correlation between different Monte Carlo steps. The results indicate that this correlation decreases exponentially as a number of steps increases. From these observed results we conclude that the sublattice structure exists in AHTL but it is short-ranged both spatially and temporally in the present DCM Monte Carlo calculation. These characteristics were observed throughout the whole temperature range ($0.02 \lesssim T \lesssim 1.5$) of our studies.

5. Discussion

In this report we introduced 'Decoupled Cell Method' Monte Carlo simulation in order to generate a canonical ensemble of quantum systems as a time sequence of a Markov chain whose transition probability W is given by (5) through (8). We applied this method to the 1D XY model and got results for internal energy, in- and out-of-plane susceptibilities and spin pair correlation functions. The results thus obtained agree well with exact solutions except at very low temperatures.

For the 2D XY model we have shown the existence of a phase transition which has close similarities to classical KT theory, through the observation of $C^x(r)$ at low and high temperature phases. We also have got a temperature dependence of a correlation length $\xi(T)$ in the high temperature region, from which the transition temperature T_c is estimated as 0.75 ± 0.03 by an extrapolation. In order to determine critical exponents of the 2D XY model we have to extend our simulation to larger systems than studied in this report. This will be the next problem.

For the spin $\frac{1}{2}$ antiferromagnetic Heisenberg model on the triangular lattice (AHTL), our numerical results suggest the existence of a sublattice structure composed of three lattices in spin ordering. It is short-ranged both spatially and temporally, and is similar to that of AITL.

The DCM Monte Carlo simulation makes it simple to observe the quantities related to microscopic thermodynamic quantities as well as macroscopic ones. Further application of this method for more general quantum many-body systems would be worthwhile.

The computations were performed at Nagoya University Computation Center, the Computer Center of the Institute of Plasma Physics of Nagoya University, and the Institute of Physical and Chemical Research. This work is partially supported by Grant-in-Aid for Scientific Research Fund of the Ministry of Education and Culture.

References

1. D. C. Handscomb, Proc.Cambridge Philos. Soc., 58 (1962),594; 60 (1964), 115
2. M. Suzuki, Prog. Theor. Phys. 56 (1976), 1454
3. M. Suzuki, Phys. Rev. B31 (1985), 2957; J. Stat. Phys. 43 (1986), 883
4. S. Chakravarty and D. B. Stein, Phys. Rev. Lett., 49 (1982), 582
5. J. W. Lyklema, Phys. Rev. Lett., 49 (1982), 88; Phys. Rev. B27 (1983),3108
6. D. H. Lee, J. D. Joannopoulos and J. W. Negele, Phys. Rev. B30 (1984),1599
7. S. Kadowaki and A. Ueda, Prog. Theor. Phys. 75 (1986), 451
8. M. Suzuki, S. Miyashita and A. Kuroda, Prog. Theor. Phys. 58 (1977), 1377
9. J. J. Cullen and D. P. Landau, Phys. Rev. B27 (1983), 297
10. H. Betsuyaku, Phys. Rev. Lett. 53 (1984), 629; Prog. Theor. Phys. 73 (1985), 319

11. T. Yokota and H. Betsuyaku, Prog. Theor. Phys. 75 (1986), 46

12. H. Takano, Prog. Theor. Phys. 73 (1985), 332

13. T. Sakaguchi, K. Kubo and S. Takada, J. Phys. Soc. Jpn. 54 (1985), 861

14. I. Satija, G. Wysin and A. R. Bishop, Phys. Rev. B31 (1985), 3205

15. E. Loh, Jr., D. J. Scalapino and P. M. Grant, Phys. Rev. B31 (1985), 4712

16. H. De Raedt, B. De Raedt and A. Lagendijk, Z. Phys. B57 (1984), 209

17. H. De Raedt and A. Lagendijk, Phys. Rep. 127 (1985), 233

18. M. Takasu, S. Miyashita and S. Suzuki, Prog. Theor. Phys. 75 (1986), 1254

19. S. Homma, H. Matsuda and N. Ogita, Prog. Theor. Phys. 72 (1984), 1245

20. S. Homma, H. Matsuda and N. Ogita, Prog. Theor. Phys. 75 (1986), 1068

21. S. Homma. K. Sano, H. Matsuda and N. Ogita, Prog. Theor. Phys. Suppl. 87

22. S. Katsura, Phys. Rev. 127 (1962), 1508

23. B. M. McCoy, Phys. Rev. 173 (1968), 531

24. J. M. Kosterlitz and D. J. Thouless, J. Phys. C6 (1973), 1181;
J. M. Kosterlitz, J. Phys. C7 (1974), 1046

25. S. Miyashita, H. Nishimori, A. Kuroda and S. Suzuki, Prog. Theor. Phys. 60 (1978), 1669

26. J. Tobochinik and G. V. Chester, Phys. Rev. B20 (1979), 3761

27. J. E. Van Himbergen and S. Chakravarty, Phys. Rev. B23 (1981), 359

28. K. Y. Szeto and G. Dresselhaus, Phys. Rev. B32 (1984), 3142

29. J. F. Fernandez, M. F. Ferreira and J. Stankiewicz, Phys. Rev. B34 (1986), 292

30. G. H. Wannier, Phys. Rev. 79 (1950), 357

31. S. Miyashita, J. Phys. Soc. Jpn. 53 (1984), 44

32. H. Nishimori and S. J. Miyake, Prog. Theor. Phys. 73 (1983), 18

33. S. J. Miyake, Prog. Theor. Phys. 74 (1984), 468

34. T. Oguchi, H. Nishimori and Y. Taguchi, J. Phys. Soc. Jpn. 55 (1986), 323

35. S. Miyashita and H. Shiba, J. Phys. Soc. Jpn. 53 (1984), 1145

36. H. Kawamura and S. Miyashita, J. Phys. Soc. Jpn. 53 (1984), 9, 4138

37. K. Hirakawa, H. Kadowaki and K. Ubukoshi, J. Phys. Soc. Jpn. 54 (1985), 3526

Quantum Monte Carlo for Spin Glasses and Related Systems

I. Morgenstern[1,2] *and D. Würtz*[1]

[1]Institut für Theoretische Physik, Ruprecht Karls Universität, Philosophenweg 19, D-6900 Heidelberg, Fed. Rep. of Germany
[2]Institute for Theoretical Physics, University of California, Santa Barbara, CA 93106, USA

We use the Suzuki-Trotter transformation to obtain $d+1$-dimensional classical representations for the corresponding quantum systems. For one-dimensional disordered quantum XY and Heisenberg chains we demonstrate the usefulness of the transfer matrix method. In two dimensions we had to rely on a Monte Carlo procedure. Results are shown for the XY and a spin-glass model. The applications of the method to fermion systems is discussed.

Introduction

The present contribution deals with a transfer of numerical methods to quantum systems; the methods were successfully applied to the spin-glass problem. First we have the "transfer matrix method" (TM) which sums numerically exact over all possible states in, e.g., a two-dimensional Ising spin glass [1]. This method can directly be applied to $1 + 1$-dimensional quantum systems using the Suzuki-Trotter transformation known as the equivalence theorem [2]. While two-dimensional spin-glasses were still in the range of the transfer matrix method, for three-dimensional systems we had to rely on the standard Monte Carlo (MC) procedure [3]. Similarly we had to apply MC for $2 + 1$-dimensional quantum systems. The reason is the storage requirement of the transfer matrix method which in two dimensions is proportional to $\exp(L)$ where L is the smaller length involved. In three dimensions the storage is proportional to $\exp(L^2)$. Therefore the limitations of the method are obvious. On the other hand, there is a philosophical difference between TM and MC. In MC we are currently looking for a quite general approach to quantum systems (with a storage more or less linearly proportional to the system size). TM is more or less restricted to one-dimensional systems or small two-dimensional lattices (maximum about 4×4, on a Cray 2, 5×5). A generalization of TM to very much larger system sizes is not possible as a certain amount of information has to be propagated, at least in the present form. Consequently we will consider the transfer matrix approach for one-dimensional XY and Heisenberg chains and their fermion analog. Here we will demonstrate the usefulness of the method originally developed for spin-glass models. The method is in the meantime quite popular in Japan and we refer the reader to the contribution of Dr. Betsayuku in these proceedings for numerous applications of the method. In the second part we will discuss a generalization of the break-up used for TM to two-dimensional systems. There we have to go back to MC.

1. Transfer Matrix Method

We restrict ourselves here to the Hamiltonian

$$\mathcal{H} = -\sum_{\langle ij\rangle} J_{ij}^x S_i^x S_j^x + J_{ij}^y S_i^y S_j^y + J_{ij}^z S_i^z S_j^z \quad . \tag{1}$$

Here the S_i^x, etc. are the Pauli spinors for the components of the spin operator at the ith site of the lattice. We confine ourselves to one-dimensional (1D) systems with nearest-neighbor interactions J_{ij}^x, etc. Two special cases are the Heisenberg chain ($J = J_{ij}^x = J_{ij}^y = J_{ij}^z$) and the XY model ($J = J_{ij}^x = J_{ij}^y$, $J_{ij}^z = 0$). In our method we do not have any restrictions considering the J_{ij}^z, etc. Therefore, we can calculate chains with arbitrary distributions of the couplings and any realizations of anisotropies. We are also able to apply magnetic fields. Thus we can provide calculations for quite a large range of 1D models.

In principle, we can apply our method to exactly the same chains (length and random configuration) as the MC method of *Hirsch* and co-workers [4, 5]. While MC provides "importance sampling", i.e., only the "important" configurations of the lattice contribute, selected by the multi-spin-flip dynamics, our method takes *all* possible spin configurations into account.

We start with the Suzuki-Trotter transformation based on the relationship for operators A and B:

$$e^{A+B} = \lim_{m\to\infty} (e^{A/m}\, e^{B/m})^m \quad . \tag{2}$$

Equation (2) is applicable to quantum-statistical-mechanical systems where the partition function is given by

$$Z = \mathrm{Tr}\, e^{-\beta\mathcal{H}} \quad . \tag{3}$$

The Hamiltonian $\mathcal{H}$ may be "decomposed" into sums of terms $\mathcal{H} = \mathcal{H}_1 + \mathcal{H}_2$ to give [4, 5]

$$Z = \lim_{m\to\infty} Z^{(m)} \quad , \tag{4}$$

where

$$Z^{(m)} = \mathrm{Tr}(e^{-\beta\mathcal{H}_1/m}\, e^{-\beta\mathcal{H}_2/m})^m \quad . \tag{5}$$

Following (4, 5) we "decompose" $\mathcal{H}$ of (1) into an odd and an even part, i.e., $\mathcal{H}_1$ contains only spins ad odd and $\mathcal{H}_2$ at even sites i. The same decomposition was used by *Hirsch* and co-workers [4, 5]. We end up with a checkerboard with four spin interactions between the resulting Ising spins surrounding the black squares. An explanation is given in Fig. 1. The weight of the corresponding spin configurations belonging to a black square are shown below the lattice.

We notice that only eight configurations contribute to the partition function. The reader should notice that our lattice in Fig. 1 is turned by 90° compared to that of *Hirsch* et al. [5]. In particular, the contributions of the configurations in Fig. 1 are the following:

$$
\begin{aligned}
E_{ij}^{(m)}(1,2) &= \exp(K\phi_{ij}) \ , \\
E_{ij}^{(m)}(3,4) &= \cosh(KP_{ij})\exp(-K\phi_{ij}) \ , \\
E_{ij}^{(m)}(5,6) &= \sinh(KP_{ij})\exp(-K\phi_{ij}) \ , \\
E_{ij}^{(m)}(7,8) &= \sinh(KM_{ij})\exp(K\phi_{ij}) \ ,
\end{aligned}
\tag{6}
$$

where the partition function is given by

$$
Z^{(m)} = \sum_{\{s\}} \prod_{\substack{ij \\ \text{black squares}}} E_{ij}^{(m)} \ . \tag{7}
$$

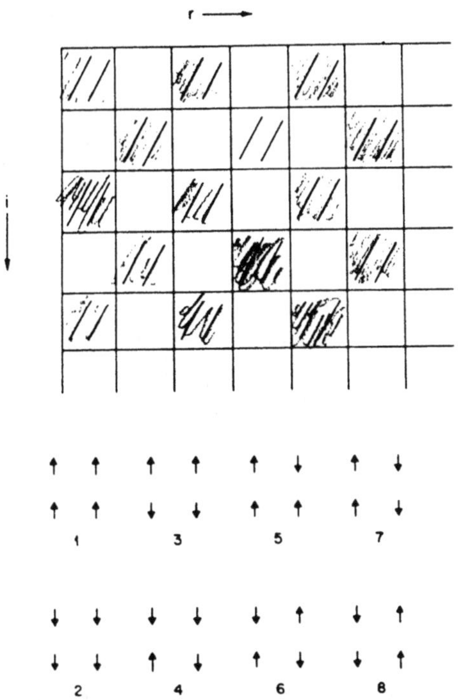

Fig. 1. Checkerboard decomposition. Columns labeled with index $r\,(r = 1, \ldots, 2m)$ correspond to additional Trotter dimension. Rows with index $i(i = 1, \ldots, N)$ correspond to sites of the chain. Four-spin interactions are indicated by squares with diagonal lines (//). *Below*: allowed four-spin configurations

$\sum_{\{s\}}$ denotes the sum over all possible (Ising) spin configurations. The product in (7) has to be taken over all sites i, j that are connected by black squares of the checkerboard. We made use of the following identities:

$$K\phi_{ij} = \beta J^z_{ij}/4m \ ,$$
$$KP_{ij} = (\beta/4m)(J^x_{ij} + J^y_{ij}) \ ,$$
$$KM_{ij} = (\beta/4m)(J^x_{ij} - J^y_{ij}) \quad ; \tag{8}$$

β is the inverse temperature. We notice that for those models where $J^x_{ij} = J^y_{ij}$ we deal with only six contributing configurations $[E^{(m)}_{ij}(7,8) = 0]$. For example, in the case of the XXZ model corresponding to the fermion model [5, 6],

$$\mathcal{H} = -t\sum_i C^+_i C_{i+1} + C^+_{i+1} C_i + V\sum_i (n_i - \tfrac{1}{2})(n_{i+1} - \tfrac{1}{2}) \quad , \tag{9}$$

the Monte Carlo procedure is quite complicated. The method of *Hirsch* et al. [5] generates closed paths along the Trotter direction to obtain a new "nonzero" configuration. Because of the complicated four-spin interactions of the checkerboard lattice it is difficult to generate paths in an unbiased and uncorrelated way.

Our method treats the same lattice as *Hirsch* and co-workers [5] by MC only turned by 90°. As we sum over *all* configurations we have avoided the problems mentioned of the MC procedure.

Description of the Method

We now turn to the checkerboard of Fig. 1. The method is restricted to free boundaries in the space direction, while in the time (m) direction periodic boundary conditions are possible. The numerical program calculates the partition function (the sum over the statistical weights of all possible configurations) in the following way.

(1) *All* possible Ising spin configurations are generated for the *first* row. The configurations are stored following a simple logical scheme; i.e., first we take all spins pointing up, then the last spin is turned and so on (binary code).

(2) Now the *first* configuration (i.e., all spins up) of the *second* row is coupled successively to *all* configurations of the *first* row. For each combination we calculate the statistical weight according to the spin arrangement around the black squares. The weights of all combinations are summed; i.e., we take the trace over the first row with the *first* configuration of the *second* row *fixed*.

(3) The same procedure is carried out successively for all configurations of the second row. The corresponding weights are stored in the order given by

the scheme. The program forgets the first row as we have already considered all the couplings connected with this row (free boundaries).

(4) All the following rows are now handled as the second; i.e., a fixed configuration of the third row is coupled to all configurations of the second row (already containing the first row).

(5) We obtain the partition function by summing over all contributions of the last row.

Because we have to store all weights corresponding to all configurations of a row, we need an array of 2^M sites where M is the number of spins in a row; i.e., in the m direction of the lattice in Fig. 1.

The calculation follows in principle the above scheme (1–5). The described scheme allows vectorization of the program using a Cray I computer. Because of the storage restriction we were only able to provide extensive calculations for $m \leq 8$. *Cullen* and *Landau* [6] also had to restrict their MC calculations to $m \leq 8$ [6], but as can be seen from their and our results, the case $m = 8$ already gives reasonable results. Only at very low temperatures are there problems with the uniform XY chain. Application of the $1/m^2$ extrapolation scheme proposed by Suzuki leads to reasonable results. The above method still holds in the case of disorder as the couplings only vary from row to row in our lattice and do not additionally affect spins in the m direction. Furthermore, our method is not restricted to the checkerboard decomposition. In principle we can apply our scheme to all kinds of different classical two-dimensional Ising systems with four-spin interactions [arising from different decomposition of the Hamiltonian (1)].

Numerical Results

Once the partition function has been obtained, we are able to calculate all interesting thermodynamic quantities by numerical differentiation. First we provide a comparison of the parallel susceptibility $\chi^{(m)}$ obtained by our method to (a) the exact solution of the pure XY chain [7, 8] and (b) to the numerical results of *Bonner* and *Fischer* [9] for the ferromagnetic Heisenberg chain (see Fig. 2). We find nice agreement at higher temperatures. At lower temperatures we see relatively large deviations in the XY case similar to the MC approach of *Cullen* and *Landau* [6]. This is caused by too-small m values. Surprisingly this is not the case for the random XY chain. Furthermore, we provide results for $\chi^{(m)}$ of the XY and the Heisenberg chain with Gaussian distributed couplings. We only consider the isotropic case ($J_{ij}^x = J_{ij}^y = J_{ij}^z$) where we expect the most problems for MC runs due to freezing effects. To obtain $\chi^{(m)}$ we first apply the Suzuki-Trotter transformation to the partition function (1) with field H applied in the z direction. It is easily found that the resulting partition function takes the form

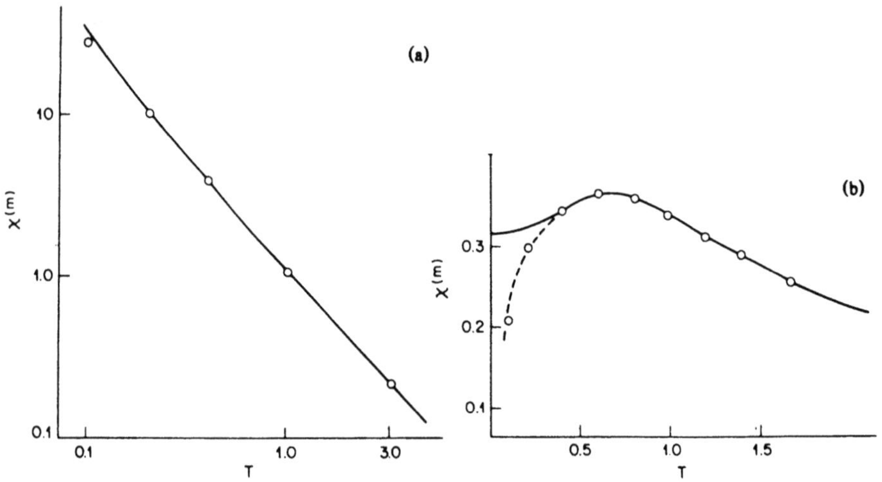

Fig. 2. **(a)** Parallel susceptibility $\chi^{(m)}$ for pure XY chain for $m = 8$ (*circles*) compared with the exact solution. **(b)** Parallel susceptibility $\chi^{(m)}$ for ferromagnetic Heisenberg chain for $m = 8$ (*circles*) compared with numerical results of Bonner and Fischer (Ref. 9) log-log plot

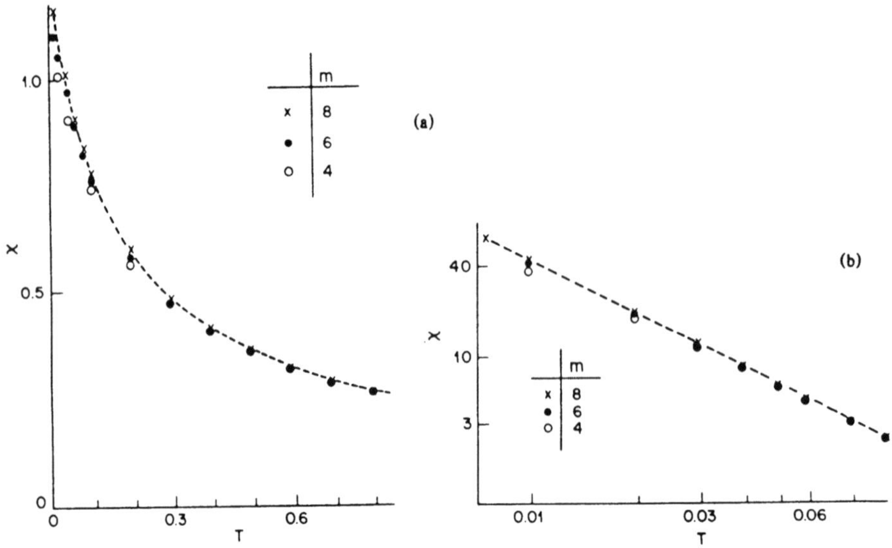

Fig. 3. **(a)** Parallel susceptibility $\chi^{(m)}$ of the isotropic XY chain with Gaussian couplings against temperature T for various Trotter lengths m. **(b)** Logarithmic plot of the parallel susceptibility $\chi^{(m)}$ of the isotropic Heisenberg chain with Gaussian couplings against temperature T. Various Trotter lengths m

$$Z_H^{(m)} = \sum_{\{s\}} \Bigg[\prod_{\substack{ij \\ \text{black squares}}} \left(E_{ij}^{(m)} \exp \frac{\beta H}{2m} \right) \sum_{\substack{\text{black} \\ \text{squares}}} S_{ir} \Bigg] \quad , \tag{10}$$

where the S_{ir} are the Ising spins belonging to the black squares of the lattice in Fig. 1. $\chi^{(m)}$ is now obtained by numerical differentiation. We provided calculations for chain lengths $N = 100$ and averaged over 20 realizations.

Figure 3(a) shows $\chi^{(m)}$ of the XY chain with Gaussian disorder against temperature T. We notice that – while in the pure model ($J^x_{ij} = J^y_{ij} = J$) $\chi^{(m)}$ does not diverge for $T \to 0$ – we find a divergence in the case of disorder. This behavior is in agreement with renormalization-group calculations of Hirsch and co-workers. Furthermore, we see that the m dependence is relatively weak even at low temperatures in contrast to the pure case. Figure 3(b) shows $\chi^{(m)}$ against T for the corresponding Heisenberg case. Here we show a logarithmic plot as we have a strong divergence for $T \to 0$. An analyis of the data leads to a value of $\gamma = \frac{5}{4}$ as we have

$$\chi \propto T^{-\gamma} \, (T \to 0) \quad . \tag{11}$$

Note that in the pure case $\gamma = \frac{7}{4}$. This means that the divergence is weakened by the disorder but the phase transition is not destroyed.

Finally, we want to add that the present method can be used to check various approaches for MC simulations of quantum systems. In spin-glasses great progress in understanding the system was achieved by just comparing TM and MC results for exactly the same lattice, leading to the detection of the famous "freezing effects" and the energy valley picture. As some MC procedures, including the one described in the next section, are limited to a constant magnetization, a careful check of these methods should be possible. The possibility that the above restriction could lead to the suppression of sizeable fluctuation near a critical temperature was thoroughly discussed at this conference considering the XY model in two dimensions. We will comment on this fact in the next section.

2. Monte Carlo Method

Here we consider again the system given by the Hamiltonian (1), but now in two dimensions. We generalize the break-up used for one-dimensional chains. We introduce a disconnected bond decomposition (DBD) which divides our system into four subsystems consisting of disconnected bonds

$$\mathcal{H} = \mathcal{H}^x_{\text{even}} + \mathcal{H}^y_{\text{even}} + \mathcal{H}^x_{\text{odd}} + \mathcal{H}^y_{\text{odd}} \quad , \tag{12}$$

where $\mathcal{H}^{x,y}_{\text{even}}$ denotes the sum over even bonds in the x- or y-direction, respectively, and $\mathcal{H}^{x,y}_{\text{odd}}$ takes care of the odd bonds. A generalization to higher dimensions is straightforward. In one dimension our DBD-scheme leads to the checkerboard decomposition discussed in Sect. 1. We introduce a complete set of eigenstates to obtain

$$\mathcal{H} = \sum_{\{u_r\}} \langle u_1 | e^{-\beta \mathcal{H}^x_{\text{even}}} | u_2 \rangle \ldots \langle u_{nm} | e^{-\beta \mathcal{H}^y_{\text{odd}}} | u_1 \rangle \quad , \tag{13}$$

where the index r labels the new "Trotter"-direction with periodic boundary conditions. Because of the local character of the DBD all the matrix elements can be easily evaluated. Finally we obtain a three-dimensional "classical system" with four-particle interactions. From the possible 16 different configurations only six contribute to the partition function given by

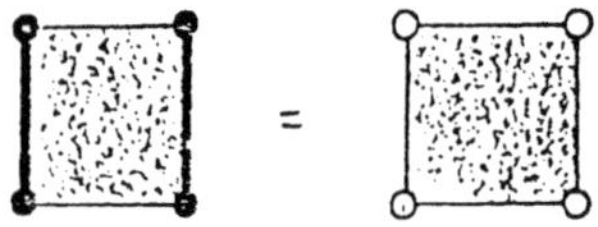

$$= e^{\beta J^z_{ij}/4m} \quad ,$$

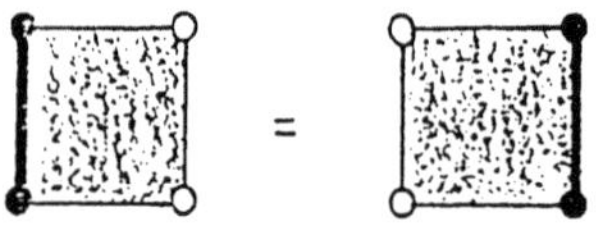

$$= \cosh\left[\frac{\beta}{4m}(J^x_{ij} + J^y_{ij})\right] e^{-\beta J^z_{ij}/4m} \quad ,$$

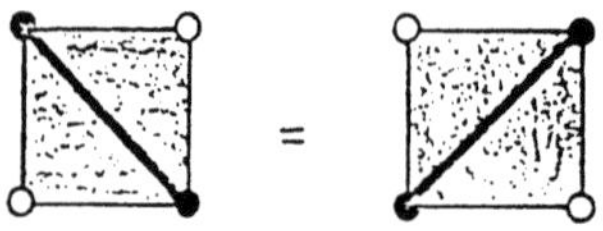

$$= \sinh\left[\frac{\beta}{nm}(J^x_{ij} + J^y_{ij})\right] e^{-\beta J^z_{ij}/4m} \quad , \tag{14}$$

where • and o denote occupied and empty sites, respectively. The unit cell of the system and a part of our "checkercube" created by the DBD are shown in Figs. 4 and 5.

There are interactions only on the shaded squares of the three-dimensional "checkercube" which means that the particles are only allowed to hop along the shaded squares. The "white" regions are forbidden. The particles now follow "world lines" [4, 5] which connect the occupied sites between different

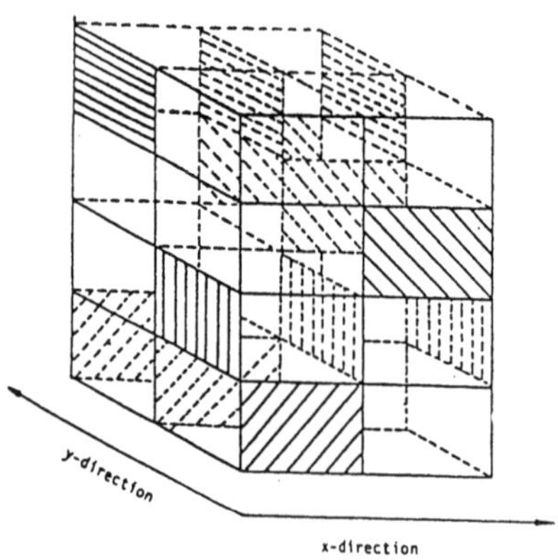

Fig. 4 Unit cell for "checkercube"

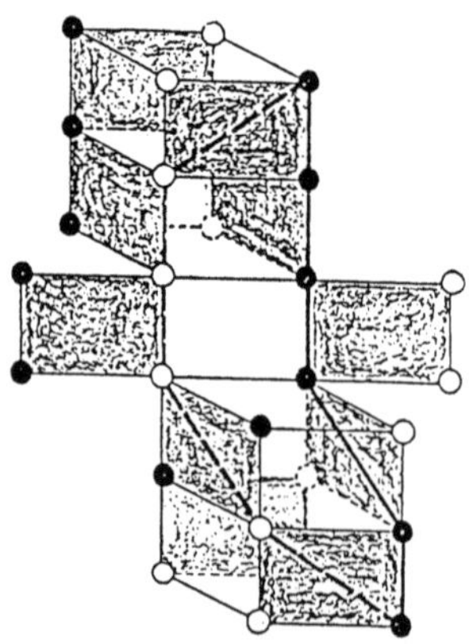

Fig. 5. Elementary MC-move for worldline

Trotter slices. These world lines can be drawn either along the vertical edges or across a shaded square. In (14) the world lines are shown by the thicker lines. They actually connect the •'s on the checkercube, see Fig. 5. Negative contributions to the partition function can only arise from the diagonal crossings. But as we have periodic boundary conditions in the Trotter direction following (13), we can avoid a negative contribution of a world line to the partition function. The negative factors just cancel out as the world line has to return to the starting point. Therefore our DBD avoids negative contributions to the partition function in all dimensions. Note that we have just the same concept as in one dimension.

The "importance sampling" of our MC simulation selects only the "important" states of the partition function instead of summing over all possible states. In our case an arrangement of world lines has to be replaced by another one according to their transition probability. To do this efficiently we introduce a local rearrangement of the world lines which avoids the forbidden configurations (Fig. 5). We consider an unshaded square of our three-dimensional checkercube. The world line may pass on the right hand side of the shaded square while the sites on the other side are empty. We now consider the continuation of this "string" up and down one step in the Trotter direction. As long as the adjacent sides are empty the world line can be moved. The connection to the rest of the world line is arranged at the top or at the bottom of the three connected strings. Obviously this rearrangement has a positive transition probability. For $d = 3$ we just have two additional planes in our four-dimensional "checkerhypercube" and therefore instead of three, five strings. The question of whether the algorithm explores the whole phase space can be answered by looking at the projection of the world line in, e.g., $d = 2$ on the $x - y$ plane. Each complicated path can be decomposed into local moves along an edge or across a square, in the latter case clockwise or anticlockwise. For these local moves one can always find the corresponding changes of the world lines following our local procedure. Here at this point we should emphasize that the dynamics of the MC simulation is in principle completely arbitrary as long as the whole phase space is explored. One has to avoid similar shortcomings as in the simulations of polymer systems [6]. The actual quantum dynamics of the system is represented by the additional Trotter direction in our DBD representation.

Numerical calculations were first performed for the two-dimensional XY model to check the reliability of our method. Figure 6 shows the result for the internal energy for 16×16 lattices. We find nice agreement with the values given by *Loh* et al. [10]. The insert shows the behavior of the maximum of the specific heat C against system size L. In contrast to *De Raedt* and *Lagendijk* [11] we find a saturation of C at about 0.65 for larger system sizes as already found by [10]. Here we would like to add that our method reproduces the results of *Loh* et al. [10] using a different break-up, but because of the "world-line" concept and the dynamics applied to our system we are restricted to

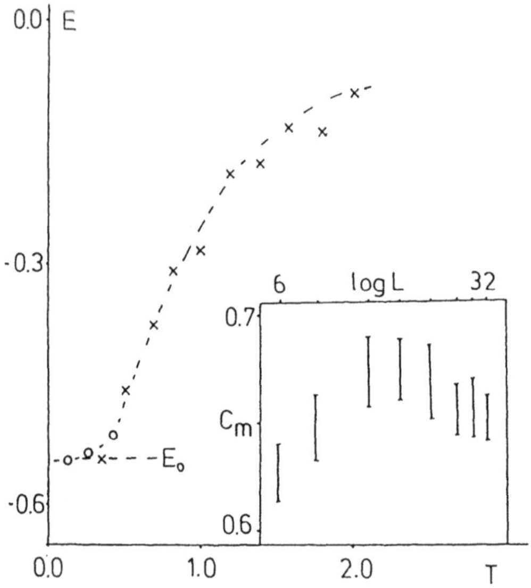

Fig. 6. Internal energy E versus temperature T, two-dimensional XY model. Crosses $m =$ 16, circles $m = 32$. Insert shows maximum of specific heat C_m for different system sizes $L \times L$

zero magnetization. The same restriction applies to Loh et al. but not to De Raedt and Lagendijk. Therefore while our results agree with those of Loh et al. They do not rule out those of De Raedt and Lagendijk as the restriction of the magnetization may suppress important fluctuations leading to a much higher peak in the specific heat. It is a result of the discussions at this conference that this point requires further very careful studies. Furthermore we report results of a quantum spin-glass model choosing $J^z = \pm J$ with $J = J^x + J^y$ in the Hamiltonian of (1). Here we publish the behavior of the linear susceptibility χ in Fig. 7.

In spin-glass theory we recognized a lively debate of the behavior of χ below the freezing temperature T where a plateau was found in field cooled experiments. While the mean-field-theory predicted the plateau it was not realized in simulations for short-range models. A possible explanation would have been the influence of quantum effects at lower tempertures, but our calculations for 16×16 lattices show a clear $1/T$-like behavior for the quantum system. Furthermore, the Edwards-Anderson order parameter tends to zero for all finite temperatures. Thus the behavior is qualitatively the same as in the two-dimensional classical $\pm J$-Ising model. We should note that the present calculations provided the fluctuations of the staggered magnetization, which in the classical case is already related to the magnetization by a gauge transformation. We agree that the staggered susceptibility is a good approximation for the nonlinear susceptibility $\chi_{\rm nl}$ ruling out a direct explanation of the plateau in terms of quantum effects.

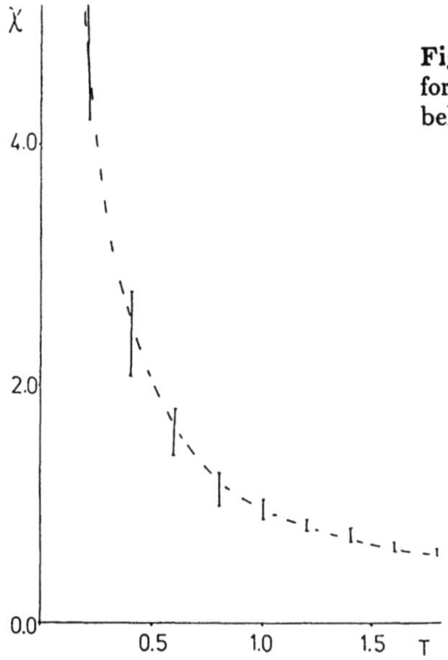

Fig. 7. Linear susceptibility versus temperature T for Heisenberg spin glass. *Broken line* denotes $1/T$ behavior

Fermion Systems

The present Monte Carlo method can also be used for fermion systems. We looked at the following Hamiltonian:

$$\mathcal{H} = -t \sum_{\langle i,j \rangle} \left(C_j^{+} C_j + \mathrm{h.c} \right) + V \sum_{\langle i,j \rangle} (n_i - \tfrac{1}{2})(n_j - \tfrac{1}{2}) \quad . \tag{15}$$

The simulation of the Hamiltonian corresponds just to that of an XXZ-Heisenberg model but we have to introduce negative transition probabilities for odd permutations of the fermions. The MC program just samples some "important" configurations for the Heisenberg as for the corresponding fermion system but for the fermions we have to take care of the additional negative signs. At lower temperatures, especially for free fermions, we encounter convergence problems for the fermion observables while the Heisenberg case is fine. This is due to the fact that we deal with ratios of the type zero/zero. The relatively large error-bars now completely destroy the ratio. This effect could be partially circumvented by introducing an additional factor α for the transition probabilities between positive and negative states and $\frac{1}{\alpha}$ respectively backwards. Calculating the average value of an observable θ we then obtain

$$\langle \theta \rangle_\alpha = \frac{\theta^{+} - \alpha \cdot \theta^{+}}{Z^{+} - \alpha \cdot Z^{-}} \quad , \tag{16}$$

where Z^{+} and Z^{-} are the contributions to the partition function of positive or negative states, respectively. $\alpha = 1$ obviously corresponds to the original case. Furthermore, it is possible to calculate values of the type

$$\langle\widetilde{\theta}\rangle_\alpha = \frac{\theta^+ - \theta^-}{Z^+ - \alpha \cdot Z^-} \quad . \tag{17}$$

Choosing different values of α it is possible to apply an extrapolation scheme to (16) and (17) leading to the *same* expectation value. The described procedure works well at higher temperatures and extends the validity of the world-line MC approach. To be precise, for the internal energy we run into the problem that our error bars still exceed 50% at lower T. But, on the other hand, calculating an order parameter defined between 0 and 1 we are not able to say the value is, e.g., 0.3 or 0.5 but we can say that it is closer to 1 or 0. Therefore we are able to provide some estimate for the boundary of the charge density wave (CDW) phase in the model (16) considering different values of the repulsive interaction V. Following usual spin-glass techniques we started at the ground state and tried to monitor the breakdown of the phase by considering the projection to the ground state as well as the Edwards-Anderson order parameter [1]. In this context we provide the phase diagram in Fig. 8.

It does not exclude the possibility of a CDW for very low or zero temperature for arbitrary small V as indicated by the dotted line. For a discussion of the phase diagram see [12]. We would like to note that the described α-trick is certainly not a solution to the minus-sign problem in the fermion case, but it extends the validity of the world-line approach to lower temperatures. Considering quantities like the order parameter may provide an insight into fermion problems at a level so far not possible. For evaluating phase diagrams within certainly substantial limitations and sizeable error bars the method seems to

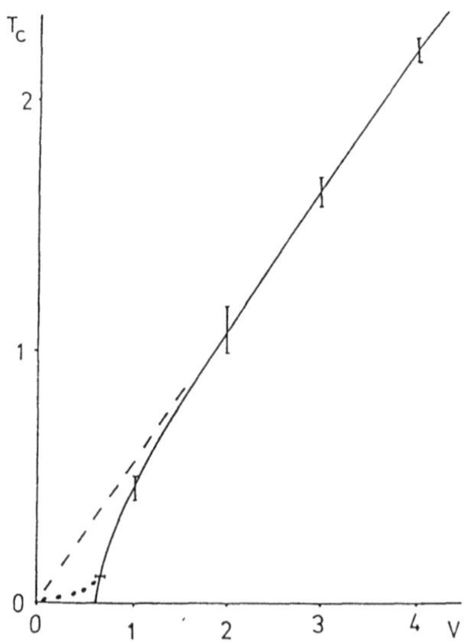

Fig. 8. Phase diagram for two-dimensional spinless fermion lattice. T_c versus n.n. interaction V. *Full line*: phase boundary. *Broken line*: extrapolation of $T_c = 0.56$ V-line (see text)

be useful. Furthermore, it shows the limitations of the world-line approach as α has to be chosen very close to 1 for lower temperatures to detect the fermionic behavior. For more substantial calculations we would like to emphasize the determinant method presented at this conference by J. Hirsch and D. Scalapino. But also here, apart from the half-filled Hubbard model we deal with negative transition probabilities where the α-trick might be useful in the future.

3. Conclusion

We conclude that the transfer matrix method is certainly the way to go for one-dimensional systems avoiding any ergodicity problems. Combined with the $1/m^2$ extrapolation scheme discussed by M. Suzuki it seems superior to the world-line approach in one dimension. Furthermore, the world-line approach has its shortcomings even in two dimensions for spin systems. The restriction to constant magnetization may cause the current problems with the XY model. Nonetheless, apart from critical behavior the method works remarkably well for spin-systems, in contrast to fermionic systems, where the application of the α-trick leads to some improvement but not to a solution of the minus-sign problem. Thus, in the view of the authors, as we are all looking for some general fermion method, the most promising way to follow clearly seems to be the determinant method, where the α-trick might also be of importance.

The authors would like to thank M. Suzuki, J. Hirsch, D. Scalapino, H. de Raedt, R. Sugar, T. Schneider and F. Klauder for helpful discussions. One of us (I.M.) acknowledges the hospitality of the Institute for Theoretical Physics at the University of California at Santa Barbara during a workshop on "Spin-Glasses, Computation and Neural Networks" where part of this work was accomplished. This research was supported in part by the National Science Foundation under Grant No. PHY82-17853, supplemented by funds from the National Aeronautics and Space Administration, at the University of California at Santa Barbara.

References

1. I. Morgenstern: In *Heidelberg Colloquium on Spin Glasses*. Lecture Notes in Physics **192** (Springer, Heidelberg 1983) and references therein
2. M. Suzuki: Prog. Theor. Phys. **56**, 1454 (1976)
3. A.T. Ogielski, I. Morgenstern: Phys. Rev. Lett. **56**, 928 (1985)
4. J.E. Hirsch, D.F. Scalapino, R.L. Sugar, R. Blenkenbecler: Phys. Rev. Lett. **47**, 1628 (1981)
5. J.E. Hirsch, D.F. Scalapino, R.L. Sugar, R. Blenkenbecler: Phys. Rev. B**26**, 5033 (1982)
6. J.J. Cullen, D.P. Landau: Phys. Rev. B**27**, 297 (1983)
7. E. Lieb, T. Schulz, D. Mattis: Ann. Phys. (N.Y.) **16**, 407 (1961)
8. S. Katsura: Phys. Rev. B**127**, 1508 (1962)
9. J.C. Bonner, M.E. Fischer: Phys. Rev. A**135**, 640 (1964)
10. E. Loh, D.F. Scalapino, P.M. Grant: Physica Scripta **32**, 327 (1985)
11. H. de Raedt, B. de Raedt, F. Fivez, A. Lagendijk: Phys. Lett. **104A**, 430 (1984)
12. J.E. Gubernatis, D.F. Scalapino, R.L. Sugar, W.D. Toussaint: Phys. Rev. B**32**, 103 (1985)

Monte Carlo Study of the Sherrington-Kirkpatrick Spin Glass Model in a Transverse Field

H. Ishii and T. Yamamoto

Department of Physics, Osaka City University,
Sumiyoshiku, Osaka 558, Japan

1. Introduction

On the basis of the Sherrington-Kirkpatrick (hereafter referred to as SK) model [1] of an Ising spin glass with infinite-range interactions, many studies have been made theoretically and by means of computer simulations [2]. The physical picture of the spin glass which has emerged is that the phase space of the spin glass is divided into many valleys separated from one another by barriers, and one of the valleys is further subdivided into two or more valleys with decreasing temperatures. This hierarchical structure manifests a succession of micro phase transitions [3].

Introduction of quantum effects into the SK model has been done by the extension to the Heisenberg spins [4],[5]. One of the motivations is to know how quantum motions resist the spin glass freezing. Bray and Moore [4] proved by the replica theory the existence of a phase transition even in the presence of the quantum motions, and calculated the depression of T_c in the static or mean field approximation.

Another quantum effect is brought by a transverse field applied to the SK model. Spin flips due to a transverse field go against the spin ordering and must decrease T_c, whose effect has been considerably studied for regular lattices [6]. The system consisting of randomly distributed Ising spins in a transverse field may be a simplified model for dipole glasses [7] in which tunneling motions of dipoles correspond to spin flips due to a transverse field.

We have studied this effect by perturbation theory [8] for the SK Hamiltonian

$$H = - \sum_{(ij)} J_{ij}\sigma_{iz}\sigma_{jz} + \Delta\sum_i \sigma_{ix}, \tag{1.1}$$

where $\vec{\sigma}_j$ are the Pauli matrices for the j'th spin and J_{ij} are independent random variables with the probability distribution $P(J_{ij})=(N/2\pi J^2)^{1/2}\exp(-NJ_{ij}^2/2J^2)$. The transverse field is represented by Δ. Since T_c is determined by $J\chi(T_c)=1$ where χ is the susceptibility, we have calculated $\chi(T)$ in the second order of Δ in the paramagnetic phase,

$$\chi(T) = (1/T)[1-2(\Delta/T)^2 \int_0^1 x(1-x)\exp[-2(J/T)^2 x(1-x)]\,dx + \ldots], \tag{1.2}$$

and from this we obtain

$$T_c = J[1-0.225(\Delta/J)^2 + \ldots]. \tag{1.3}$$

The contributions of the second and fourth order of Δ tend to T and T^3 for $T\to0$, respectively. If all the coefficients for Δ^{2n} vanish similarly for $T\to0$, we consider that the paramagnetic susceptibility $\chi(T)$ would diverge at zero temperature. Then $J\chi(T)=1$ has always a solution, so that we stated in [8] the conjecture that T_c always exists irrespective of Δ. Recently Fedorov and Shender [9] have studied this problem by the replica theory following [4]. They have shown that (1.3) can be derived again, however for $\Delta\gg J$ they pointed out that our conjecture stated above is made beyond the radius of convergence of the series expansion in Δ, and that there should be an upper bound of Δ for the spin glass transition. Thereupon we have examined the problem again and have obtained the result in the static approximation just as used in [4] that T_c is given by the solution of

$$\int_0^\infty (x^2-2)\exp(-x^2/2)\cosh[Jx^2/T_c+(\Delta/T_c)^2]^{1/2}\,dx=0. \qquad (1.4)$$

When (1.4) is expanded in Δ, (1.3) is reproduced and it is shown that T_c tends to 0 for $\Delta=2J$. Equation (1.4) is numerically solved and the result is shown in Fig. 1. Its detail, together with the result for some thermodynamic quantities, will be reported elsewhere [10]. Because of the static approximation, T_c thus obtained may give an upper bound. Therefore our conjecture should be withdrawn.

In the spin glass phase, too, the transverse field must bring quantum motions in quenched spins. We expect that they tend to decrease the Edwards-Anderson order parameter and increase the ac (nonequilibrium) susceptibility due to the quantum fluctuation, in contrast to the behaviour in the paramagnetic phase. However, the effect of the transverse field in the spin glass phase has not been studied so far. The purpose of the present work is, by means of the Monte Carlo (MC) simulations, to study the low-temperature properties of the present system as well as to examine T_c obtained theoretically.

The MC simulations in the SK model [11]-[13] have played a role which is usually done by experiments and also have clarified some important points, e.g., the structure of the phase space. In the present problem, as a spin flip term due to the transverse field does not commute with the SK Hamiltonian, we use a quantum MC method proposed by Suzuki [14]. This method is based on the fact that a quantum system can be mapped onto a classical system with an additional dimension through the use of the generalized Trotter formula. The mapping is exact when the lattice thickness n (Trotter number) of the additional dimension (Trotter direction) is infinite. It has been proved by Suzuki [15] that the correction to the n-th approximant for the thermodynamic quantities corresponding to the finite n is of the order n^{-2}. This yields the extrapolation method to predict the correct value at the limit of $n\to\infty$.

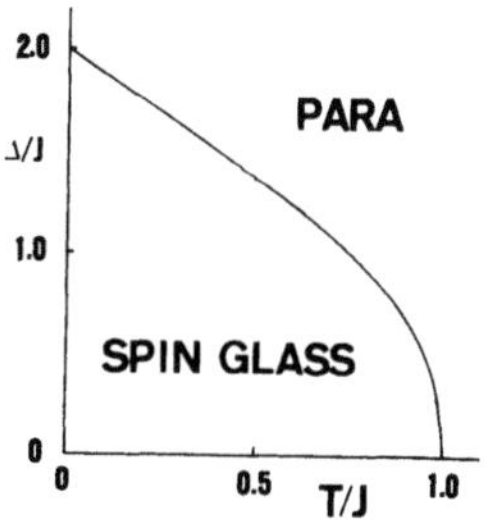

Fig. 1 Phase diagram of the SK model obtained from (1.4) (Δ, transverse field; T, temperature)

When the MC simulations are performed to obtain the thermodynamic quantities in the present problem, the following steps are to be taken; firstly several samples are prepared for a fixed spin number N and the values of n are suitably selected, secondly for each sample with n the thermodynamic quantities are calculated by the MC simulations, thirdly the extrapolation to $n \to \infty$ due to the n^{-2}-law is made to obtain them for each sample, fourthly these are averaged over samples. Furthermore the N-dependence is desired to be studied for the correct value in the thermodynamic limit. In comparison with the simulation in the classical spins on regular lattices, the number of systems to be computed is multiplied by the number of samples and by the number of n. This yields surely a difficult problem in real computation.

Although the approximation becomes better with increasing n as mentioned above, Wiesler [16] found in carrying out a simulation of one-dimensional transverse Ising model that for large n the MC simulation consisting of a sequence of one-spin-flip process ceases to work as an importance sampling. This is due to the singular behaviour of the coupling constant between spins along the Trotter direction for large n. In addition to this upper limit for n, it becomes evident in the present paper that the lower limit emerges from the condition of the importance sampling for the ordered phase of the system which undergoes a phase transition. To find a suitable range of n for the SK model, we describe in Sec. 2.1 simulations for the Ising model with ferromagnetic infinite-range interactions. The reason why we study this system is that its ordered phase is evidently characterized by two distinct minima in the phase space and it is known [17] that the mean field theory gives the exact solution at least for $\Delta=0$, in addition that the interactions in both systems are of infinite range.

On the basis of the analysis for n in Sec. 2.1, we study the SK model in Sec. 2.2. As T_c decreases with increasing Δ, the condition that the importance sampling works reasonably becomes severe and demands large n's. This forces us to limit the study in the range $\Delta \leq 1.5J$. Moreover, in the vicinity of T_c a problem arises from the finiteness of the system. Even if a macroscopic system undergoes the phase transition, the finite system in the actual MC simulation remains ergodic. It migrates within a finite time τ_{eg} between two time-reversed states [13]. Just below T_c, τ_{eg} is not so long that the observation (MC simulation) time exceeds it. This prevents the evaluation of the thermal average. Furthermore, when we pass through T_c, the false ergodic reverse in the ordered phase changes to the true one characteristic of the paramagnetic phase. To discriminate between them we ought to have studied the N-dependence. Because of the computation time we had to renounce the investigation of the N-dependence in this work. Instead of this, to see the feature near T_c, we introduce the function P(t) defined as the projection of the state realized at the t-th MC step onto the starting state which has the lowest energy as far as we have searched. At the instant when an occurrence of the ergodic reverse can be recognized from the time-evolution of P(t), we reverse all the spin directions to cancel out the ergodic reverse. The thermodynamic quantities are averaged in the time evolution corrected in this way. We calculate the energy, order parameter and susceptibility, mainly in the case $\Delta=J$. The last section presents a discussion of the results.

2. Simulations of the Ising Model in a Transverse Field

Suzuki [14] showed that the partition function of the quantal spin system (1.1) is written in the form

$$Z = \lim_{n\to\infty} [(\sinh 2\Delta/nT)/2]^{nN/2} \sum_{\sigma_{jk}=\pm 1} \exp(-H_{eff}'/T), \tag{2.1}$$

$$H_{eff}' = - \sum_{k=1}^{n} [\sum_{(ij)} (J_{ij}/n)\sigma_{ik}\sigma_{jk} + \gamma \sum_j \sigma_{jk}\sigma_{j\,k+1}], \tag{2.2}$$

where σ_{jk} denotes the Ising spin defined on the lattice site (j,k), j the position in the original SK model and k the coordinate along the Trotter direction, and $\gamma=(T/2)\ln[\coth(\Delta/nT)]$. The thermal average of the energy E is expressed as

$$E = \lim_{n\to\infty} [-N\Delta \coth(2\Delta/nT) - \langle \sum_{k=1}^{n} [\sum_{(ij)} (J_{ij}/n)\sigma_{ik}\sigma_{jk} - (\Delta/n)\mathrm{cosech}(2\Delta/nT) \sum_j \sigma_{jk}\sigma_{j\,k+1}]\rangle], \tag{2.3}$$

where $\langle\ \rangle$ stands for the thermal average governed by (2.1) and (2.2).

By the Metropolis method [11] we calculate the MC average $\langle Q\rangle_n$ of a thermodynamic quantity Q for the classical spin system (2.2) with finite n. Then we extrapolate them to $n\to\infty$ by the n^{-2}-correction law [15],

$$\langle Q\rangle_{exact} = \langle Q\rangle_n + O(n^{-2}). \tag{2.4}$$

In doing it, as pointed out in [16], we need to restrict n within a suitable range in which the importance sampling is expected to work reasonably.

We notice from (2.2) that with increasing n the interaction γ increases logarithmically whereas J_{ij}/n decreases. Therefore above certain n, spin flips for the MC sampling are mainly governed by γ, where the parallel spin alignment along the Trotter axis is favoured. For such a case, independently of whether J_{ij} is defined in the SK model, the situation may be similar to that in the system with $J_{ij}=0$, namely the linear chains along the Trotter axis. The probability to flip one of spins aligned in parallel is given by $\exp(-4\gamma/T)=[\tanh(\Delta/nT)]^2$ in the Metropolis method, as is seen from (2.2). If $n \gg \Delta/T$, spin-flips can scarcely occur and the importance sampling breaks down. To know the upper bound for n in the SK model we present the MC study first for the Ising model with ferromagnetic infinite-range interactions ($J_{ij}=J/N$, $J>0$), which is much simpler than the SK model and for which the mean field theory can be used to examine the MC result.

2.1 Ising Model with Ferromagnetic Infinite-Range Interactions

We have studied the case $\Delta=J/2$ for which the mean field theory predicts that a ferromagnetic state is realized at low temperatures. MC simulations were performed for $N=100$ with several values of n at temperatures $T=0.1J\sim0.5J$. At each temperature, by starting the system in the ferromagnetic state ($\sigma_{jk}=1$ for all j,k) and letting it evolve 500 MC steps per spin (MCS) as a relaxation process, the next 1000 MCS were used for taking the thermal averages of the various quantities.

In Fig. 2 we represent the data for the energy E per spin obtained by the MC simulations of (2.3). At each temperature it can be seen that the data points vary linearly with n^{-2} as predicted by (2.4). However, deviations occur for large n and with lowering temperature the point n_d at which the deviation sets up shifts toward large n. This suggests the occurrence of the singular behaviour of γ, which is connected with the decrease of Δ/nT as noted before. We estimated the ratio of n_d to Δ/T for five cases of T to find the relation $n_d \sim 5\Delta/T$ roughly holds. On the other hand, the data for small n show also deviations from the straight line (these points lie in the out side of the region in Fig. 2). For example, $E=-0.814J$ is obtained for $n=3$ at $T=0.2J$, whereas the extrapolation of the straight line in Fig. 2 gives $-0.845J$. This is because these n are too small to satisfy the inequality $J/nT<1$ which should hold for $n\to\infty$ where (2.1) and (2.2) are correct. This condition yields lower limit for n. The magnetization data also show similar behaviour in the n^{-2}-plot to that of the energy. At low temperatures there is a range between these two limits for the plot. However, above $T=0.8J$ its width narrows so much that the extrapolation turns impossible. Figure 3 shows the energy and magnetization obtained by the extrapolation procedure for $T\leq 0.7J$ and those obtained directly at $n=2$ for $T\geq 0.8J$. Agreement of the results with those of the mean field theory is fairly good.

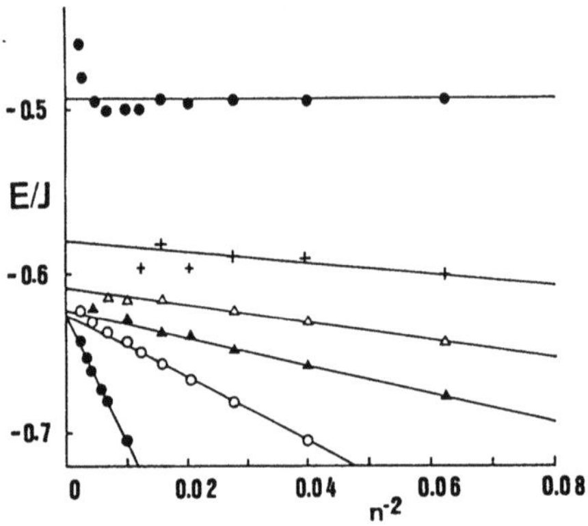

Fig. 2 Energies (per spin) vs n^{-2} in the ferromagnetic infinite range model (●:T=0.1J, ○:T=0.2J, ▲:T=0.3J, Δ:T=0.4J, +:T=0.5J). The solid circles in the upper part represent E for the linear chain (Δ=0.5J, T=0.2J) whose exact solution is shown by the horizontal line.

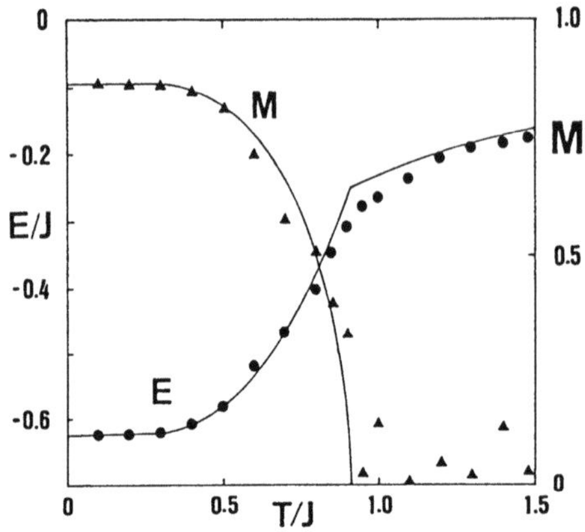

Fig. 3 Temperature dependence of the energy and magnetization (per spin) in the ferromagnetic infinite-range model. The lines represent the results of the mean field theory.

For comparison, the data for linear chains (namely $J_{ij}=0$ in (2.2)) are shown also in the upper side of Fig. 2, where the straight line represents the exact result $E=-\Delta\tanh(\Delta/T)$ which is independent of n. The data lie on this line, however with increasing n they deviate downward at first and then turn upward. Similar feature is seen for the paramagnetic phase ($\Delta=2J$). As for the lower limit of n it is found in the case of $n=3$ at $T=0.2J$ that the magnetization appears in the paramagnetic phase. This is due to too strong ferromagnetic couplings used beyond the limit $J/n<T$.

From these studies it may be allowed to deduce the range of n suitable for the importance sampling,

$$J/T < n \lesssim 5\Delta/T. \qquad (2.5)$$

2.2 SK Model

In this subsection we present the MC simulations for the SK model. Following the Gaussian distribution law for J_{ij} given in Sec. 1, we prepared three samples consisting of 80 spins. By replacing J with that defined in the SK model, (2.5) was used to determine the Trotter number. It is noticed that the ground or metastable state $\{\sigma_{jk}\}$ for (2.2) is obtained as a result of positive sign of γ from that $\{\sigma_j\}$ found in the SK model (Δ=0) by putting $\sigma_{jk}=\sigma_j$ (all spins on each Trotter axis are parallel). Thus we searched the lowest energy state in the SK model in the following way. For each sample, several random starting arrangements of spins were constructed, and for each starting configuration a procedure was followed to reach a state which was stable under any one-spin flip. Next, this state was slightly disturbed (warming the state to a temperature 3 times as large as J) and then the procedure to reach a state stable under one-spin flip was followed again. This repetition was made 20 times and the lowest known energy state was obtained. The energies per spin of these states for three samples (a)-(c) were -0.741J, -0.704J, -0.700J, respectively. These are in fact shallow in comparison with -0.76J, the more elaborate MC result obtained in [12].

On the basis of the Metropolis method we have performed MC simulations on these samples mainly for the case Δ=J. Calculation was carried out by starting each system in its lowest known energy state and letting it evolve for 1500MCS at an ascending series of temperatures. The last 1000MCS at each temperature were prepared to take the averages of the thermodynamic quantities.

In our simulation the order parameter describing spin glass freezing can be defined in the following two ways,

$$q = (1/N) \sum_j <(1/n) \sum_k \sigma_{jk}>^2, \tag{3.1}$$

$$q' = (1/nN) \sum_j \sum_k <\sigma_{jk}>^2. \tag{3.2}$$

The order parameter q describes the degree of freezing for each Trotter direction consisting of n spins and q' represents freezing of each spin. Although q is smaller than q' by definition, the two values are not so different in our results. Therefore we describe the result of q in the following.

For the study near T_c as mentioned in the Introduction, we define the function P(t) by

$$P(t) = (1/nN) \sum_j \sum_k \sigma_{jk}(t)\, \sigma_{jk}(0), \tag{3.3}$$

where $\sigma_{jk}(t)$ denotes the spin state realized at the t-th step. For the case Δ=J, at T=0.2J for all the three samples, P(t) always keeps a positive sign with fluctuations of small amplitude, showing that the state is in the spin glass phase. However, at T=0.4J in sample (c) with n=3, and at T=0.6J in the sample (a) with n=3 and also in the sample (b) with n=5, P(t) shows a reverse of the sign. Figure 4 shows two typical time-evolutions of P(t) at T=0.6J and J on one sample (a) (T_c=0.78J is given for Δ=J in Fig. 1). Since the difference between them is clearly seen, we dare to regard the reverse of P(t) in Fig. 4a as the size-effect and that in Fig. 4b as the thermal fluctuation [18]. For the former at the instant when the reverse of the sign occurs we reverse all the spin directions to cancel out the ergodic reverse. In this way we have corrected the time-evolutions in the three cases of T=0.4J and

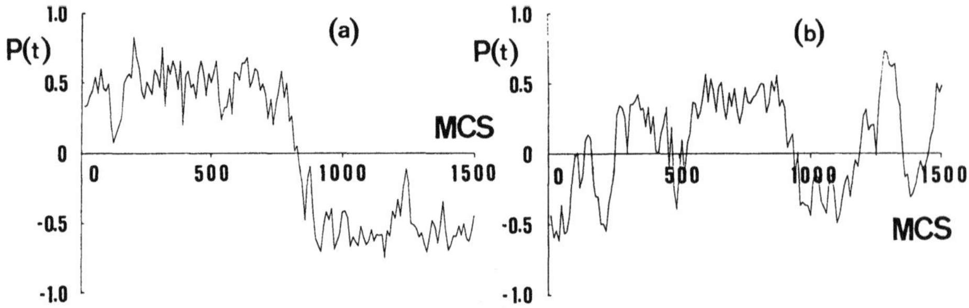

Fig. 4a-b Examples of the time-evolution of P(t), the projection of the state realized at the t-th MCS onto the initial state for Δ=J, n=3.
a) T=0.6J, b) T=1.0J

0.6J for which P(t) shows once a reverse of the sign. At T=0.8J the difference between the finite-size effect and the thermal fluctuation can not be discriminated, so we do not dare to correct the time-evolutions. With the time-evolutions corrected in this way the thermodynamic quantities were computed.

Above T=0.6J, n is limited by (2.5) in the narrow range of small value and the n^{-2}-extrapolation is not possible in fact. Thus we have done the n^{-2}-plot at T=0.2J, 0.4J and 0.6J. The results for the energy E per spin and the order parameter q on one sample are shown in Fig. 5. The other samples show the similar behaviour as in Fig. 5. We extrapolate these data to $n\to\infty$ according to (2.4) for each sample, and then average them over samples. The results

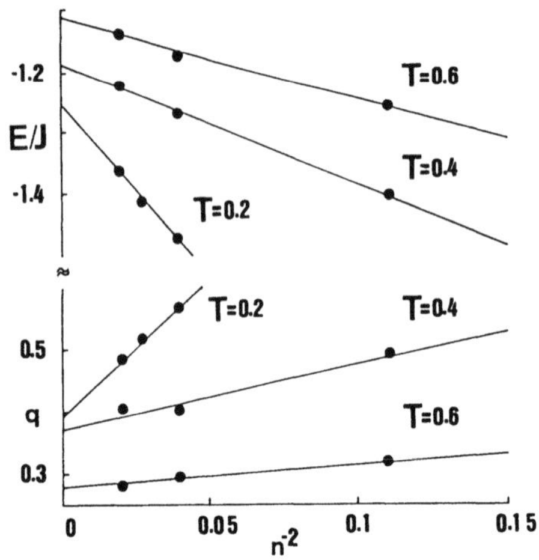

Fig. 5 Energies (per spin) and order parameters vs n^{-2} for Δ=J in the SK model

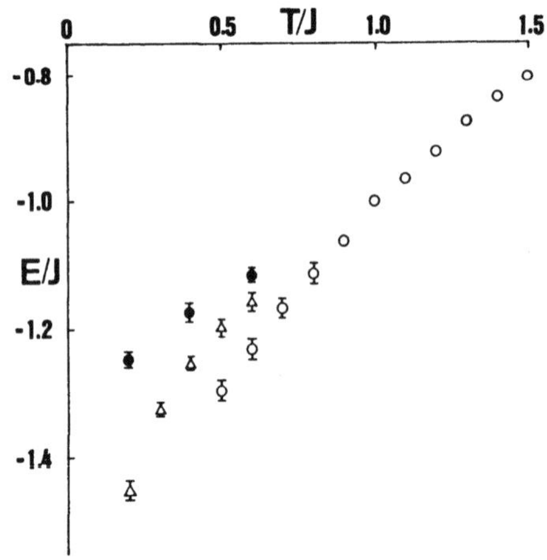

Fig. 6 Temperature dependence of the energy (per spin) for Δ=J in the SK model. Symbol ● shows the sample average of the extrapolated values. Symbols o and Δ show the sample average at n=3 and 5 respectively by cooling from high temperatures. Error bars indicate rms deviations over samples.

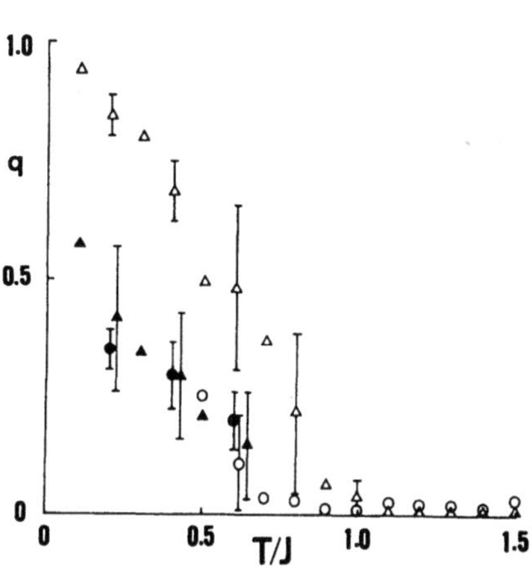

Fig. 7 Temperature dependence of the order parameter for Δ=J in the SK model. Symbol ● shows the sample average of the extrapolated values. Symbols O and ▲ show the sample average at n=3 and 5 respectively by cooling from high temperatures. Symbol Δ shows the average for Δ=0.

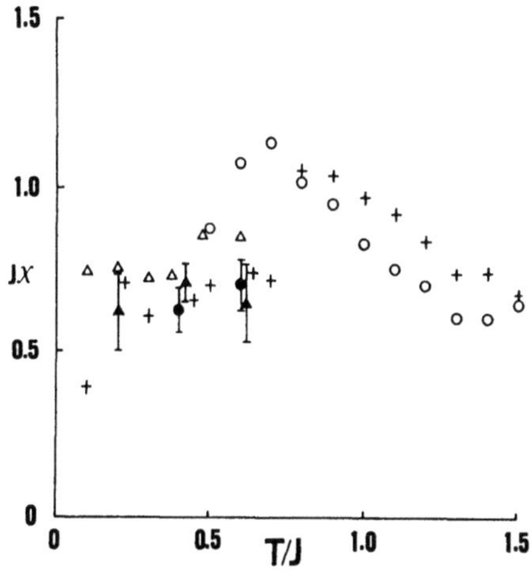

Fig. 8 Temperature dependence of the susceptibility (per spin) for Δ=J in the SK model. Symbols ● and ▲ show the sample averages at n=3 and 5 respectively by heating from low temperatures. Error bars indicate rms deviations over samples. Symbols O and Δ show the average at n=3 and 5 respectively by cooling from high temperatures. Symbol + shows the average for Δ=0.

for E and q are shown in Figs. 6 and 7, where to supplement three points thus obtained we add the data of n=3 and 5 averaged over samples by cooling gradually from high temperatures. For the sake of comparison we add in Fig. 7 the result for Δ=0 which agrees considerably with the result of [12]. The data for the susceptibility χ scatter not to show clearly the n^{-2}-dependence in our calculation. The same is seen in the MC simulation in 1-dimensional antiferromagnetic Heisenberg model [19]. Thus we show the data averaged over samples for various n in Fig. 8, which is again supplemented by the data of n=3 and 5 averaged over samples by cooling gradually from high temperatures.

In Figs. 7 and 8 it is found that χ(T) has a maximum at $T\sim0.7J$ and q(T) nearly disappears above this temperature. This might allow us to conclude $T_c\simeq0.7J$ for Δ=J, which agrees roughly with the theoretical result of (1.3) and Fig. 1. In Fig. 8 the decrease of q(T) due to Δ can be seen. In particular the extrapolation of q(T) to T=0 shows deviation from 1. This spin contraction manifests the quantum fluctuation due to Δ, by which it may be interpreted that the system goes back and forth in the vicinity of the metastable state. The same effect also brings about the increase of χ(T) below T_c and the decrease from that given by the Curie law above T_c as shown in Fig. 8.

To examine the theoretical result in Fig. 1, we have tried to do MC simulation for Δ=1.5J. At T=0.1J we choose $n=10\sim15$ which lie in the region $(10<n<75)$ determined by (2.5). We have found that two samples show through the behaviour of P(t), q and χ that the state is in the spin glass phase, whereas in one sample we could not find certain evidence for it. It is necessary for the definitive answer to increase the number of samples with large N.

3. Summary and Discussion

We have applied the quantum MC method to the simulation of the SK Ising spin glass in the transverse field Δ. This has been possible since the quantum effect due to Δ is local, whereas the quantum MC method would be inapplicable for the extension of the SK model to the Heisenberg spin, since the infinite-range nature of the interactions itself manifests the non-locality of the quantum effect. However, in the present simulations for random spin systems, the number of systems to be calculated is multiplied by the number of samples and by the number of cases necessary for the n^{-2}-extrapolation as well as the increase of the spin number multiplied by n. In other words, in limited computation, this confines the accuracy of our simulations to some extent, as will be stated below.

After studying the range of n for the importance sampling, we have been able to calculate E and q through the extrapolation law at three points below T_c. At high temperatures this extrapolation turns impossible on account of the narrow range of n. On the other hand the data of χ have scattered not to show the n^{-2}-dependence. Thus these data have been plotted without extrapolation. The quantum motion due to Δ is manifested in the temperature dependence of q and χ, from which T_c is estimated as $\sim 0.7J$ for $\Delta = J$. This agrees roughly with the theoretical result in Fig. 1.

We have noticed certain difficulties in our simulations. The first is the problem of the finite-size effect. To remedy it the size dependence should be investigated. However, in the SK model the increase of N in one sample means to prepare the other sample. This is because the lowest energy (or metastable) state in a system with N is not guaranteed to be again the lowest energy state in the system with N+N' even if the same J_{ij} are taken for the bonds common to both systems. Thus the N-dependence is to be considered after taking the sample average. The second difficulty is in the scattering of the data of χ mentioned above which hinders the n^{-2}-extrapolation. Moreover, the data for the nonlinear susceptibility have fluctuated so much in a single 1500MCS that we could not take its MC average. Because of this we could not determine T_c by finding its singular point [20],[8]. These shortcomings may be improved by 1) finding a better starting state having lower energy, 2) increasing the number of MC steps and 3) studying the size-dependence to discriminate the ergodic reverse.

In addition to the study of thermodynamic quantities intended in the present work, it will be interesting to study how the system goes back and forth around a metastable state (valley) in the phase space by the quantum effect of Δ. To see the nature of this localization it will be helpful if the Suzuki-Trotter formula can be applied to the time-evolution operator exp(-itH) as proposed by de Raedt [21].

Acknowledgements

The authors would like to thank Professor M. Suzuki for valuable discussions and encouragement.

References

1. D. Sherrington, S. Kirkpatrick: Phys. Rev. Lett. 35, 1792 (1975)
2. For reviews, see K. H. Fischer: Phys. Status Solidi B116, 357 (1983); 130, 13 (1985)
3. M. Mézard, G. Parisi, N. Sourlas, G. Toulouse, M. Virasoro: Phys. Rev. Lett. 52, 1156 (1984); J. Phys (Paris) 45, 843 (1984)

4. A. J. Bray, M. A. Moore: J. Phys. C 13, L655 (1980)
5. H. -J. Sommers: J. Magn. Magn. Mat. 22, 267 (1981)
6. R. J. Elliott, C. Wood: J. Phys. C 4, 2359 (1971)
P. Pfeuty, R. J. Elliott: J. Phys. C 4, 2370 (1971)
A. Yanase, Y. Takeshige, M. Suzuki: J. Phys. Soc. Japan 41, 1108 (1976)
7. A. T. Fiory: Phys. Rev. B4, 614 (1971)
8. H. Ishii, T. Yamamoto: J. Phys. C 18, 6225 (1985)
9. Ya. V. Fedorov, E. F. Shender: Pis'ma. Zh. Eksp. Teor. Fiz. 34, 526 (1986)
10. T. Yamamoto, H. Ishii: (unpublished)
11. For reviews, see K. Binder, D. Stauffer: In Applications of the Monte Carlo Method in Statistical Physics, Topics Current Phys., Vol 36, ed. by K. Binder (Springer, Berlin, Heidelberg 1984), p.1, p.241
12. S. Kirkpatrick, D. Sherrington: Phys. Rev. B 17, 4384 (1978)
13. N. D. Mackenzie, A. P. Young: J. Phys. C 16, 5321 (1983)
14. M. Suzuki: Prog. Theor. Phys. 56, 1454 (1976)
15. M. Suzuki: Phys. Rev. B 31, 2957 (1985)
16. A. Wiesler: Phys. Lett. 89A, 359 (1982)
17. See, e.g., H. E. Stanley: Phase Transitions and Critical Phenomena (Oxford Univ. Press, London, 1971), p.91
18. A weak support to such a judgement may be given by the similarity of the shape of P(t) in Fig. 4a to the correspondent obtained below T_c in the ferromagnetic model dealt in Sec. 2.1. In the latter the reverse found for N=100 disappears for N=200, showing the finite-size effect.
19. J. J. Cullen, D. P. Landau: Phys. Rev. B 27, 297 (1983)
20. M. Suzuki: Prog. Theor. Phys. 58, 1151 (1977)
21. H. De Raedt: preprint

Note added

After the symposium was finished we became aware that the phase diagram identical with Fig. 1 together with the correct conclusions in the static approximation concerning an upper bound of Δ for the spin glass transition was already obtained by a method different from the replica theory in the following paper, K. D. Usadel: Solid State Commun. 58, 629 (1986). We thank Prof. K. D. Usadel for pointing out this fact to us.

Part III

Fermion Systems and Nuclear Systems

Computer Simulation of Polaron and Bipolaron Systems

H. De Raedt[1], *A. Lagendijk*[2], *and P. de Vries*[2]

[1]Physics Department, University of Antwerp, Universiteitsplein 1, B-2610 Wilrijk, Belgium

[2]Natuurkundig Laboratorium, University of Amsterdam, Valckenierstraat 65, NL-1018 XE Amsterdam, The Netherlands

I. INTRODUCTION

Thermodynamic behavior of classical many-particle systems such as Ising models, hard sphere gases, fluids etc. can be studied by means of analytical, e.g. high-temperature series, perturbation expansions and renormalization group methods, and computer simulation techniques. There is no doubt that the latter have contributed a lot to the present understanding of the behavior of classical many-body systems [1,2]. Extending the techniques, originally developed for classical-mechanical models, to quantum systems is not straightforward. In this paper a Monte Carlo approach for quantum models at non-zero temperatures ($T \neq 0$) is discussed. The technique is complementary to the $T = 0$ Monte Carlo methods, see for instance CEPERLEY and KALOS [3]. In the following section an outline of the basic concepts and mathematical justification of the method are given. The general methodology is illustrated in subsequent sections by considering explicit examples. Topics treated include the one-, two- and three-dimensional Molecular Crystal Model, a two-level system with non-linear coupling and bipolaron formation.

II. QUANTUM STATISTICAL MECHANICS

The physical properties of a quantum statistical system in thermal equilibrium with its surroundings follow from the knowledge of the expectation value of physical observables. Quantum statistical mechanics postulates that the expectation value of a physical observable A is defined by

$$< A >= \mathbf{Tr}\ Ae^{-\beta H} / \mathbf{Tr}\ e^{-\beta H} \quad , \tag{1}$$

where H is the Hamiltonian of the system, β is proportional to the inverse of the temperature T, and **Tr** denotes the sum of all diagonal matrix elements. Evaluation of matrix elements of the exponent of an operator generally requires diagonalization of the operator in question and for (1) this it tantamount to finding the exact solution. A possible solution to this fundamental problem is the use of path-integral methods. As demonstrated in [4], the Trotter product formula

$$e^{A+B} = \lim_{m \to \infty} \left(e^{A/m} e^{B/m} \right)^m \tag{2}$$

and its generalizations [5,6] provide a convenient starting point to derive useful path-integral representations. The key point is to write the Hamiltonian as a sum of two (or more) operators, each of them being sufficiently simple so that they are easily diagonalized separately. In most cases there is a lot of freedom in choosing a decomposition of the Hamiltonian. For particles moving in continuum space, the Hamiltonian is the sum of kinetic and potential energy. Adopting this self-suggesting break-up yields the well-known Feynman path-integral [7]. In the case of spin-Hamiltonians such a break-up does not seem to be applicable but then one can use a real-space approach in which the Hamiltonian is written as a sum of pair interactions [4,8]. Just as for the Feynman path-integral, the "path-integral" representation of the quantum spin system is obtained by introducing appropriate resolutions of the identity operator and working out all matrix elements. In a shorthand notation the process of deriving a path-integral representation can be summarized as follows:

$$\begin{aligned} Z &= \mathbf{Tr}\, e^{-\beta H} \quad , \\ &= \lim_{m\to\infty} \mathbf{Tr}\,(\text{product of operators}) \quad , \\ &= \lim_{m\to\infty} \sum_{\psi_1} \dots \sum_{\psi_m} (\text{product of matrix elements}) \quad , \end{aligned}$$

where the summations are over a complete set of states. Similar results for expectation values can be derived following the same route.

In practice one can rarely take the limit $m \to \infty$ and therefore it is necessary to make one, but only one, approximation: One has to keep m relatively small. Making the finite-m approximation is not an uncontrolled approximation since convergence of Trotter formulae is guaranteed (see [4,6]) under conditions which in most numerical work are automatically fulfilled. In numerical work, we adopt a very pragmatic strategy and start by taking $m = 1$. The value of m is increased until the systematic errors due to the finite-m approximation, become "small" enough. As most numerical results are obtained from Monte Carlo simulations, "small" usually means that the systematic errors disappear in the statistical noise of the data.

Calculating quantities such as (3), each quantity being a sum of a very large number of terms, is a difficult task. The basis idea of performing computer simulations is to extract the relevant physics by sampling a small number of terms contributing to (3). This will make sense if most of these terms do not contribute significantly to (3) because then one can use importance sampling techniques, in particular the Metropolis method (see [1,2,9]), to generate the most "important" (dominant) configurations. This tremendous simplification can, in principle, always be made if one can attach a positive weight to each configuration. In classical statistical physics this weight is nothing but the Boltzmann factor $e^{-\beta H}$. In the case of a quantum system, getting rid of the operators requires the use of Trotter formulae, and this gives rise to some complications. Owing to symmetry properties of the quantum system (such as conservation of magnetization for instance), many matrix elements appearing in (3) are zero. This makes it difficult, if not impossible, to define transition probabilities between states of different symmetry, a problem which is not often encountered in Monte Carlo simulations of classical systems. A direct consequence of this is that it becomes hard to satisfy one of the important mathematical conditions necessary to justify the applicability of the Metropolis Monte Carlo method, namely the ergodicity requirement [9]. Non-ergodic algorithms only sample a subset of the important contributions. An even more serious problem is the possibility that there are negative contributions to (3). This would preclude the use of importance sampling techniques at all because some of the transition probabilities become negative. Negative contributions still constitute a largely unsolved problem. To set up a Monte Carlo simulation of a given quantum model we proceed as follows.

1) Choose an appropriate decomposition of the Hamiltonian. Remember that in practice any break-up will converge if $m \to \infty$.
2) Determine the matrix elements. If there are many zero or negative matrix elements try to find a transformation to remove them. If no such transformation can be found, choose another model.
3) Perform exact calculations for small systems, i.e. diagonalize the Hamiltonian. Compute physical quantities in the weak or strong coupling limit.
4) Perform rigorous finite-m summations and study the convergence as a function of m using the results of 3 as reference data.
5) Construct a Monte Carlo algorithm that satisfies the conditions stated in [9]. Check the algorithm by comparing the results with exact ones (see 3 and 4).
6) Perform Monte Carlo simulations of larger systems than those studied in 3 and 4. Take into account that statistical errors obtained from a single long run usually underestimate the true statistical error because of large correlations among states generated by the Metropolis Monte Carlo method (this problem is much more severe for quantum systems than for classical systems). It is safer to make a number of shorter runs and consider the outcome of each run as an independent measurement.

III. MOLECULAR CRYSTAL MODEL

In this section we consider the simplest model of an electron interacting with lattice vibrations. The Hamiltonian is given by [10]

$$H = t\sum_{l\in\Lambda}\sum_{\delta}\left(c_l^+ c_{l+\delta} + c_{l+\delta}^+ c_l\right) + \lambda\sum_{l\in\Lambda} x_l\, n_l + \frac{1}{2}\sum_{l\in\Lambda}\left(p_l^2 + \Omega^2 x_l^2\right) \quad , \tag{4}$$

where the first term represents the electron kinetic energy, the second term describes the on-site interaction between the electron and the lattice vibrations and the last term models the Einstein oscillators at each site. The sum over l runs over all points of the hyper-cubic lattice Λ and the sum over δ goes over all basis vectors of the lattice.

One of the most important questions in the study of polaron models is whether or not there exists some kind of transition from a state where the electron moves through the system as an almost free particle to a state where the electron is tightly bound to the phonon cloud surrounding the electron. The latter state is usually referred to as the self-trapped state. Note that there is an essential difference between a self-trapped state and a localized state. The former is translational invariant whereas the localized state is not.

Most of the work on this problem starts from a continuum description. For continuum polarons the Feynman path-integral approach supplemented with a variational ansatz has proven to be the most powerful tool [11,12]. Its power stems from the fact that the Hamiltonian is a quadratic function of the phonon coordinates. This feature can be exploited to integrate out analytically the phonon degrees of freedom. Most (but not all) of such calculations yield a genuine first-order ground-state phase transition [11,12].

Turning back to the lattice polaron problem, the main question is whether a behavior different from the one described by continuum models is to be expected. This will depend on the range over which the self trapping is effective. If it extends over many lattice sites, the continuum approximation to the lattice problem must be adequate. If this range is of the order of a lattice spacing, the continuum approximation breaks down.

It is remarkable that as far as we know, nobody has given a path-integral formalism for the lattice polaron model, probably because traditionally one starts from a classical Lagrangian which in this case is hard to write down. However, using the Trotter-Suzuki approach a discrete version of the Feynman path-integral can be derived directly [13]. Just as in the continuum case the phonons can be eliminated analytically. For example, in the adiabatic limit the m-th approximant to the partition function of the 1D model reads

$$Z_m = Z_m^{(0)} \sum_{\{x_j\}} \prod_{j=1}^{m} I_{x_j - x_{j+1}}(2\beta t/m) \exp\left(\frac{\beta C}{m^2} \sum_{i,j=1}^{m} \delta_{x_i,x_j}\right) \quad , \tag{5}$$

where $Z_m^{(0)}$ is the m-th approximant to the partition function of the free-phonon system, $I_n(z)$ is the modified Bessel function, and C is a measure for the strength of the effective, phonon-mediated self-interaction of the electron. Generalizing (5) to the d-dimensional or non-adiabatic case is straightforward [13]. In (5) each x_j runs over all lattice sites.

To study the behavior of systems described by discrete path-integral representations such as (5), we have carried out extensive Monte Carlo simulations and also exact numerical calculations (for references see [13]). As already mentioned, one of the main issues is whether there is a difference between lattice and continuum models. The ultimate non-continuum model is definitely a lattice of only two sites. This model is equivalent to a spin-boson model and has the appealing feature that in the adiabatic limit it can be solved rigorously. The most important result is that in the ground state it undergoes a genuine phase transition [14]. The second derivative of the free energy with respect to the coupling C is discontinuous. For C smaller than its critical value, the particle is free whereas for large C the particle is self trapped. There is however a subtle point here. In the adiabatic limit there is breaking of translational invariance, or in other words the particle is localized. Adding the slightest non-adiabatic correction will smear out this phase transition.

Except for the two-site and adiabatic limit the transition from quasi-free to self-trapped state is not a genuine phase transition but rather reflects a dramatic change in behavior of the particle as the coupling energy approaches its "critical" value. Our simulation data showed that there is no qualitative difference between the behavior of the two-site, 1D, 2D or 3D model [13]. In all cases there is a smooth transition from the quasi-free to self-trapped state. This observation is further supported by a variational calculation that takes into account all effects of lattice structure of the theory [15]. In this manner full agreement between Monte Carlo simulations and variational calculations has been obtained.

It is worthwhile mentioning that variational work based on a continuum model with wave vector cut-off reproduced some but not all of the features observed in the simulations [16]. The reason for this is to be found in the failure of the continuum theory to account for the boundedness of the kinetic energy of the electron on the lattice.

Additional simulation evidence that the continuum approximation breaks down is found by computing the "size" of the polaron. The simulations show that for C larger than a certain value C^* (the actual value of which depends on the dimensionality and other model parameters) the size of the polaron goes to zero [13].

IV. TWO-LEVEL SYSTEM WITH QUADRATIC COUPLING

In this section we present some results of a numerical study of a two-level system coupled quadratically to a harmonic oscillator. The Hamiltonian reads

$$H = -h\sigma^x + p^2/2M + M\Omega^2 x^2/2 + \lambda\sigma^z x^2 \quad , \tag{6}$$

where h is the tunnel energy of the free spin, M and Ω are the mass and frequency of the free oscillator and λ is the coupling constant, which can always be chosen to be non-negative and must not be larger than $M\Omega^2/2$ so that (6) is bounded from below. This upper-bound for λ signals the presence of a phase transition as a function of the coupling strength. Model (6) might be relevant for the behavior of quadratically coupled polarons. Gutfreund and Weger have suggested that quadratic coupling of electrons to librons in TTF-TCNQ type compounds is important [17]. However the coupling they propose is more complicated than that of (6).

We have studied this problem in the adiabatic limit ($M \to \infty$, $M\Omega^2$ finite) [18]. An advantage of working in this limit is that all thermodynamic and ground-state properties can be computed numerically to any desired precision, for any allowed value of λ. Therefore the accuracy and reliability of calculations based on path-integral representations can be examined in detail. Such a study yields useful information because this method is applied to many systems for which the exact solution is not known.

From the exact $T = 0$ properties it follows that model (6) does not exhibit a ground state transition as a function of λ. There is no self-trapping for $T = 0$. To study the system at $T > 0$ exact numerical calculations and computations based on a path-integral representation have been performed. In the adiabatic limit the m-th approximant to the partition function is given by

$$Z_m/Z_{OSC} = \sum_{\{S_i=\pm 1\}} \left(1 + \frac{C}{m}\sum_{i=1}^{m} S_i\right)^{-1/2} \exp\left(J\sum_{i=1}^{m} S_i S_{i+1}\right) \quad , \tag{7}$$

where Z_{OSC} is the free-oscillator partition function, $2J = \ln\coth(\beta h/m)$ and $0 \leq C = 2\lambda/M\Omega^2 \leq 1$. Clearly (7) resembles the partition function of an Ising system. However the spin-spin interaction J is size(m)-dependent. The sum over all spin states can easily be done exactly for $m \leq 20$. For $m > 20$ Monte Carlo simulations have been performed.

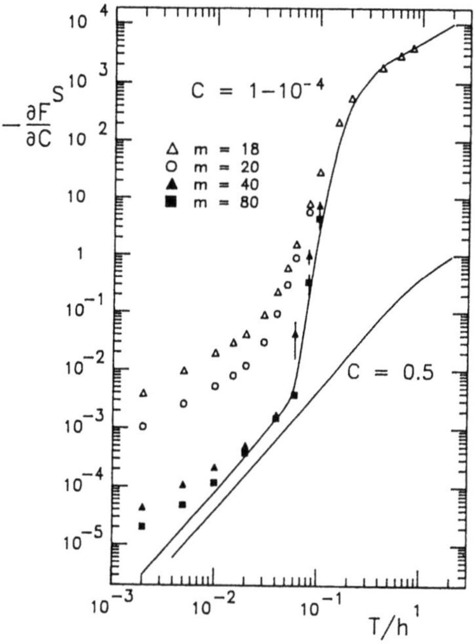

Fig.1. Derivative of the free energy with respect to the coupling C for $C = 0.5$ and $C = 0.9999$ as a function of T/h. Symbols are results for path-integral calculations for $C = 0.9999$. Open symbols: exact results for $m = 18, 20$. Solid symbols: Monte Carlo data for $m = 40, 80$ (error bars not shown have the size of the symbols). Full lines: exact results

In Fig.1 the coupling energy $\partial F/\partial C$ is shown as a function of temperature for $C = 0.5$ and $C = 1 - 10^{-4}$. From comparison with perturbation theory we conclude that for all C there is weak-coupling behavior if the temperature is sufficiently low. For C not too close to 1 (see the $C = 0.5$ curve in Fig.1) the behavior changes smoothly to that of the high-temperature regime. As $C \to 1$ the spin becomes polarized ($< \sigma^z > \to -1$) and the oscillator mode becomes soft. Then at a certain temperature (which decreases if C approaches 1) the derivative of the free energy begins to increase rapidly. Clearly there is an instability for $C = 1$. In this intermediate regime the derivative shows a divergence as a function of C. From a comparison of Monte Carlo data with exact results it follows that Monte Carlo simulations can be used to study properties of models such as (6) in an accurate and systematic manner.

V. BIPOLARON FORMATION

The ultimate goal of studying polarons is undoubtedly the complete understanding of the many-polaron system. An important step towards this ambitious goal is the investigation of bipolaron formation. A bipolaron is a self-trapped entity consisting of two polarons. In a two-polaron system one can have in addition two separate self-trapped polarons and also two (quasi) free carriers. Unraveling the zero-temperature phase diagram with respect to the existence of these quasi-particles is an important aspect of the study of the bipolaron system. Bipolaron formation in continuum space has been the subject of recent investigations [19,20]. Here we focus on bipolaron formation on a 3D lattice.

The generic Hamiltonian of the model reads

$$H = \sum_{\mathbf{k},s} \epsilon(\mathbf{k}) c^{+}_{\mathbf{k},s} c_{\mathbf{k},s} + \sum_{\mathbf{k}} V(\mathbf{k}) \rho_{\mathbf{k}} \rho_{-\mathbf{k}} + \sum_{\mathbf{k}} Q(\mathbf{k}) (a^{+}_{-\mathbf{k}} + a_{\mathbf{k}}) \rho_{\mathbf{k}} + \sum_{\mathbf{k}} \omega(\mathbf{k}) a^{+}_{\mathbf{k}} a_{\mathbf{k}} \quad . \tag{8}$$

The operator $a^{+}_{\mathbf{k}}$ ($a_{\mathbf{k}}$) creates (annihilates) a phonon excitation of wave vector $\mathbf{k}$, $c^{+}_{\mathbf{k},s}$ ($c_{\mathbf{k},s}$) creates (annihilates) an electron with spin s and $\rho_{\mathbf{k}}$ is the Fourier transformed electron density. In this paper we will, for the sake of simplicity, assume a simple cubic lattice and as usual in the study of bipolaron formation [19,20] confine ourselves to the case where one electron has spin up and the other has spin down. As (8) has a quadratic form in the phonon coordinates, path-integral concepts are ideally suited to study this model because in such an approach the phonon degrees of freedom can be eliminated analytically [7]. This combination of analytical and numerical techniques makes the Monte Carlo simulation very efficient. First of all the phonon contributions have been factored out and because of this the (few) electron properties can be calculated directly since the polaron contributions are not hidden in the statistical noise that would have been present if the phonons degrees of freedom were also treated by means of sampling methods.

Elimination of the phonons leads to an effective electron-electron interaction which in the adiabatic limit depends only on $Q^2(\mathbf{k})/\omega(\mathbf{k})$ (see [21] for details) showing that in the adiabatic limit, models with different $Q(\mathbf{k})$ and $\omega(\mathbf{k})$ can lead to the same effective interaction. We have carried out Monte Carlo simulations for (8) in the adiabatic limit and confined ourselves to the case where $Q^2(\mathbf{k})/\omega(\mathbf{k})$ does not depend on $\mathbf{k}$. Here we only discuss results for an on-site repulsive interaction $V(\mathbf{x}) = U\delta_{\mathbf{x},0}$. Calculations using a long-range interaction have been performed and show the same features [21].

In Fig.2 results are shown for the coupling-energy susceptibility $\chi^E_m = -\partial^2 F^E_m/\partial C^2$ as a function of C for a $32 \times 32 \times 32$ model with $U = 15$. The occurrence of two pronounced maxima is clear. Simulation data for the RMS distance D_m between the electrons (normalized to the free-electron value) can be found in Fig.3. The sharp drop in distance is connected to the largest peak in Fig.2. This demonstrates that this peak signals condensation into the bipolaron state. Comparison with single-polaron simulations [13,15] shows that the smallest peak is due to the self-trapping of the two individual polarons. Also shown in Fig.3 are the kinetic energy and the coupling energy. It is seen that the kinetic energy drops quickly when the two independent polarons are formed but keeps changing smoothly when the transition to the bipolaron state takes place. The results for the coupling energy confirm the behavior of the susceptibility since the latter is the derivative of the first and both quantities have been measured independently. The regions in (C,U)-parameter space where the bipolaron or two-polaron state is stable can be found by inspecting Fig.4.

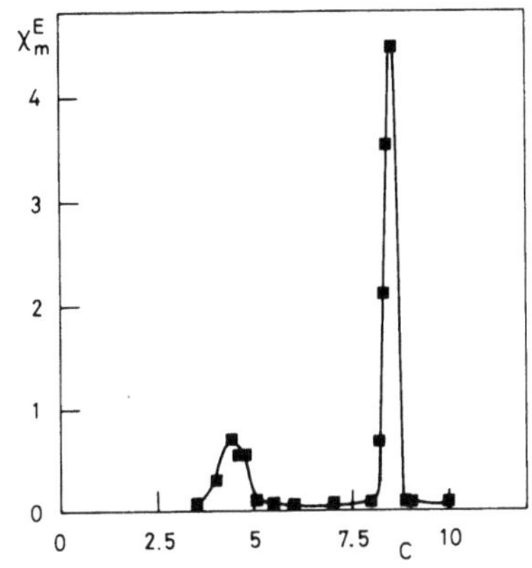

Fig.2. Monte Carlo simulation results for the coupling-energy susceptibility as a function of C for $U = 15$

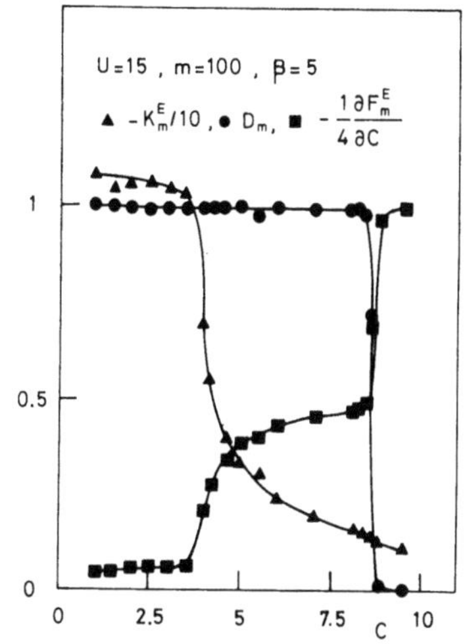

Fig.3. Simulation data for the kinetic energy (triangles), the normalized distance between the electrons (dots), and the coupling energy (squares) as a function of C

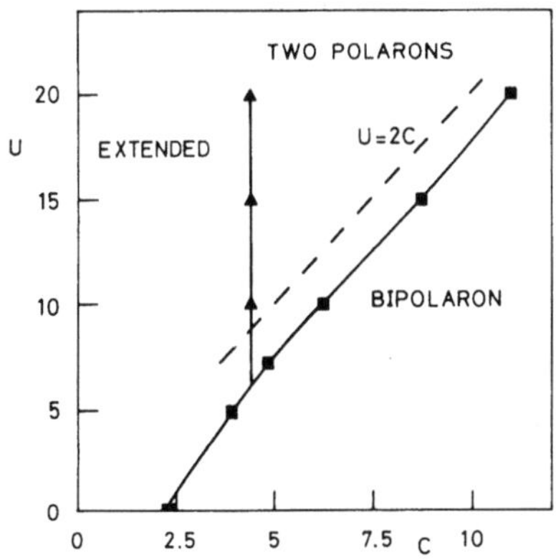

Fig.4. U–C phase diagram for an on-site Coulomb repulsion. The dashed line is the prediction of a simple strong-coupling calculation. Solid lines are guide to the eyes

VI. CONCLUSION

We have discussed the application of the Monte Carlo method to quantum statistical systems. Simulation of quantum spin models requires rather sophisticated and dedicated algorithms in order to achieve ergodicity and efficient use of computer time. Spin-boson and fermion-boson systems constitute a class of models that are well-suited to a path-integral treatment. Implementation of Monte Carlo algorithms for path-integral representations is not straightforward and there are still a number of fundamental problems to be solved. However it is clear that the combination of Monte Carlo simulations and Trotter-Suzuki mappings can be a fruitful approach for studying quantum systems.

Acknowledgement

This work is partially supported by the "Supercomputer Project" of the Belgian National Science Foundation (N.F.W.O.), the research program of the Dutch Stichting voor Fundamenteel Onderzoek der Materie (FOM) which is financially supported by the "Nederlandse Organizatie voor Zuiver-Wetenschappelijk Onderzoek" (ZWO), and NATO grant no. RG.86/0026. One of us (H.D.R) thanks the N.F.W.O. for financial support.

REFERENCES

1. K. Binder, and D. Stauffer: in Applications of the Monte Carlo Method in Statistical Physics, ed. by K. Binder, Topics in Current Physics, Vol. 36 (Springer, New York 1984)
2. K. Binder: in Monte Carlo Methods in Statistical Physics, ed. by K. Binder, Topics in Current Physics, Vol. 7 (Springer, New York 1979)
3. D.M. Ceperley, and M.H. Kalos: in Monte Carlo Methods in Statistical Physics, ed. by K. Binder, Topics in Current Physics, Vol. 7 (Springer, New York 1979)
4. M. Suzuki: Prog. Theor. Phys. 56, 1454, (1976)
   M. Suzuki, S. Miyashita, and A. Kuroda: Prog. Theor. Phys. 58, 1377, (1977)
5. H. De Raedt, and B. De Raedt: Phys. Rev. A 28, 3575, (1983)
6. M. Suzuki: J. Math. Phys. 26, 601, (1985)
7. R.P. Feynman, and A.R. Hibbs: Quantum Mechanics and Path Integrals (McGraw-Hill, New York 1965)
8. H. De Raedt, and A. Lagendijk: Phys. Rep. 127, 233, (1985)
9. J.M. Hammersley, and D.C. Handscomb: Monte Carlo Methods (Methuen, London 1964)
10. T. Holstein: Ann. Phys. (N.Y.) 8, 325, (1959)
11. A. Sumi, and Y. Toyozawa: J. Phys. Soc. Jpn. 35, 137, (1973)
12. F.M. Peeters, and J.T. Devreese: Phys. Stat. Sol. 112, 219, (1982)
13. see ref. 8 chapter 6.
14. R. Beck, W. Götze and P. Prelovšek: Phys. Rev. A 20, 1140, (1979)
15. A. Lagendijk, and H. De Raedt: Phys. Lett. 108A, 91, (1985)
16. S. Das Sarma: Solid State Commun. 54, 1067, (1985)
17. H. Gutfreund, and M. Weger: Phys. Rev. B 16, 1753, (1977)
18. P. de Vries, A. Lagendijk, and H. De Raedt: Phys. Rev. B (in press)
19. M.H Cohen, E.N. Economou, and C.M. Soukoulis: Phys. Rev. B 29, 4496 (1985)
20. H. Hiramoto, and Y. Toyozawa: J. Phys. Soc. Jpn. 54, 245 (1985)
21. H. De Raedt, and A. Lagendijk: Z. Phys. B 65, 43 (1986)

A Stochastic Algorithm for Many-Fermion Systems

D.J. Scalapino

Department of Physics, University of California,
Santa Barbara, CA93106, USA

Abstract

A new algorithm for simulating many-fermion systems is discussed in the context of the 3-D Hubbard model.

Numerical simulation of the many-fermion problem is difficult because of the anticommuting nature of the basic fields. Typically one introduces a Hubbard-Stratonovich field, or there is an intrinsic phonon field, which reduces the Hamiltonian to a form that is bilinear in the fermion fields and then traces out the fermion fields. This leads to an effective action for the Hubbard-Stratonovich field which is nonlocal, and it is a challenging problem to develop efficient algorithms for a stochastic sampling of the Hubbard-Stratonovich field. That is, algorithms whose running time scales linearly with the space-time volume. Thus, for example, one technique[1–3] which has been applied to a variety of condensed matter fermion problems scales as the third power of the spatial volume times the number of imaginary time slices. This severely restricts the size of the lattices which one can study. For example, this approach has been used to study 2-D spinless[3] fermions on spatial lattices as large as 12×12, the 2-D Hubbard model on 8×8 and the 3-D Hubbard model on $6 \times 6 \times 6$ spatial lattices.[2]

While the lattice size required to meaningfully scale onto the bulk limit clearly depends upon the specific problem of interest, there are various questions about the Hubbard model, as well as other many-fermion models, which make it highly desirable to extend the linear size of the lattices which can be simulated. Hirsch[2] used a self-consistently calculated boundary field to study the phase transition of the 3-D Hubbard model on 4^3 and 6^3 lattices. It will be interesting to compare this with results on periodic 4^3, 6^3, 8^3 and 10^3 lattices, particularly to further explore the weakly first order transition found in reference 2. An interesting problem also arises when the para-ordering of a strongly interacting system gives rise to collective excitations which may be exchanged, leading to a weak interaction capable of producing a new low temperature phase. For example, can paramagnon antiferromagnetic spin fluctuations in a nearly magnetically ordered Hubbard model give rise to singlet d-wave superconductivity? Here again one sees the desirability of going to larger lattices. Within an RPA approximation of the paramagnon exchange, one indeed finds[4] an attractive pairing interaction in the d-wave channel. In this case, it is not that one is seeking to simulate the superconducting transition but rather to obtain lattices of sufficient size that the space-time structure of the *pairing interaction* in the nearly ordered, para–antiferromagnetic state, can be adequately represented. The role of the simulation is to provide input for an effective interaction (replacing the RPA approximation) that can then be used in an analytic calculation of the low-temperature superconducting transition. Since the structure of the spin fluctuations is set on a space-time scale an order of magnitude smaller than the scale of the superconducting pairing correlations, this can be a useful approach.

A variety of ways to construct more efficient algorithms have been suggested.[5–7] At UCSB, Scalettar, Sugar and I have investigated several approaches.[8] Here I will describe an algorithm that the three of us with Toussaint from UCSD have been recently studying. It is based on a hybrid approach suggested by the work of Duane and Kogut.[7] In this scheme, a combination of heat bath and deterministic molecular dynamic moves are used to produce a rapid evolution of the Hubbard–Stratonovich field through configuration space. We find that a "leapfrog" solution of the discrete molecular dynamic equations provides a stable and accurate method for carrying out the molecular dynamics. We also introduce a new method for calculating fermion Green's functions which makes this part of the simulation much more efficient than previous approaches. This algorithm has recently been applied to SU(3) lattice gauge theory with dynamic quarks,[9] and is currently being used to study the 3-D Hubbard model.[10] Here, in this latter context, I'll describe the method and discuss how it scales with the volume of the system.

The Hamiltonian describing the particle-hole symmetric (half-filled)[11] Hubbard model is

$$H = -t \sum_{\substack{\langle ij \rangle \\ \sigma}} (C_{i\sigma}^{+} C_{j\sigma} + C_{j\sigma}^{+} C_{i\sigma}) + U \sum_{i} (n_{i\uparrow} - \tfrac{1}{2})(n_{i\downarrow} - \tfrac{1}{2}) . \qquad (1)$$

Here $C_{i\sigma}^{+}$ is the usual fermion creation operator for an electron in a Wannier state of spin σ at site i and $n_{i\uparrow} = C_{i\uparrow}^{+} C_{i\uparrow}$. In the usual way,[12] the partition function is expanded so that

$$Z = Tr\ e^{-\beta H} \cong Tr\ (e^{-\Delta\tau K} e^{-\Delta\tau V})^L \tag{2}$$

with K the kinetic hopping term in Eq. (1) and V the potential. Then a Hubbard-Stratonovich field is introduced at each of the L time slices so that

$$e^{-\Delta\tau U(n_\uparrow - \frac{1}{2})(n_\downarrow - \frac{1}{2})} = \left(\frac{\Delta\tau}{\pi}\right)^{\frac{1}{2}} e^{-\frac{\Delta\tau U}{4}} \int_{-\infty}^{\infty} dx e^{-\Delta\tau[x^2 + \sqrt{2U}x(n_\uparrow - n_\downarrow)]} \tag{3}$$

and the fermion degrees of freedom are traced out leaving

$$Z = \int \delta x e^{-S_0[x]} \det M_\uparrow \det M_\downarrow \ . \tag{4}$$

Here

$$S_0[x] = \Delta\tau \sum_{il} x_i^2(\tau_l), \tag{5}$$

$$\begin{aligned} \det M_\uparrow[x(\tau)] = \det M_\downarrow[-x(\tau)] &= Tr\left(\prod_{l=1}^{L} e^{\Delta\tau \sum_{ij} C_{i\uparrow}^+ t_{ij} C_{j\uparrow}} e^{\Delta\tau \sum_i \sqrt{2U} x_i(\tau_l) n_{i\uparrow}}\right) \\ &= \det(1 + B_L \cdots B_1) \end{aligned} \tag{6}$$

and

$$B_l = e^{\Delta\tau t} e^{\Delta\tau V(\tau_l)} \tag{7}$$

with $t = (t_{ij})$ and $V(\tau_l) = \delta_{ij}\sqrt{2U}x_i(\tau_l)$. In our work, we further truncated $e^{\Delta\tau t}$ so that it contains only nearest-neighbor terms and is sparse.

Equation (6) formed the basis for the exact updating scheme of reference 1 and had the advantage that the B_l matrices are *only* $N \times N$ matrices with N the number of spatial sites rather than $NL \times NL$ matrices. However, for the approach discussed here, we return to the $NL \times NL$ matrix form in which

$$\det M_\uparrow = \det \begin{pmatrix} B_1 & -1 & & & \\ & B_2 & -1 & & \\ & & B_3 & -1 & \\ & & & \ddots & \\ 1 & & & & B_L \end{pmatrix} . \tag{8}$$

For the particle-hole symmetric case $\det M_\downarrow[x(\tau)] = e^{-\Delta\tau \sum_{il} \sqrt{2u} x_i(\tau_l)} \det M_\uparrow[x(\tau)]$ so that $(\det M_\uparrow \det M_\downarrow)$ is positive definite, and we can replace $\det M_\uparrow$ by

$$\sqrt{\det(M_\uparrow^T M_\uparrow)} = \int \delta\phi_\uparrow e^{-\phi_\uparrow 0_\uparrow^{-1} \phi_\uparrow} \tag{9}$$

with $0_\uparrow = M_\uparrow^T M_\uparrow$. Proceeding in the same way for $\det M_\downarrow$, the simulation can be viewed as a stochastic calculation involving three real classical fields x, $\phi_\uparrow$ and $\phi_\downarrow$.

$$Z = \int \delta x \delta\phi_\uparrow \delta\phi_\downarrow e^{-S_{eff}[x,\phi_\uparrow,\phi_\downarrow]} \tag{10}$$

with an effective action

$$\tilde{S}_{eff} = \Delta\tau\Sigma x_{i\tau}^2 + \phi_\uparrow (M_\uparrow^T M_\uparrow)^{-1}\phi_\uparrow + \phi_\downarrow (M_\downarrow^T M_\downarrow)^{-1}\phi_\downarrow \ . \tag{11}$$

The updating of the ϕ-field involves constructing two vectors of Gaussianly distributed random numbers $R_\uparrow$ and $R_\downarrow$ such that $P(R) \sim e^{-R^T R}$. Then setting

$$\phi_\uparrow = M_\uparrow^T(x)R_\uparrow \ , \quad \phi_\downarrow = M_\downarrow^T(x)R_\downarrow \tag{12}$$

generates a heat bath updating of the ϕ-fields in which successive configurations are completely independent. Since M is a sparse matrix, this provides for a very efficient updating of the ϕ fields.

The x-field is now updated using molecular dynamics. Adopting a notation in which $x_i(\tau_l) = x_{i\tau}$, a conjugate momentum $p_{i\tau}$ is introduced, and a kinetic energy term is added to Eq. (11)

$$S[p, x, \phi_\uparrow, \phi_\downarrow] = \tilde{S} + \sum_{i\tau} p_{i\tau}^2/2 \quad . \tag{13}$$

Then, to within a constant normalization factor, the partition function is simply

$$Z = \int \delta_p \delta x \delta\phi_\uparrow \delta\phi_\downarrow e^{-S} \ . \tag{14}$$

The momentum field $p_{i\tau}$, like the ϕ fields, plays no role in the physical expectation values but simply serves to provide a mechanism for ultimately generating the desired distribution of the Hubbard–Stratonovich field x required by Eq. (4). A gaussian distribution of the $p_{i\tau}$ variables is easily generated using a heat bath algorithm, and this is done at the same time as the ϕ–fields are updated. Then we generate a new x configuration using molecular dynamics. For fixed ϕ, S given by Eq. (13) can be interpreted as a *classical* Hamiltonian. Introducing a "molecular dynamics" time t, Hamilton's equations of motion give

$$\begin{aligned} \frac{dx_{i\tau}}{dt} &= \frac{\partial S}{\partial p_{i\tau}} = p_{i\tau} \ , \\ \frac{dp_{i\tau}}{dt} &= -\frac{\partial S}{\partial x_{i\tau}} = -2\Delta\tau x_{i\tau} + \phi_\uparrow 0_\uparrow^{-1}\frac{\partial 0_\uparrow}{\partial x_{i\tau}} 0_\uparrow^{-1}\phi_\uparrow + \phi_\downarrow 0_\downarrow^{-1}\frac{\partial 0_\downarrow}{\partial x_{i\tau}} 0_\downarrow^{-1}\phi_\downarrow \ . \end{aligned} \tag{15}$$

Integrating these equations forward in time generates a new configuration $(x_{i\tau}(t), p_{i\tau}(t))$. According to Lionville's theorem, the differential volume in configuration space as well as S are constants of the motion. Therefore, following the usual Monte Carlo importance sampling approach, we always accept the new "molecular dynamics" generated configuration $(x_{i\tau}(t), p_{i\tau}(t))$. Typically, after a time t of order 1, we find that the new configuration is independent of the previous one.

The equations of motion, Eq. (15), are solved on a discrete t-lattice in which the x variables and p variables occupy alternate sites separated by $\Delta t/2$. We find that this "leapfrog" method, used in classical molecular dynamics, gives a stable numerical procedure. Furthermore, since the error in each step is of order Δt^3, the error after $(1/\Delta t)$ steps (e.g. $t = 1$) is only of order $(\Delta t)^2$. This violation of detailed balance leads to errors of this same order in the measured quantities. We have typically used Δt values less than 0.1.

After each Δt step it is necessary to recompute $0_{\uparrow}^{-1}\phi_{\uparrow}$ and $0_{\downarrow}^{-1}\phi_{\downarrow}$. A conjugate gradient routine is used and it is here that the calculational bottleneck arises. The iterates in the conjugate gradient evaluation of $0^{-1}\phi$ involve several matrix multiplications by M, and the number of iterations needed to obtain a given accuracy grows as β or U increases. To minimize the number of these iterations during the integration of the equations of motion (15), we note that if $y(t) = 0^{-1}(t)\phi$ evolves continuously, then $y(t+\Delta t) \simeq y(t) + y'(t)\cdot\Delta t$. Since $y'(t) \cong \frac{y(t)-y(t-\Delta t)}{\Delta t}$, we can obtain an approximate for $y(t+\Delta t)$ from the *two previous* conjugate gradient steps $y(t+\Delta t) \cong 2y(t) - y(t-\Delta t)$. Using this interpolation to start the conjugate gradient calculation of $0^{-1}(t+\Delta t)\phi$ typically reduces the number of conjugate gradient iterations by a factor of two to three over those needed immediately after the heat bath updating of p and ϕ when no good starting point is available.

To summarize, the algorithm we use provides a stochastic procedure for generating independent sets of configurations $(x, p, \phi_{\uparrow}, \phi_{\downarrow})$ which are distributed according to a probability S with

$$S = \sum \frac{p_{i\tau}^2}{2} + \sum x_{i\tau}^2 + \phi_{\uparrow}(M_{\uparrow}^T M_{\uparrow})^{-1}\phi_{\uparrow} + \phi_{\downarrow}(M_{\downarrow}^T M_{\downarrow})^{-1}\phi_{\downarrow}\,. \tag{16}$$

The ϕ and p fields are generated using heat baths and x and p are swept through phase space for fixed ϕ according to the molecular dynamics equations of motion Eq. (15).

Typical quantities which are calculated are the kinetic and potential energy per site

$$K = -4t\langle C_{i\uparrow}C_{j\uparrow}^{+}\rangle \tag{17}$$

$$V = U\langle(n_{i\uparrow} - \tfrac{1}{2})(n_{i\downarrow} - \tfrac{1}{2})\rangle \tag{18}$$

and the q-dependent magnetic susceptibility

$$\chi(q) = \Delta\tau \sum_{\tau}\sum_{l} e^{i\vec{q}\cdot\vec{l}} \langle (C^+_{i+l\uparrow}(\tau)C_{i+l\downarrow}(\tau) + C^+_{i+l\downarrow}(\tau)C_{i+l\uparrow}(\tau)) \cdot (C^+_{i\uparrow}(0)C_{i\downarrow}(0) + C^+_{i\downarrow}(0)C_{i\uparrow}(0))\rangle . \tag{19}$$

Here we find as discussed in reference 2 that the estimator for χ_{xx} has less noise than that for χ_{zz}. From V we can also obtain the mean square site moment $\langle (n_{i\uparrow} - n_{i\downarrow})^2 \rangle$.

Now, ordinarily it takes a conjugate gradient calculation to obtain each row of M^{-1} so that a great deal of time is required to calculate the Green's functions needed in Eqs. (17)-(19). One could, of course, run a separate Monte Carlo calculation to obtain the Green's functions, as in the traditional pseudo-fermion method,[5] but this will have significant noise as the sites are separated. The algorithm we have described above allows for a much more efficient method.[8] After a ϕ update given by $\phi = M^T R$,

$$0^{-1}\phi = (M^T M)^{-1}\phi = M^{-1}R \tag{20}$$

so that

$$\langle M^{-1}_{ij} \rangle = 2\langle (0^{-1}\phi)_i R_j \rangle \tag{21}$$

since R is a Gaussian random vector such that $\langle R_i R_j \rangle = \frac{1}{2}\delta_{ij}$. Measurements of products of Green's function such as occur in evaluating χ requires the generation of an additional random R' and one additional conjugate gradient to obtain $0^{-1}\phi'$.

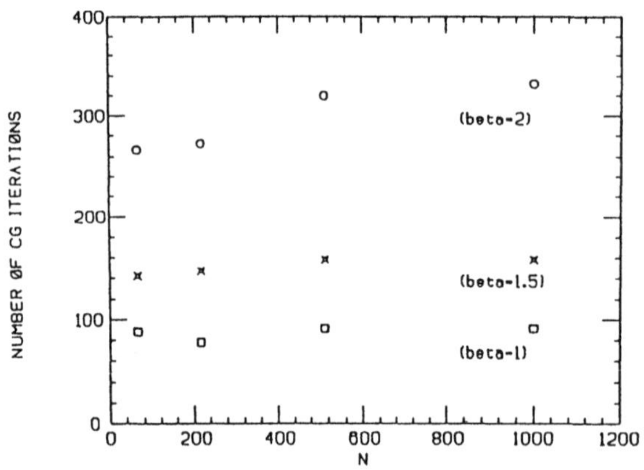

Fig. 1. The average number of conjugate gradient iterations following a molecular dynamics step versus the lattice size N for $U = 6$, $dt = .04$ for various values of β. According to Ref. 2, β_c is of order 2.

Initial runs for $U = 6$ and various temperatures[13] have been carried out on lattices ranging in spatial size from 4^3 to 10^3. We have taken $\Delta\tau = 0.1$ and used a molecular dynamics time step $dt = 0.04$. At $\beta = 1.6$, the generation of 100 independent configurations on a $10^3 \times 16$ lattice required several hours of Cray XMP time. At fixed β we find, as shown in Figure 1, that the number of conjugate gradient iterations required in the $0^{-1}\phi$ calculation is essentially independent of the spatial lattice size as N varies from 4^3 to 10^3. However, as shown in Figure 1, this number increases as the transition is approached. The number of iterations also depends upon U and with the other parameters fixed drops a factor of order 2 for $U = 4$. There are a variety of methods for reducing the number of conjugate gradient iterations which involve conditioning the matrix M, and we are presently investigating this aspect of the problem.

Acknowledgments

This review covers work done with R.T. Scalettar, R.L. Sugar at UCSB and W.D. Toussaint at UCSD. The UCSB work was supported in part by the National Science Foundation under Grant DMR83-20481 and the Department of Energy under Grant No. DE-FG03-85ER45197. An ST-100 Array Processor was purchased under NSF-DMR83-20423. We gratefully acknowledge E.I. du Pont de Nemours and Company and the Xerox Corporation for their support.

REFERENCES

1. R. Blankenbecler, D.J. Scalapino and R.L. Sugar, *Phys. Rev. D* **24**, 2278, (1981); D.J. Scalapino and R.L. Sugar, *Phys. Rev. B* **24**, 4295, (1981).
2. J.E.. Hirsch, *Phys. Rev. B* **31**, 4403, (1985), and *Phys. Rev.*, to be published.
3. J.E. Gubernatis, D.J. Scalapino, R.L. Sugar and W.D. Toussaint, *Phys. Rev.* **32**, 103, (1985).
4. D.J. Scalapino, E. Loh and J.E. Hirsch, *Phys. Rev.*, to be published.
5. F. Fucito, E. Marinari, G. Parisi and C. Rebbi, *Nucl. Phys. B* **180**, 369, (1981).
6. G.G. Batrouni, G.R. Katz, A.S. Kronfeld, G.P. Lepage, B. Svetitsky and K.G. Wilson, *Phys. Rev. D* **32**, 2736, (1985).

7. S. Duane, *Nucl. Phys. B* **257**, 652, (1985); S. Duane and J. Kogut, *Phys. Rev. Lett.* **55**, 2774, (1985).

8. R.T. Scalettar, D.J. Scalapino and R.L. Sugar, *Phys. Rev. B*, to be published.

9. S.A. Gottlieb, W. Liu, D. Toussaint, R.L. Renken and R.L. Sugar, *Phys. Rev. D*, to be published.

10. R.T. Scalettar, D.J. Scalapino and R.L. Sugar, to be published.

11. The overall fermion determinate is positive definite for the $U > 0$ particle-hole Hubbard model. It is also positive definite for any filling with $U < 0$ as well as for the electron-phonon model in which further potential terms are added to the x-field action.

12. R.L. Stratonovitch, *Dokl. Akad. Nauk SSSR* **115**, 1097, (1957); J. Hubbard, *Phys. Rev. Lett.* **3**, 77, (1959); S.Q. Wang, W.E. Evanson and J.R. Schrieffer, *Phys. Rev. Lett.* **23**, 92, (1969).

13. Here we measure energies in units of the transfer integral t in Eq. (1) so, for example, the bandwidth is 12.

Simulation of Magnetic Impurities in Metals

J.E. Hirsch

Department of Physics, University of California,
San Diego, La Jolla, CA 92037, USA

Abstract: I discuss a Monte Carlo method to study properties of a few magnetic impurities in an electron gas. Results for the magnetic susceptibility of a single impurity show the expected universal Kondo behavior at low temperatures. Other properties of a single impurity as well as properties of two magnetic impurities are discussed.

I. Introduction

As it will be evident to any reader of these proceedings, there are almost as many different quantum Monte Carlo techniques as there are problems one wants to solve. Much more so than in classical Monte Carlo simulations, the nature of the quantum problem dictates which technique has to be used to solve it. Attempting to use another technique, even though in principle possible, will frequently lead to a failure to reach the interesting physical regime of the problem of interest. As an example, using a "world-line algorithm" [1] to solve the Hubbard model in more than one dimension is in principle possible but leads to unmanageable minus-sign problems, except at high temperatures where the physics is uninteresting.

The other well-known point I would like to mention is that what distinguishes quantum from classical Monte Carlo simulations is the importance of the extra dimension (imaginary time, or temperature) relative to the spatial dimension. Some problems are more "quantum" than others, and some are "almost classical" in the sense that the spatial dimension is the dominant one and the time dimension introduces only small perturbations; an example of the latter would be a half-filled spinless fermion model with a nearest-neighbor repulsion that is much larger than the hopping.

The problem I would like to discuss here, that of magnetic impurities in metals, is certainly the most properly "quantum" I have encountered, in that the time-dimension is all-important in determining the behavior of the system while the spatial dimensions play only a minor role. Our technique for studying this problem [2] takes this into account by analytically eliminat- the spatial degrees of freedom at the outset and focusing on the time-like direction. This appears to be the ideal technique for this problem, in that the resulting algorithm is simple and stable, and can reach the interesting low-temperature regime without difficulty. In contrast, an algorithm that does not eliminate the spatial degrees of freedom cannot reach the low-temperature regime of interest due to instabilities [3] .

An attractive feature of our method is that it allows for the first time to study an infinite sea of conduction electrons with a sharp Fermi surface, rather than a finite system which necessarily has discrete energy levels.

The conduction electrons enter the problem only through their Green's function at the impurity sites, which is easily calculable if the conduction electrons are assumed to be non-interacting. It is thus straightforward to study arbitrarily complicated band structures in any dimension.

In the next section I give some technical details of the method. Section III discusses various results for a single magnetic impurity, including correlations of the impurity with the surrounding conduction electrons and dynamic properties. Section IV discusses properties of two impurities, and I conclude in Section V with a discussion. Different portions of the work discussed here were done in collaboration with R. M. Fye, H. Q. Lin, D. J. Scalapino and J. E. Gubernatis.

II. The Method

We consider the single impurity Anderson model:

$$H = \sum_k \epsilon_k c^\dagger_{k\sigma} c_{k\sigma} + \sum_{k,\sigma} V_k (c^\dagger_{k\sigma} d_\sigma + \text{h.c.}) + \epsilon_d \sum_\sigma n_{d\sigma} + U n_{d\uparrow} n_{d\downarrow}$$

$$\equiv H_0 + H_1 . \qquad (1)$$

A similar development applies to the Kondo Hamiltonian. In a discrete path-integral formulation the partition function is:

$$Z = \mathrm{Tr}\, e^{-\beta H} = \mathrm{Tr} \prod_{\ell=1}^{L} e^{-\Delta\tau H} \cong \mathrm{Tr} \prod_{\ell=1}^{L} e^{-\Delta\tau H_0} e^{-\Delta\tau H_1} \qquad (2)$$

with $\beta = L\Delta\tau$. We decouple the interaction part of the Hamiltonian by introducing auxiliary Ising variables [5]:

$$\exp(-\Delta\tau H_1) = \exp\left\{-\Delta\tau U\left[n_{d\uparrow} n_{d\downarrow} - \tfrac{1}{2}(n_{d\uparrow} + n_{d\downarrow})\right]\right\} = \tfrac{1}{2}\,\mathrm{Tr}_\sigma \exp[\lambda\sigma(n_{d\uparrow} - n_{d\downarrow})] \quad (3)$$

with $\cosh\lambda = \exp(\Delta\tau U/2)$, and take the trace over fermion degrees of freedom. The result can be written as:

$$Z = \mathrm{Tr}_\sigma \prod_{\mu=\pm 1} \det{}_{N,L}\, O_\mu(\{\sigma_\ell\}) , \qquad (4)$$

where O_μ is an $NL \times NL$ matrix, with N the number of spatial sites (or k-vectors) for the conduction electrons plus 1 (the impurity orbital), with matrix elements in the time indices:

$$(O_\mu)_{\ell,\ell} = 1 ,$$

$$(O_\mu)_{\ell,\ell-1} = -e^{(-\Delta\tau K)} e^{V_{\ell-1}} (1 - 2\delta_{\ell,1}) , \qquad (5)$$

and $(O_\mu)_{\ell,m} = 0$ otherwise. K is an $N\times N$ matrix corresponding to the bilinear part of the Hamiltonian, and $V^\mu_\ell = \lambda_\mu \sigma(\ell) |d\rangle\langle d|$ is a time-dependent potential acting only at the impurity site.

The Green's function $g_\mu = O_\mu^{-1}$ is found to obey the Dyson equation [2]

$$g'_\mu = g_\mu + (g_\mu - 1)\left(e^{V'^\mu - V^\mu} - 1\right) g'_\mu \tag{6a}$$

or its transpose

$$g'_\mu = g_\mu + (g'_\mu - 1)\left(1 - e^{V^\mu - V'^\mu}\right) g_\mu \tag{6b}$$

relating any two spin configurations. From these equations we can obtain arbitrary correlation functions for both impurity and conduction electron properties, as well as the rules for the Monte Carlo updating. For the updating only the time components of the Green's function at the impurity site enter. The spin at time slice ℓ is flipped if the ratio of determinants

$$R = \prod_\mu \left[1 + (1 - g_{dd}^\mu(\ell,\ell))\left(e^{V_\ell'^\mu - V_\ell^\mu} - 1\right)\right] \tag{7}$$

is larger than a random number between zero and one, in which case all time components of the d-Green's function $f^\mu(\ell_1,\ell_2) \equiv g_{dd}^\mu(\ell_1,\ell_2)$ are updated through the relation:

$$f'^\mu(\ell_1,\ell_2) = f^\mu(\ell_1,\ell_2) + (f^\mu(\ell_1,\ell) - \delta(\ell_1,\ell)) \times$$

$$\times \left[e^{V_\ell'^\mu - V_\ell^\mu} - 1\right]\left\{1 + [1 - f^\mu(\ell,\ell)]\left[e^{V_\ell'^\mu - V_\ell^\mu}\right]\right\}^{-1} f^\mu(\ell,\ell_2). \tag{8}$$

The procedure is similar to the one introduced by Blankenbecler, Scalapino and Sugar [6] to study fermion lattice systems, the difference being that here time components of the Green's function rather than space components are updated.

Initially, an $L \times L$ matrix is inverted (L^3 operations) to obtain the impurity Green's function for the initial field configuration from the Green's function for the fields set equal to zero:

$$g_{dd}^o(\ell,\ell') = T \sum_n e^{-i\omega_n \Delta\tau(\ell - \ell')} g_{dd}^o(i\omega_n) \tag{9}$$

$$g_{dd}^o(i\omega_n) = -\left[i\omega_n - \left(\epsilon_d + \frac{U}{2}\right) - \sum_k \frac{|V_k|^2}{i\omega_n - \epsilon_k}\right]^{-1} \tag{10}$$

with $\omega_n = (2n+1)\pi/\beta$. Then, g_{dd} is updated in L^2 operations every time the move is accepted, so that the number of operations per sweep is of order L^3 ($(nL)^3$ for n impurities). A typical run of 6,000 sweeps for two impurities with $L = 64$ takes 16 min on a Cray X-MP. The procedure is found to be stable, and g does not have to be recomputed from the matrix inversion for at least 50 sweeps in all cases studied. It is also possible to go to very low temperatures without encountering any instabilities in the algorithm, in contrast to what happens in some lattice problems [7].

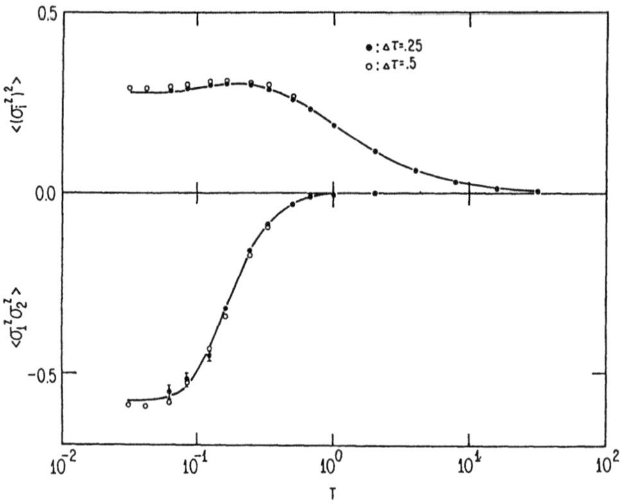

Fig. 1. Comparison of Monte Carlo (points) and exact diagonalization results (full lines) for local moment $\langle(\sigma_z^i)^2\rangle$ and spin-spin correlations $\langle\sigma_1^z\,\sigma_2^z\rangle$ for a two-impurity two-site conduction electron lattice. U = 2, $\Delta = \pi V^2 \rho(\epsilon_F) = 0.5$.

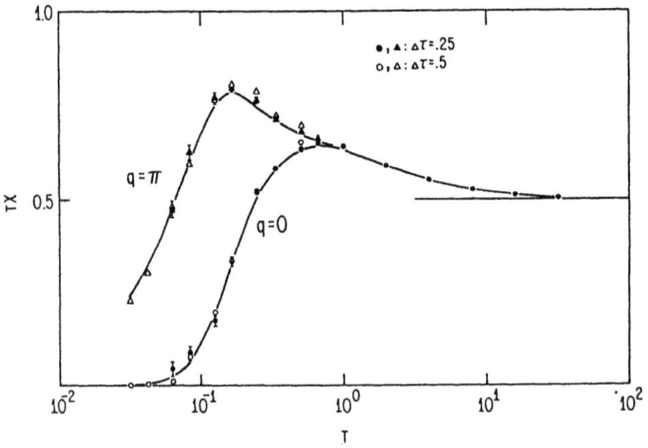

Fig. 2. Same as Fig. 1 for uniform (circles) and staggered (triangles) susceptibilities.

To demonstrate the accuracy of the method, Figs. 1 and 2 show Monte Carlo results for two Anderson impurities in a two-site conduction electron "lattice"; this Hamiltonian can be easily diagonalized, and the exact results (full line) and the Monte Carlo results (points) are found to be in good agreement. We expect similarly accurate results for any conduction electron band structure.

III. A Single Symmetric Anderson Impurity ($\epsilon_d = -U/2$)

Figure 3 shows results for the local moment $\langle \sigma_z^2 \rangle$ and for $T\chi$ (χ = magnetic susceptibility) plotted versus log T for U = 1, 2, 3, 4 and hybridization width $\Delta = \pi V^2 \rho(\epsilon_F) = 0.5$ ($\rho(\epsilon_F)$ = density of states at the Fermi energy). The local moment builds up around $T \sim \Delta$ and stays constant at lower T, while $T\chi$ first increases as T decreases and then reaches a maximum and decreases as correlations between the impurity spin and the conduction electrons start to build up. The curves for $T\chi$ are found to follow the universal Kondo behavior [8] (dashed curves), with Kondo temperatures in agreement with those obtained from a Bethe Ansatz solution [9]. Figure 4 shows the behavior of the impurity spin-conduction electron-spin correlation function $S(r) = \langle \sigma_d \sigma_c(r) \rangle$ for three values of the temperature (β = 8, 16, 32) and U = 0 and U = 4 [10]. These correlations are not easily obtained by other methods. Note that for U = 0 only antiferromagnetic correlations exist (due to the Fermi hole created around the impurity when the impurity orbital is occupied) and for U = 4 both ferromagnetic and antiferromagnetic correlations exist. The temperature dependence is stronger for U = 4 than for U = 0 because the relevant temperature scale is T_K rather than Δ, although surprisingly, the ferromagnetic correlations are temperature independent. The integral of this correlation function should exactly compensate the local moment, and it is found that this compensation occurs over very large distances. A detailed discussion of these results is given in Ref. 10. Similarly, one can obtain results for charge oscillations around the impurity, Knight shift, etc. [11]. By putting in band structures of specific solids in g_o one should obtain results that are directly comparable with experiments.

Another set of important questions in these systems concern dynamic properties. As is well known, it is very difficult to obtain dynamic properties from quantum Monte Carlo simulations, since the results obtained directly are in imaginary time or frequency and they have to be "analyti-

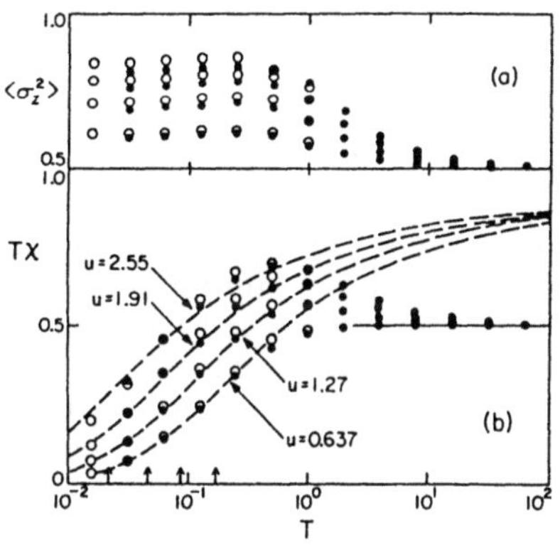

Fig. 3. (a) Local moment $\langle \sigma_z^2 \rangle$ and (b) $T\chi$ (spin susceptibility) for a single Anderson impurity; $\Delta = 0.5$ and several values of $u = U/\pi\Delta$. The closed and open circles correspond to $\Delta\tau = 0.25$ and $\Delta\tau = 0.5$, respectively. The dashed lines are the universal Kondo susceptibility for the Kondo temperature T_K obtained from Bethe Ansatz (denoted by arrows).

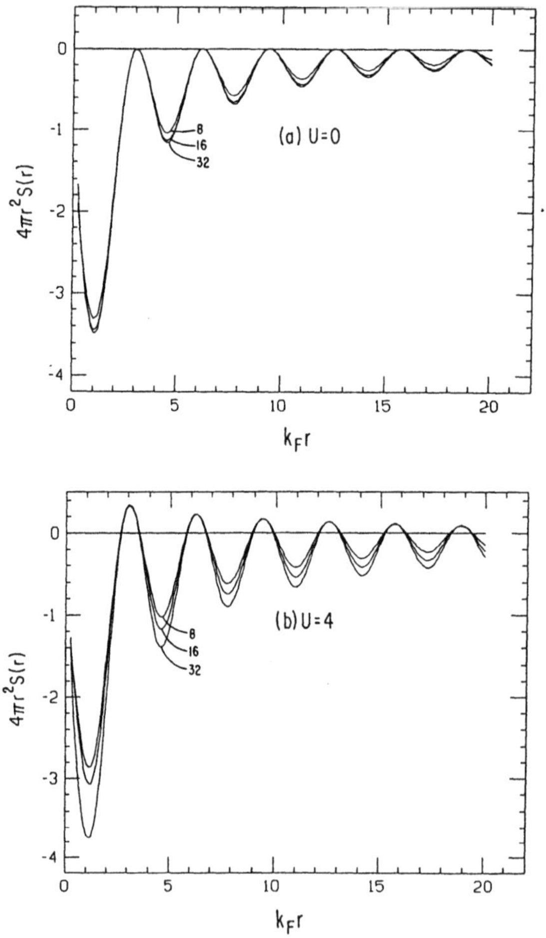

Fig. 4. Spatial dependence of impurity spin-conduction electron-spin correlations multiplied by $4\pi r^2$. $\Delta = 0.5$, $\beta = 8$, 16 and 32.

cally" continued to the real axis. We have succeeded in doing this for this problem [11] using N-point Padé approximants [12]. Figure 5 shows the impurity Green's function as a function of imaginary time and frequency, for $U = 0$ and $U = 4$, as they are directly obtained from the Monte Carlo simulation. As can be seen, the curves for both cases are quite featureless. After performing the analytic continuation to the real axis, however, the difference is quite striking, as shown in Fig. 6. The non-interacting case shows a lorentzian shape of width Δ, while the interacting case shows a narrow resonance at the Fermi level (the "Kondo resonance") of width $\sim T_K$ and two side-lobes at positions $-U/2$ and $U/2$, corresponding to the energies of the single and doubly occupied d-orbitals. The results are found to be very sensitive to statistical errors in the imaginary time data, and very long Monte Carlo runs are needed. Nevertheless, this is to our knowledge the only reliable technique to obtain dynamic properties beyond the range of perturbation theory even for just the single impurity.

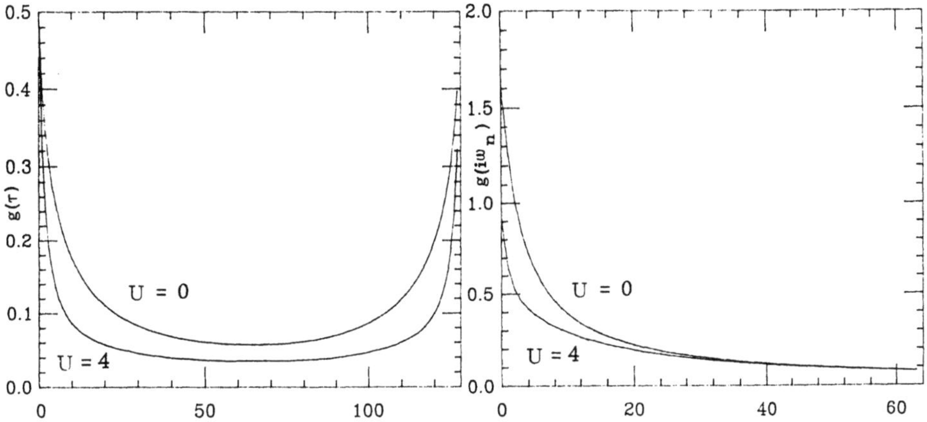

Fig. 5. Imaginary time and imaginary frequency dependence of the impurity Green's function for $U=0$ and $U=4$, $\beta = 32$, $\Delta = 0.5$.

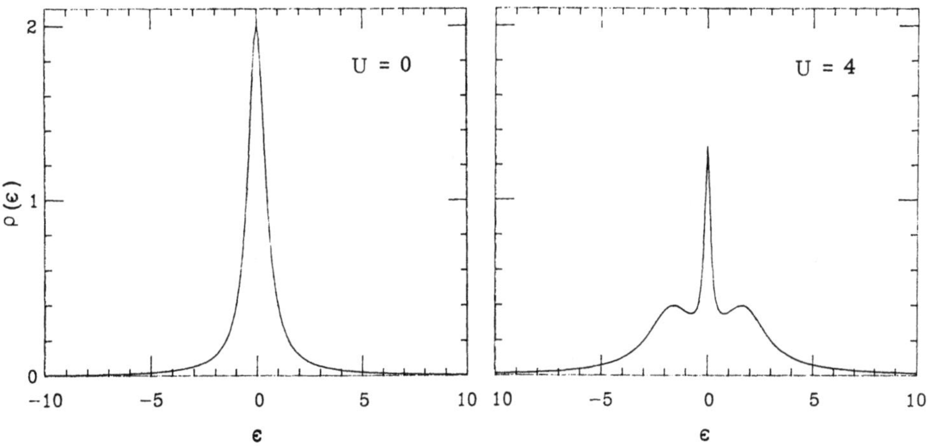

Fig. 6. Spectral density $\rho(\varepsilon) = -1/\pi \text{ Im } g_{dd} (\varepsilon + i0)$ obtained from "analytic" continuation of the imaginary frequency data of Fig. 5 using N-point Padé approximants.

IV. Two Impurities

There are several questions regarding interactions between two magnetic impurities that have been controversial, and Monte Carlo simulations offer the possibility to settle these questions. One such question concerns the nature of the spin-spin correlations (whether ferromagnetic or antiferromagnetic) for two impurities coupled by a direct hopping term V_{12}, as a function of the position of the d-level with respect to the Fermi level [13] (Alexander-Anderson model). Both Alexander and Anderson [14], using Hartree-Fock and Yamada [15], using perturbation theory in U, found that for large U correlations change from antiferromagnetic to ferromagnetic as the d-level approaches the Fermi level. In contrast, Gottlieb and Suhl [16], by perturbation theory in Δ, found that correlations are always antiferromagnetic.

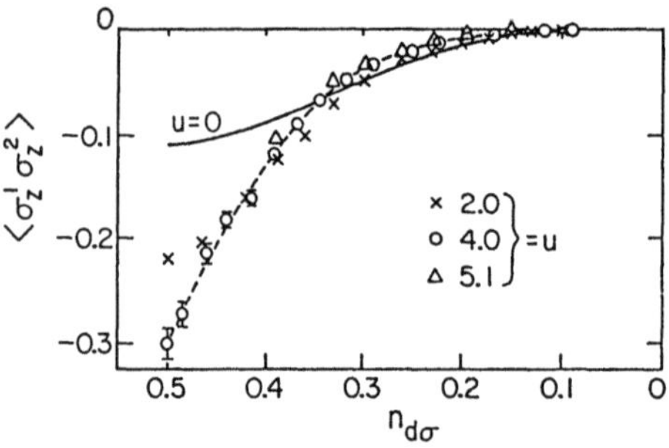

Fig. 7. Spin-spin correlations for two magnetic impurities coupled by a direct hopping term $V_{12} = 0.5$ as a function of the occupation of the d-level. $\Delta = 0.5$, $\beta = 8$, $u = U/\pi\Delta$.

Figure 7 shows spin-spin correlations for two impurities for $V_{12} = 0.5$, $\Delta = 0.5$, and three values of $u = U/\pi\Delta$ as a function of the occupation of the d-level (which is a function of ϵ_d). It can be seen that correlations are always antiferromagnetic, in contrast with Yamada's results, who finds ferromagnetism for $u \geq 3$ and $n_{d\sigma}$ around 0.25, and Hartree-Fock which predicts ferromagnetism for $u \gtrsim 1.6$. We have explored the parameter space for this model quite extensively [13], and conclude that correlations are always antiferromagnetic. Thus, the model cannot be used to explain the sign and magnitude of the magnetic interactions in transition metals [17], and a proper treatment of a model that includes band degeneracy is necessary.

Another controversial question concerns the interplay between the Kondo effect and the indirect exchange interaction (RKKY) [18] between two impurities. The most commonly held view is that if the Kondo temperature T_K is larger than the RKKY exchange interaction J, the impurities will be individually Kondo-quenched; while if T_K is smaller than J, a collective state of the two impurities will be formed and then quenched by the conduction electrons [19, 20]. Recently, however, it has been suggested that the RKKY interaction plays a dominant role even for $J \ll T_K$, based on perturbation theory [21] and renormalization-group calculations [22] that suggest that the effective RKKY interaction becomes very strong as the temperature is lowered even if it is small at high temperatures.

The Monte Carlo simulation results [23] do show residual correlations between impurities, even for $J < T_K$, but they are indeed quite small if J is small compared to the Kondo temperature. Figures 8 and 9 show results for various properties of two impurities for the case $U = 2$, $\Delta = 0.5$, and separation between impurities $k_F R = 1.25$ and $k_F R = 2$, respectively. The upper part of the figure shows the local moment and the spin-spin correlation function. The dashed line denotes the spin-spin correlations that would be obtained from a temperature-independent RKKY exchange J, which is $J = 0.039$ and $J = 0.008$ for $k_F R = 1.25$ and 2, respectively. The single impurity Kondo temperature for this case is $T_K = 0.088$. Note that the spin-spin correlations follow the RKKY predictions down to roughly the Kondo temperature and then they level off. The susceptibility is shown on the bottom part of the figures; at high temperatures it coincides with the

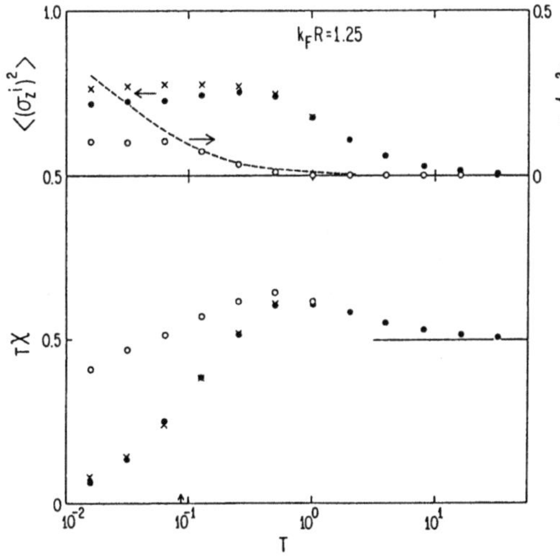

Fig. 8. Top part: spin correlations (open circles) and local moment (crosses) for two impurities versus temperature. The full circles show the local moment for $k_FR = \infty$. Bottom part: $T\chi$ susceptibility. Open circles: uniform, crosses: staggered susceptibility. Full circles: susceptibility for $k_FR = \infty$.

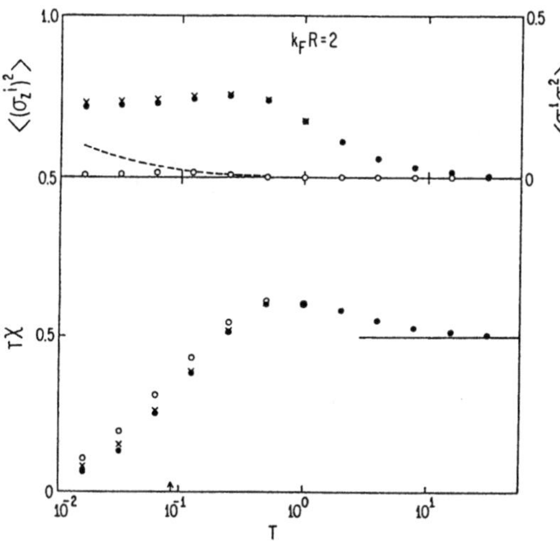

Fig. 9. Same as Fig. 8 for $k_FR = 2$.

single impurity values but at low temperatures it decays slower due to the ferromagnetic correlations between the impurities. For $k_FR = 2$, the RKKY interaction is very small (close to a node) and the properties of the two-impurity system are very close to the ones of the single impurity even at low temperatures. These results show that the size of the RKKY interaction at high temperatures found from perturbation theory does determine the low-temperature behavior, contrary to the recent suggestions.

V. Discussion

We have discussed a Monte Carlo algorithm to study properties of a few magnetic impurities in an electron gas. The computer time in this algorithm goes as $(nL)^3$, with n the number of impurities and L the number of time slices. The number of conduction electron degrees of freedom does not affect the computer time, so that one can study systems with an infinite number of conduction electrons with a sharp Fermi surface, and with arbitrary band structures. The algorithm does not suffer from instabilities at low temperatures, and one can easily reach the interesting temperature regime.

In this paper we discussed only Anderson impurities. Application of this method to Kondo impurities and to impurities with degenerate orbitals is similar in principle although more complicated, and will be reported elsewhere [4, 24].

As we have attempted to show in this paper, simulations can provide quantitative answers on magnetic impurity systems that are not obtainable by any other known method. Hopefully, these results will be of use in obtaining a better understanding of the effects of electron correlations in various solid state systems such as dilute magnetic alloys, transition metal compounds and heavy fermion compounds.

Acknowledgments: This work was supported in part by NSF Grants No. DMR-86-17756 and PHY-82-17853, supplemented by NASA, as well as by contributions from AT& T Bell Laboratories and Cray Research Corporation.

References

1. J. E. Hirsch, D. J. Scalapino, R. L. Sugar and R. Blankenbecler, Phys. Rev. Lett. 47, 1628 (1981); Phys. Rev. B26, 5033 (1982).
2. J. E. Hirsch and R. M. Fye, Phys. Rev. Lett. 56, 2521 (1986).
3. J. E. Gubernatis, T. C. Olson, D. J. Scalapino and R. L. Sugar, J. Stat. Phys. 43, 831 (1986).
4. R. M. Fye and J. E. Hirsch, unpublished.
5. Functional integral formulations of the Anderson model using Gaussian auxiliary fields were introduced in S. Q. Wang, W. E. Evenson and J. R. Schrieffer, Phys. Rev. Lett. 23, 92 (1969) and D. R. Hamman, Phys. Rev. Lett. 23, 95 (1969); Ising auxiliary fields are discussed in J. E. Hirsch, Phys. Rev. B28, 4059 (1983) and Phys. Rev. B34, 3216 (1986).
6. R. Blankenbecler, D. J. Scalapino and R. L. Sugar, Phys. Rev. D24, 2278 (1981).
7. J. E. Hirsch, J. Stat. Phys. 43, 841 (1986).
8. K. G. Wilson, Rev. Mod. Phys. 47, 773 (1973); H. R. Krishna-Murthy, J. W. Wilkins and K. G. Wilson, Phys. Rev. B21, 1003 (1980).
9. P. B. Wiegmann and A. M. Tsvelick, J. Phys. C16, 2281 (1983); B. Horvatic and V. Zlatic, J. Phys. (Paris) 46, 1459 (1985).
10. J. E. Gubernatis, J. E. Hirsch and D. J. Scalapino, to be published.
11. J. E. Hirsch and D. J. Scalapino, unpublished.
12. H. J. Vidberg and J. W. Serene, J. Low Temp. Phys. 29, 179 (1977).
13. J. E. Hirsch and H. Q. Lin, to be published.
14. S. Alexander and P. W. Anderson, Phys. Rev. 133A, 1594 (1964).

15. K. Yamada, Prog. Theor. Phys. 62, 901 (1979).
16. P. Gottlieb and H. Suhl, Phys. Rev. 134A, 1586 (1964).
17. S.H. Liu, Phys. Rev. 163, 472 (1967); T. Moriya, Prog. Theor. Phys. 33, 157 (1965).
18. M.A. Rudermann and C. Kittel, Phys. Rev. 96, 99 (1954).
19. C. Jayaprakash, H.R. Krishnamurthy and J.W. Wilkins, Phys. Rev. Lett. 47, 737 (1981).
20. S. Chakraverty and J.E. Hirsch, Phys. Rev. B25, 3273 (1982).
21. E. Abrahams and C.M. Varma, preprint.
22. B. A. Jones and C.M. Varma, preprint.
23. R.M. Fye, J.E. Hirsch and D.J. Scalapino, to be published.
24. H.Q. Lin and J.E. Hirsch, unpublished.

The Spatial Dependence of Spin and Charge Correlations in a One-Dimensional, Single Impurity, Anderson Model

J.E. Gubernatis

Theoretical Division, Los Alamos National Laboratory, Los Alamos, NM 87545, USA

1. Introduction

Recent advances in Monte Carlo algorithms for systems of interacting electrons have made possible the detailed study of the spatial dependence of spin and charge correlations in the Anderson single impurity model. The hamiltonian for this model [1]

$$H = \sum_{ks} \varepsilon_k n_{ks} + \sum_k (V_{kd} c^+_{ks} c_{ds} + V^*_{dk} c^+_{ds} c_{ks})$$

$$+ \varepsilon_d \sum_s n_{ds} + U n_{d\uparrow} n_{d\downarrow} \tag{1}$$

has a non-degenerate impurity state with energy ε_d hybridized with strength V_{kd} to energy states ε_k of some conduction band. In addition it has a Coulomb interaction U between electrons with oppositely paired spins that try to occupy the impurity state.

As a function of interaction parameters and temperature, the model exhibits a variety of interesting many-body phenomena [2,3]. One of these phenomena is the existence of a local magnet moment at the impurity and the compensation of this moment by a spin cloud that leaves the system in a singlet state as $T \to 0$. Prior to recent Monte Carlo simulations the structure of this spin compensation cloud was accessible only through perturbative theories that led to somewhat controversial or incomplete results [4].

In investigations of the spatial dependence of this spin compensation, the first [5] of a series of Monte Carlo studies found RKKY and Friedel-like oscillations with a strong negative (antiferromagnetic) correlation between electrons at the impurity site and those at the lattice site with which it interacts. With a wave number of $2k_F$, these oscillations alternated in sign and decreased in amplitude as the distance from the impurity site increased. The principal limitation of this study was its inability to probe temperatures well below the Kondo temperature T_K. Subsequently proposed was another approach [6] that reached these low temperatures, at the expense of losing the information about the spatial dependence of correlations, and had the additional advantage of embedding the impurity in an infinite system. The most recent study [7] showed within this new approach how to recover the spatial dependence and presented low-temperature results that had several surprising features.

Perhaps the most surprising feature was the possible extension of the spin compensation cloud to distances beyond the apparent natural length v_F/T_K, where v_F is the fermi velocity. If the compensation in physical systems is this long-ranged, then the single impurity results would most likely have applicability only to very dilute alloys, and the quantitative comparison between theory and experiment, notwithstanding the neglect of orbital degeneracy, would be difficult. Along with a closer examination of this possible long-ranged correlation, several other interesting features were found that await a clearer understanding. These novel features included a non-monotonic variation of the lattice site to impurity site spin correlation as a function of the Coulomb interaction.

The principal purpose of this short note is to present corroborative results for most of these features. A secondary purpose is to highlight several additional systematic, but small, features that become apparent by going to lower temperatures. Whereas the original study assumed the computationally convenient model of the conduction band as a broad flat density of states and a free-electron dispersion, here we assume a one-dimensional, tight-binding band

$$\varepsilon_k = -2\cos k \quad . \tag{2}$$

In the next section we summarize the Monte Carlo procedure. Then in the third section we present the results of our calculation. In the fourth and final section we summarize our findings and comment on further applications of the procedure to the study of quantum many-body phenomena.

2. Procedure

The procedure we use is that of [7] and starts with a discrete, path integral approach [8] to the evaluation of the partition function in which the temperature axis is divided into L steps of size $\tau = \beta/L$. At each step the following approximation is made

$$e^{-\tau H} \simeq e^{-\tau H_0}\, e^{-\tau H_I} \tag{3}$$

and then the interaction term in the hamiltonian is eliminated by introducing auxiliary Ising variables σ_ℓ [9], one for each step,

$$e^{-\tau\frac{U}{2}(n_{d\uparrow}-n_{d\downarrow})^2} = \frac{1}{2}\sum_{\sigma_\ell=\pm1} e^{-\tau J\sigma_\ell(n_{d\uparrow}-n_{d\downarrow})}$$

with $\cosh \tau J = \exp(-\tau U/2)$. After this transformation the partition function becomes

$$Z = \mathrm{Tr}_\sigma \exp(-\sum_s \mathrm{Tr}\,\ell n\, g^s[\sigma]) \, , \tag{4}$$

where the matrix $g^s[\sigma]$ is the equal-time thermodynamic Green's function whose elements are defined by

$$g^s_{ij}(\ell\ell') = \langle T c^+_{is}(\ell)\, c_{js}(\ell')\rangle \, . \tag{5}$$

The Monte Carlo part of the procedure performs the trace over the σ configurations and consists of successively flipping the σ_ℓ and accepting or rejecting the flip on the basis of whether the quantity $R/(1+R)$ is greater than or less than a random number between 0 and 1 [5]. R is the ratio of the partition function for the new configuration to partition function for the old configuration and equals $R_\uparrow R_\downarrow$ with

$$R_s = 1 + (1 - g^s_{dd}(\ell\ell))(\Delta^s_d(\ell) - 1), \tag{6}$$

where $\Delta^s_d(\ell)$ is the change in energy caused by the change in the field.

As pointed out by Hirsch and Fye [6], within the Trotter approximation there is a general relation between the spatial and time elements of the Green's functions for two σ configurations differing at time step ℓ''. For each component of electron spin

$$g'_{ij}(\ell\ell') = g_{ij}(\ell\ell') + (\delta_{id}\delta_{\ell\ell''} - g_{id}(\ell\ell''))$$
$$\times (\Delta_d(\ell'') - 1)\, g'_{dj}(\ell''\ell'). \tag{7}$$

In our Monte Carlo procedure we use this relation for three different purposes [7]. First to initialize the calculation, we chose the σ_ℓ randomly. Having computed the $g_{dd}(\ell\ell')$ for the non-interacting problem, we use (7) iteratively to turn on the σ_ℓ one at a time at each step of the iteration to obtain $g'_{dd}(\ell\ell')$ for the complete configuration: With the unperturbed $g_{dd}(\ell\ell')$ we use (7) to find $g'_{dd}(\ell\ell')$ for system with just σ_1. We then use this $g_{dd}(\ell\ell')$ on the right side of (7) to find $g'_{dd}(\ell\ell')$ for a system with σ_1 and σ_2. We continue turning on the spins this way until we construct the Green's function for all the spin variables. With the $g_{dd}(\ell\ell')$ we can begin the Monte Carlo steps and evaluate (6). If we accept the flip, we use (7) again to compute the $g_{dd}(\ell\ell)$ for the new configuration [10],

$$g'_{dd}(\ell\ell) = g_{dd}(\ell\ell) - \frac{(g_{dd}(\ell\ell)-1)(\Delta_d(\ell)-1)g_{dd}(\ell\ell)}{1+(1-g_{dd}(\ell\ell))(\Delta_d(\ell)-1)} \; . \tag{8}$$

The third use of (5) is to recover the spatial information about the Green's functions, since only the impurity elements $g_{dd}(\ell\ell')$ are the natural product from the Monte Carlo steps. Here we are interested in determining $g_{dj}(LL)$, $g_{id}(LL)$ and $g_{ij}(LL)$. From (7) we have

$$g'_{dd}(L\ell') = g_{dd}(L\ell') + (\delta_{L\ell''} - g_{dd}(L\ell''))(\Delta_d(\ell'') - 1)g'_{dd}(\ell''\ell'), \tag{9a}$$

$$g'_{dj}(L\ell') = g_{dj}(L\ell') + (\delta_{L\ell''} - g_{dd}(L\ell''))(\Delta_d(\ell'') - 1)g'_{dj}(\ell''\ell'), \tag{9b}$$

where j is a lattice site. We use (9) the same way as we used (7) to get $g_{dd}(\ell\ell')$: Starting with the non-interacting Green's functions $g_{dd}(L\ell')$ and $g_{dj}(L\ell')$, we iterate (9) and at each step turn on a new spin until $g'_{dj}(L\ell')$ for the given configuration is constructed.

To find $g_{dj}(LL)$ and $g_{ij}(LL)$, we use a similar procedure except we rewrite (7) by interchanging the labels of the old and new configurations

$$g'_{ij}(\ell\ell') = g_{ij}(\ell\ell') - (\delta_{id}\delta_{\ell\ell''} - g'_{id}(\ell\ell''))$$
$$\times (\Delta^{-1}_d(\ell'') - 1)\, g_{dj}(\ell''\ell') \, . \tag{10}$$

This leads to

$$g'_{id}(LL) = g_{id}(LL) + g'_{dd}(L\ell'')(\Delta^{-1}_d(\ell'')-1)g_{dd}(\ell''L) \tag{11a}$$

$$g'_{ij}(LL) = g_{ij}(LL) + g'_{id}(L\ell'')(\Delta^{-1}_d(\ell'')-1)g_{dd}(\ell''L) \tag{11b}$$

with i and j being lattice sites. Hence starting with $g'_{dd}(LL)$ and $g'_{id}(LL)$ found from equation (8) and with the non-interacting $g_{dj}(LL)$ and $g_{ij}(LL)$, we use the iteration-turn-on technique to find $g'_{dj}(LL)$ and $g'_{ij}(LL)$. An important point to appreciate about the procedures just outlined is we can start with the Green's function for an infinite, non-interacting system and obtain the Green's function of the infinite interacting system at a finite number of states of our choosing.

With the Green's functions one can compute the thermodynamics quantities of interest. In this paper we are mainly interested in the impurity susceptibility

$$T\chi_d = \int_0^\beta d\tau \langle \sigma_d(\tau)\sigma_d(L)\rangle , \tag{12a}$$

the impurity moment

$$S_d = \langle (\sigma_d(L))^2 \rangle \tag{12b}$$

and the correlation between the spin at the impurity and the spin at lattice sites

$$S(i) = \langle \sigma_d(L)\sigma_i(L)\rangle , \tag{12c}$$

where $\sigma_d = n_{d\uparrow} - n_{d\downarrow}$ and $\sigma_i = n_{i\uparrow} - n_{i\downarrow}$.

The averages over the products of these spins are related to the Green's functions by use of Wick's theorem. The application of Wick's theorem simplifies somewhat because the introduction of the auxiliary fields leaves the Green's function in a representation block-diagonal in the electron spin. As an example, we can express (12c) as

$$S(i) = \langle (g^\uparrow_{dd} - g^\downarrow_{dd})(g^\uparrow_{ii} - g^\downarrow_{ii})\rangle - \langle g^\uparrow_{id} g^\uparrow_{di} + g^\downarrow_{id} g^\downarrow_{di}\rangle , \tag{13}$$

where the average is over σ configurations.

We will also present results showing the nature of charge compensation. Defining $n_d = n_{d\uparrow}+n_{d\downarrow}$ and $n_i = n_{i\uparrow}+n_{i\downarrow}$, we have the following quantities in analogy to equation (12):

$$T\,\Pi_d = \int_0^\beta d\tau \langle n_d(\tau)n_d(L)\rangle , \tag{14a}$$

$$C_d = \langle (n_d(L))^2 \rangle , \tag{14b}$$

$$C(i) = \langle n_d(L)\, n_i(L)\rangle - \langle n_d(L)\rangle\langle n_i(L)\rangle . \tag{14c}$$

To compute the averages over the σ configurations, we generate 1024 configurations to equilibrate the system, and then generate an additional 5120 configurations from which we compute the averages. These additional configurations are split into 20 groups on which we do coarse grain-averaging and obtain estimates of statistical error [11]. In general, our statistical error is no more than several percent and is larger than error we make using the Trotter approximation. This latter error limits the accuracy of our Green's functions to about the fourth decimal place and is noticeable only at low temperatures when some of our measured quantities become comparably sized.

3. Results

Here we report results for the particle-hole symmetric case where $\varepsilon_d = -U/2$ and $\langle n_d \rangle = 1$. For all our calculations $\tau = 0.25$, and the fermi energy $\mu = 0$. At $T = U = 0$ this choice of μ corresponds to a half-filled band. We considered cases for $V = 1$ and $1/\sqrt{2}$, so that at $U = 0$ the full width of the impurity resonance, $2\Gamma = \pi\rho_F V^2$, is 1 and 1/2. For $\Gamma = 1/2$, we computed cases for $U = 1/4$, 1/2, 1, 2, 3, and 4; for $\Gamma = 1/4$, $U = 0$, 1, 2, 4. A summary of our results will now be given.

Three-dimensional plots of the spatial and temperature dependence of the spin-correlation (12c) and charge correlation (14c) functions for $U = 0$ and 2 are shown in Fig. 1. Apart from the overall scale, the two U cases are remarkably similar: There is a strong negative correlation between the impurity and the central lattice site, and the amplitude decreases as it spreads to other sites in the lattice. For $\beta < 10$ only a few nearest neighbors are correlated to the impurity, while for $\beta < 10$, the correlation spreads more rapidly outward and the amplitudes at the impurity and the closest neighbors are becoming somewhat independent of β.

To see specific effects of $U \neq 0$ on these correlations, we have to take a closer look at their U and β dependence. We find the principal effect of $U \neq 0$ is the induction of ferromagnetic spin correlations at the sites between the antiferromagnetically correlated sites. These correlations increase with increasing U. On the other hand, the correlations at the antiferromagnetic sites are often suppressed if U becomes large enough. The number of sites at which this suppression occurs is reduced to just the impurity site if β is well above β_K. The importance of being at low temperatures to reduce the suppression of the antiferromagnetic correlations was underscored by comparing the results for $\Gamma = 1/2$ and 1/4. For most of our U values the lower Γ pushes β_K outside the range of our simulation and the suppression then occurs at most antiferromagnetic sites studied.

The Coulomb interaction also induces a negative charge correlation at sites where the ferromagnetic spin correlation was induced. These correlations show a non-monotonic variation with U -- first increasing, then decreasing. The correlations at the antiferromagnetically correlated spin sites are suppressed as U is increased. We note that our charge correlation function (14c) differs from that discussed in [7] since at each site we subtract from $\langle n_d n_i \rangle$ the decorrelated value $\langle n_d \rangle \langle n_i \rangle$ for that site as opposed to the decorrelated valued for a site infinitely far away. The definition also allows (16), defined below, to express charge conversation at $T = 0$ as $T\,\Pi_d \to 1$ for the symmetric model and makes $S(i) = C(i)$ for $U = 0$ as can be seen from Fig. 1. Hence the $U = 2$ plots in this figure dramatizes the different consequences a non-zero interaction has on the spin and charge correlations. Overall increasing U enhances spin correlations and suppresses charge correlations, but specific details show a range of behavior. Some of the behavior just summarized is illustrated in Fig. 2. We point out that most of the effects are quite small.

For a given U, the effect of increasing β is to increase the antiferromagnetic spin correlations. At large β the rate of increase slows to the

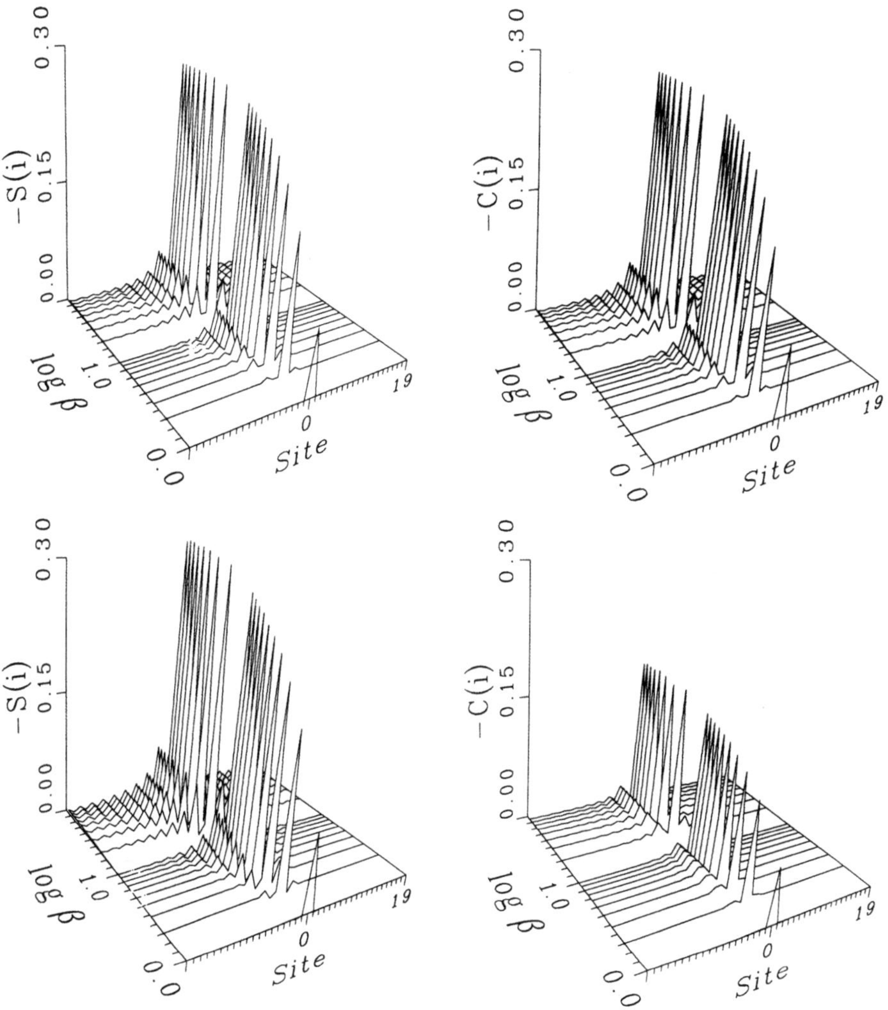

Fig 1 In the top row are the spin and charge correlation functions for V = 1 and U = 0 as a function of position and β. In the bottom row are the same correlation functions but with U = 2.

point where in some cases the correlations at the impurity and several nearby sites appear to be saturating. On the other hand, the ferromagnetic spin correlations at first increase and then become suppressed as β is increased. Only for β above β_K do the trends become most apparent. The charge correlations behave similarly. Some of this behavior is depicted in Fig. 3. Again the effects are generally small.

In [7] the following sum rule, representing the spin compensation in the same spirit as the Clogston-Anderson compensation theorem [12], was found to hold

$$\sum_i \langle\sigma_d(L)\,\sigma_i(L)\rangle = T\chi_d - \langle\sigma_d(L)^2\rangle\,. \tag{15}$$

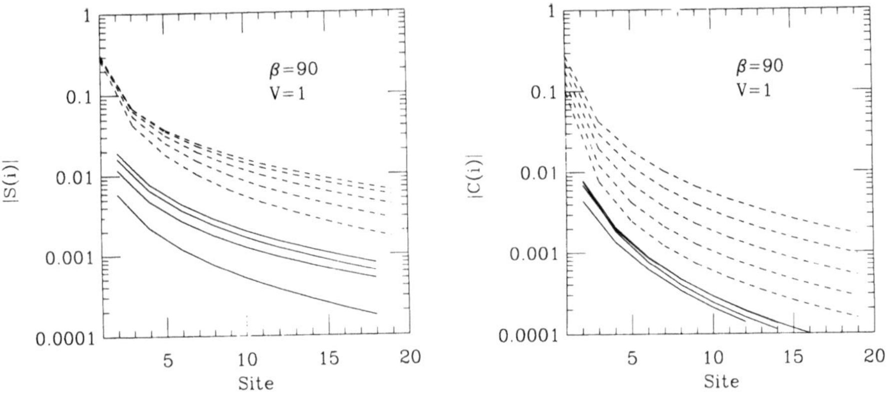

Fig. 2 Variation of the spin and charge correlations as a function of position and U. The solid lines connect the values of the function at odd lattice sites, which include the impurity site, and the dashed lines, the even sites. For the spin correlation the upper line of each type corresponds to U = 4. For the charge correlations, the lowest solid line corresponds to U = 0, while lowest dashed line corresponds to U = 0. Plotted are cases for U = 0, 1, 2, 3, and 4.

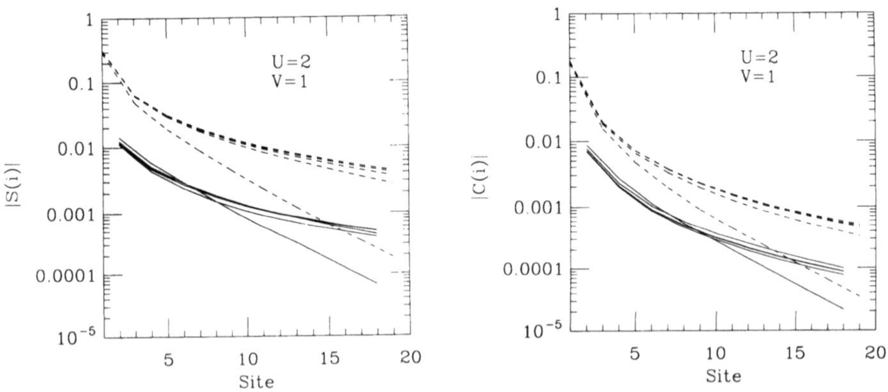

Fig. 3 Variation of the spin and charge correlations as a function of position and β. The line types are described in Fig. 2. In both figures the lowest lying solid and dashed lines, which for the most part are straight lines, correspond to β = 10. Plotted are the cases for β = 10, 30, 50, 70, and 90.

We also find this relation to hold, as well as the analogous relation for charge compensation

$$\sum_i \langle n_d(L)\, n_i(L)\rangle = T\,\Pi_d - \langle n_d(L)^2\rangle \,. \tag{16}$$

At low temperatures we find the correlation functions decay as $i^{-\alpha}$. For the spin compensation sum rule to hold, we had to correct the sum over $1 \le i \le |20|$ by fitting the ferromagnetic and antiferromagnetic peaks separately to the form $A\, i^{-\alpha}$ and then summing the resulting expression over a large number of additional sites. For charge compensation this procedure, for all prac-

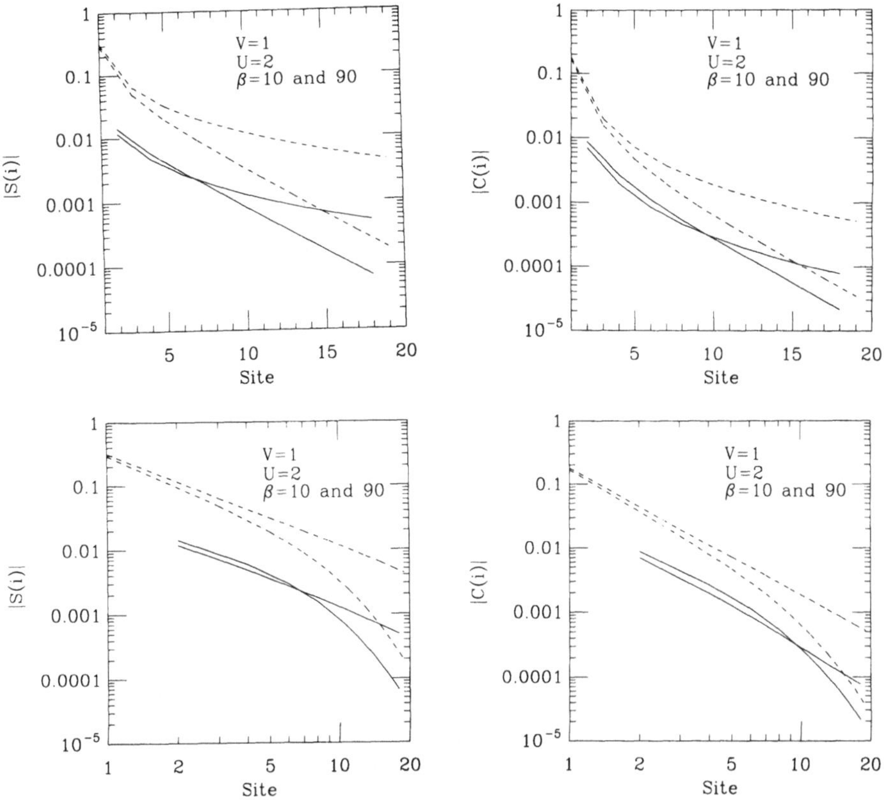

Fig. 4 The functional form of the spatial dependence of the spin and charge correlations at high and lowest temperatures. In the top row the straight lines, indicative of an exponential, are for $\beta = 10$; in the bottom row the straight lines, indicative of a power law, $\beta = 90$. The line types are defined in Fig. 2.

ticle purposes, was unnecessary. At high temperatures we found an exponential decay instead of the power law behavior. These functional dependences on i were also found for the U = 0 case. The different behaviors are illustrated in Fig. 4.

4. Summary

We summarized the results of a series of quantum Monte Carlo calculations of the spatial dependence of spin and charge correlations in a one-dimensional, single impurity, symmetric Anderson model. We corroborated several features of [7], and because we achieved lower temperatures, we were able to identify several additional unusual features in the behavior of the correlations as functions of U and β. We also showed the existence of a charge compensation sum rule and found a power law decay of the correlations at low temperatures.

To corroborate some of the low-temperature behavior discussed here, carrying the calculation to lower temperatures would be desirable but unfortunately would also be costly. The procedure described scales in computer time as L^3, and on a Cray X-MP/48 for the number of sweeps described, the 360 steps to achieve $\beta = 90$ took around 145 minutes. The vectorization achieved on a Cray computer reduces power in L^3 to something slightly less than 3. Although slightly less, the reduction in time is noticeable at low temperatures. Notwithstanding this, doubling β would be expensive to do and most likely be just partly helpful.

Within the parameter and temperature range discussed here, there are other interesting aspects of the Anderson model that can be explored. For the asymmetric Anderson model, valence fluctuation phenomena occur and can easily be studied,and have been observed with the same program used for the symmetric model [13]. Orbital degeneracy effects and magnetic fields can be added to the model, and as pointed out elsewhere [6,7], more than one impurity can be treated and realistic band structures can be used. Some of this work is currently underway.

5. Acknowledgements

Support for the calculations was provided by the Center for Materials Science at the Los Alamos National Laboratory.

6. References

1. P. W. Anderson: Phys. Rev. 124, 41 (1965).
2. F. D. M. Haldane: Phys. Rev. Lett. 40, 416 (1978).
3. H. R. Krishna-Murthy, J. W. Wilkins and K. G. Wilkins: Phys. Rev. B 21, 1003 (1980); 21, 1044 (1980).
4. For example, see G. Grüner and A. Zawadowski: Rep. Prog. Phys. 37, 1497 (1974). Recently, a new approach combining perturbative scaling and non-perturbative renormalization-group methods has been applied to a related problem by K. Chen, C. Jayaprakresh and H. R. Krishna-Murthy in the preprint "Spatial Correlations around a Kondo Impurity."
5. J. E. Gubernatis, T. C. Olson, D. J. Scalapino, and R. L. Sugar: J. Stat. Phys. 43, 831 (1986); J. E. Gubernatis, Los Alamos Technical Report LA-UR-86-2748.
6. J. E. Hirsch and R. M. Fye: Phys. Rev. Lett. 56, 2521 (1986).
7. J. E. Gubernatis, J. E. Hirsch, and D. J. Scalapino: Los Alamos Technical Report LA-UR-86-3510.
8. R. Blankenbecler, D. J. Scalapino, and R. L. Sugar: Phys. Rev. D 24, 2278 (1981). For related, earlier work see: M. Suzuki: Prog. Theor. Phys. 56, 1454 (1976); Comm. Math. Phys. 51, 183 (1976).
9. J. E. Hirsch: Phys. Rev. B 28, 4059 (1983).
10. Historically, this equation was first derived in Ref. 8 and is the basis for other finite temperature quantum Monte Carlo methods for the simulations of fermion systems. For examples of prior use, see Ref. 5 or 9.
11. K. Binder: in Applications of the Monte Carlo Method in Statistical Mechanics, ed. by K. Binder. Topics in Current Physics, Vol. 36, (Springer, New York, 1984) p. 1.
12. A. M. Clogston and P. W. Anderson: Bull. Am. Phys. Soc. 6, 124 (1961).
13. J. E. Gubernatis, unpublished.

Added Note

For $U = 0$, the spin and charge correlation functions were just calculated for 100 sites from the impurity. At large distances the correlations appear to decay exponentially, while close to the impurity the decay appears to follow a power law that depends on Γ. As β is increased, the exponentials seem to asymptote to the power law, that is, the regions over which the power law is exhibited grows. For $\beta \geq 30$, more than 20 sites are needed to see these features, and because of the similarity of the dependences of the $U = 0$ and $U \neq 0$ correlation functions on distance, it is possible for $U \neq 0$ that 20 sites are insufficient to see the asymptotic spatial dependence of the correlation functions. The functional form of the decay of the correlation functions is being investigated further.

Quantum Monte Carlo Studies of Nuclear Systems

J.W. Negele

Center for Theoretical Physics, Laboratory for Nuclear Science
and Department of Physics, Massachusetts Institute of Technology,
Cambridge, MA 02139, USA

I. INTRODUCTION

Quantum Monte Carlo methods provide a powerful tool for exploring diverse quantum systems possessing large numbers of degrees of freedom, ranging from condensed matter, spin systems, atoms and molecules to nuclei and quantum chromodynamics. This talk will survey a range of problems in nuclear systems in which we are attempting to obtain new insight into the physics of the diverse nuclear and sub-nuclear degrees of freedom which arise in nuclear physics.

The appropriate degrees of freedom for low energy nuclear physics are nucleons interacting with static potentials, and in this context stochastic methods provide a new opportunity to study tunneling and fission. Experimental studies of meson exchange currents in nuclei motivate the exact solution of a model meson-nucleon field theory. With the knowledge that nucleons are extended objects comprised of quarks, it is instructive to see how a confining quark model manifests both hadronic degrees of freedom at low energy and observable effects of quark substructure at high energies. Finally, since the ultimate understanding of nucleon structure and interactions must be understood in terms of quantum chromodynamics, it is interesting to see how the quantum Monte Carlo methods familiar from many-body problems presents insights complementary to those obtained with conventional lattice gauge theory techniques.

Path integrals provide a useful physical perspective from which to approach Monte Carlo studies of quantum systems with many degrees of freedom, and this approach is close in spirit to the original method pioneered by Suzuki[1] based on the equivalence theorem. This approach retains the physical insight provided by summing over time histories, and the Monte Carlo method may be viewed simply as an alternative to the stationary-phase approximation for summing all the fluctuations around the stationary paths. It has the advantage that one may directly exploit the usual freedom in formulating path integrals for quantal systems and utilize one's physical understanding of stationary solutions such as instantons. It therefore provides a natural perspective from which to develop stochastic methods which incorporate as much physics as possible. In practical calculations, the path integral Monte Carlo approach is roughly comparable to the Green's Function Monte Carlo Method.[2]

To distinguish the nuclear physics problems discussed in this work from applications in other fields, it is useful to recall the diverse physical scales involved in the nuclear many-body problem. The nucleon-nucleon interaction is strongly repulsive at separations less than 0.5 fm and strongly attractive in the region of 1 fm so that the range of rapid variation of the potential is characterized by $\Delta x \sim 0.2$ fm. Since a heavy nucleus has a radius of the order of 7 fm and a diffuse surface, it must be calculated in a box of linear dimension at least $L \sim 20$ fm on a side. These values of Δx and L imply that if the nuclear many-body problem were to be formulated on a lattice, for example using a functional integral over an auxiliary field, the lattice must have $(100)^3$ spatial points. Similarly, the time mesh size must be small enough that $\Delta\tau < \frac{1}{2}m(\Delta x)^2 \sim 5\times 10^{-4}\,\mathrm{MeV}^{-1}$ which requires of the order of 10^3 time steps if the energy of the first excited state is of the order of 1 MeV. The practical problems associated with solving for a single nucleon in lattice QCD are equally severe. Thus, for the present, Monte Carlo studies in nuclear physics must concentrate on pedagogical models rather than definitive realistic calculations. In the remainder of this talk, I will explain how our understanding of some of the underlying physics is incorporated in stochastic calculations and discuss what is being learned from studies of models.

II. STOCHASTIC METHODS

The principal intellectual challenge is to exploit the freedom in formulating Monte Carlo methods and to utilize one's physical insight to incorporate as much of the essential physics of the problem as possible into the stochastic algorithm.

There are three primary categories of choices which will be discussed here: the physical quantity to calculate, the functional integral representation used to formulate the path integral, and the choice between sampling the global action or solving an initial-value diffusion problem.

The first question is what physical quantity to evaluate. One option is the trace of the evolution operator $\sum_n \langle n|e^{-\beta H}|n\rangle$ which provides all thermodynamic information at inverse temperature β. Whereas this is obviously useful for a system which is in thermodynamic equilibrium, it precludes incorporating knowledge of the structure of a specific state. A second option treated in detail below is to choose a specific matrix element of $e^{-\beta H}$ which is optimal for calculation of a particular observable.[3–6] For example, in the limit of large β, the ground state energy may be written

$$E = \frac{\langle\phi|He^{-\beta H}|\phi\rangle}{\langle\phi|e^{-\beta H}|\phi\rangle} \;, \tag{2.1}$$

where $|\phi\rangle$ is an approximate wave function. The expectation value of an operator $\mathcal{O}$ may be written exactly as

$$\begin{aligned}\langle\mathcal{O}\rangle &= \frac{\langle\phi|e^{-\beta H}\mathcal{O}e^{-\beta H}|\phi\rangle}{\langle\phi|e^{-2\beta H}|\phi\rangle} \\ &\approx 2\frac{\langle\phi|\mathcal{O}e^{-\beta H}|\phi\rangle}{\langle|e^{-\beta h}|\phi\rangle} - \frac{\langle\phi|\mathcal{O}|\phi\rangle}{\langle\phi|\phi\rangle} \;.\end{aligned} \tag{2.2}$$

The phase shift for the scattering of two composite particles at energies below the excitation threshold for the first excited state is obtained by imposing the boundary condition that the wave functions vanish at a specified fragment separation distance and evaluating the energy E of the total system as a function of this distance.[7] The fission lifetime may be calculated exactly below the threshold for fragment excitation by calculating the scattering phase shift and approximately for any high barrier by calculating the overlap matrix element V between interior and exterior states [8,9]

$$\frac{\langle\phi_{\mathrm{int}}|e^{\beta H}|\phi_{\mathrm{ext}}\rangle}{\left[\langle\phi_{\mathrm{int}}|e^{-\beta H}|\phi_{\mathrm{int}}\rangle\langle\phi_{\mathrm{ext}}|e^{-\beta H}|\phi_{\mathrm{ext}}\rangle\right]^{\frac{1}{2}}} = \tanh(V\beta). \tag{2.3}$$

A second choice is the functional integral representation.[10] Different, yet exact, functional integrals may be obtained by inserting alternative resolutions of unity at each time step: integrals over coordinate states, alternate integrals over coordinate and momentum states, boson coherent states, overcomplete sets of Slater Determinants, and Grassmann coherent states to name a few. Alternatively, by applying the Hubbard Stratonovich transformation, the many-body evolution operator may be expressed as an integral over an auxiliary field of a one-body evolution operator. Whereas each form has its distinct advantages and limitations, in view of the exceedingly large mesh size necessitated by the behavior of the nuclear force for any functional integral over fields, the Feynman path integral in coordinate representation is most suitable for our present purposes.

Finally, the last question is whether to repeatedly sample the action for the system, thereby generating a sequence of complete time histories, or to formulate an initial value problem in which an initial ensemble of points is evolved in time to generate an ensemble of time histories. Each approach offers distinct advantages which I will briefly summarize.

The initial value diffusion problem may be formulated as a guided random walk,[11] which is written here for simplicity for the case of quantum mechanics in one dimension. Instead of evolving the solution to the imaginary time-dependent Schrödinger equation

$$-\dot{\psi}(x,\tau) = \left[-\frac{1}{2m}\frac{d^2}{dx^2} + V(x)\right]\psi(x,\tau) \tag{2.4}$$

one evolves the product of ψ with a trial function $\phi(x,\tau)$

$$f(x,\tau) = \phi(x,\tau)\psi(x,\tau) \quad . \tag{2.5}$$

The equation of motion for $f(x,\tau)$ is

$$-\dot{f}(x,\tau) = -\frac{1}{2m}\frac{\partial^2}{\partial x^2} f + \frac{1}{m}\frac{\partial}{\partial x}\left(\frac{\phi'}{\phi} f\right) + \frac{-\frac{1}{2m}\phi'' + V\phi - \dot{\phi}}{\phi} f \; , \tag{2.6}$$

which includes a source sink term $\frac{\left(-\frac{1}{2m}\frac{\partial^2}{\partial x^2} + V - \frac{\partial}{\partial \tau}\right)\phi}{\phi}$ and a drift term $\frac{1}{m}\frac{\partial}{\partial x}\left(\frac{\phi'}{\phi} f\right)$. The infinitesimal evolution operator contributing to a path integral may be written

$$\left\langle x_n \middle| \phi(\hat{x},\tau+\epsilon) e^{-\epsilon\hat{H}} \frac{1}{\phi(\hat{x},\tau)} \middle| x_{n-1}\right\rangle = \sqrt{\frac{m}{2\pi\hat{\epsilon}}}\, e^{-\frac{m}{2\hat{\epsilon}}\left(x_n - x_{n-1} - \frac{\epsilon}{m}\frac{\phi'}{\phi}\right)^2 - \epsilon\frac{\left(H-\frac{\partial}{\partial\tau}\right)\phi}{\phi}} \; , \tag{2.7}$$

where

$$\hat{\epsilon} = \hat{\epsilon}(x_{n-1}) = \epsilon\left(1 + \frac{\epsilon}{m}\frac{d^2}{dx^2}\ln\phi\right)\bigg|_{x_{n-1}}$$

and ϕ, ϕ', and $\left(H - \frac{\partial}{\partial\tau}\right)\phi$ are evaluated at x_{n-1}. A matrix element of $e^{-\beta H}$ is then evaluated by chaining infinitesimal matrix elements of the form (2.7) in the usual way

$$\left\langle\phi_a\middle|e^{-\beta H}\middle|\phi_b\right\rangle = \int dx_1\cdots dx_n \left\langle\phi_a\frac{1}{\phi}\middle|x_n\right\rangle \prod_{m=2}^{n}\left\langle x_m\middle|\phi e^{-\epsilon H}\frac{1}{\phi}\middle|x_{m-1}\right\rangle\left\langle x_1\middle|\phi\phi_b\right\rangle \tag{2.8}$$

gives rise to a branched random walk. First, x, is randomly sampled according to $\phi(x)\psi_b(x)$. Given a set of points $\{x_{m-1}\}$, the next set is generated by Gaussian diffusion $e^{-\frac{m}{2}\left(x_m - x_{m-1} - \frac{\epsilon}{m}\frac{\phi'}{\phi}\right)^2}$ and replication or deletion with probability $e^{-\epsilon\frac{\left(H-\frac{\partial}{\partial\tau}\right)\phi}{\phi}}$. The final generation is used to sample $\int dx_n \phi_a(x_n)\frac{1}{\phi(x_n)}$ and thus evaluate the matrix element. In the special case in which the trial function ϕ is 1, this corresponds to path integral Monte Carlo in which diffusion accounts for the kinematic energy and replication and deletion account for the potential energy. In the case of nuclear forces, this replication and deletion introduces such a large variance that quantitative calculation become difficult or impossible. In the opposite extreme in which $\phi(x,\tau)$ is an exact solution to the time reversed evolution equation, the source-sink term is a constant contributing nothing to the variance, and all the physics is contained in the drift term which guides the evolution of the population.To the extent to which an intelligent approximation to ϕ incorporates much of the essential physics, the stochastic calculation only has to account for the residual error.

The many-Fermion problem is treated analogously. The matrix element of the evolution operator between antisymmetrized states is

$$\begin{aligned}&\langle x_1^n \dots x_A^n | \phi e^{-\epsilon(\hat{H}-E)} \frac{1}{\phi} | x_1^{n-1} \dots x_A^{n-1} \rangle \\ &\quad = \sum_P (-1)^P \left[\tfrac{m}{2\pi\epsilon}\right]^{\frac{A}{2}} e^{-\frac{m}{2\epsilon}\left(x_{Pi}^n - x_i^{n-1} - D_i(x^{n-1})\right)^2 - \epsilon S(x^{n-1})} ,\end{aligned} \tag{2.9}$$

where the drift term is $D_i(x^{n-1}) = \frac{\epsilon}{m} \frac{d}{dx_i^{n-1}} ln\phi(x_1^{n-1} \dots x_A^{n-1})$ and the source term is $S(x^{n-1}) = \frac{(\hat{H}-\frac{\partial}{\partial\tau})\phi(x^{n-1})}{\phi(x^{n-1})}$. In more than one spatial dimension, interference between positive and negative contributions to the functional integral degrades the statistical accuracy such that very good trial functions and exceedingly large ensembles are required to obtain useful results.[12] In one dimension, however, antisymmetry completely specifies the nodal points and a positive definite result may be obtained by simply evolving the wavefunction in the subspace $x_1 < x_2 < x_3 \dots x_A$. As described in Ref. [4], it is useful to approximate the determinant by

$$\begin{aligned}&\sum_P (-1) P e^{-\frac{m}{2\epsilon}(x_{Pi}^n - x_i^{n-1} - D_i)^2} \\ &\quad = e^{-\frac{m}{2\epsilon}\sum_i (x_i^n - x_i^{n-1} - D_i)^2} \det \left| e^{-\frac{m}{2\epsilon}\left[(x_i^n - x_j^{n-1} - D_j)^2 - (x_i^n - x_i^{n-1} - D_i)^2\right]} \right| \\ &\quad \approx e^{-\frac{m}{2\epsilon}\sum_i (x_i^n - x_i^{n-1} - D_i)^2} \prod_{i=2}^{A} \left(1 - e^{-\frac{m}{\epsilon}(x_i^n - x_{i-1}^n)(x_i^{n-1} - D_i - x_{i-1}^{n-1} + D_{i-1})}\right),\end{aligned} \tag{2.10}$$

which includes the effect of the first images surrounding each node.

Sampling the global action using the Metropolis algorithm or some variant is particularly well-suited to problems in collective motion such as nuclear fission. The essential issue may be seen in the familiar case of a double well potential in one dimension. We know from applying the stationary phase approximation to the Feynman path integral that for a high barrier, the dominant configurations for tunneling processes are given by instantons. Nevertheless, a naive application of the Metropolis algorithm with uncorrelated updates or application of the Langevian equation gives rise to trajectories which become trapped in one of the two wells.

Once the essential problem with sequential updates has been recognized, it is possible to introduce generalizations to include instanton physics. One alternative is to define the following correlated additive trial move for the entire trajectory

$$x_i^T = x_i + \sigma f_i , \quad \sigma = \pm 1 . \tag{2.11}$$

Here, f_i is some global change defined on all the time slices and σ is a random variable with value ± 1. In order to make the move reversible, f must be randomly added and subtracted, with the disadvantage that half the moves are thereby sure to be rejected. A physical choice for f is the semiclassical two-instanton trajectory calculated in the stationary phase approximation. The cost in action for such a global change is distributed around the classical result and assures that the equilibrium distribution of instantons is built up. Note that if the positions are also picked randomly, subsequent updates will change the instanton number by ± 2 and distribute the instantons randomly.

In addition to the introduction of collective moves, one always has the option of using importance sampling. We define a function $N(\vec{x})$ which counts the number of instantons in a configuration and write the tautology

$$\int d\vec{x}\ \mathcal{O}(\vec{x})e^{-S(\vec{x})} = \int d\vec{x} \left[\mathcal{O}(\vec{x})e^{-\alpha N(\vec{x})}\right] e^{-(S(x)-\alpha N(\vec{x}))}\ . \qquad (2.12)$$

By adjusting α, one can make the acceptance of instantons in the Metropolis updates arbitrarily desirable and thus control the average number of instantons. Although the extraneous factor is exactly divided out in evaluating observables, if the quantity of interest is dominated by multi-instanton configurations, the technique lives up to its name by preferentially sampling the space in the "important" regime. Note also that the definition of an instanton doesn't need to be absolutely precise: as long as $N(\vec{x})$ counts configurations in the desired sector, it will introduce configurations in this sector which subsequently properly equilibrate. Using these techniques, it is straightforward to calculate the energy splitting in a double well and the lifetime of a metastable state, and detailed results may be found in Refs. [8] and [9].

III. MANY-NUCLEON PROBLEM WITH STATIC INTERACTIONS

Even at the level of non-relativistic quantum mechanics with static interactions, the nuclear many-body problem is extremely difficult to solve, so it is useful to study model systems stochastically to assess the validity of conventional approximations. The model one-dimensional nuclear system described in Refs. [4] and [12] provides a simple case in which to study fission. A potential with strong short-range repulsion, intermediate range attraction and a long-range Coulomb-like force force gives the ground state densities shown in Fig. 1 and produces a 16-particle bound state which is unstable with respect to fission into two eight-particle daughters. The essential element in calculating fission for such a model is introducing collective updates which correspond to instantons in the

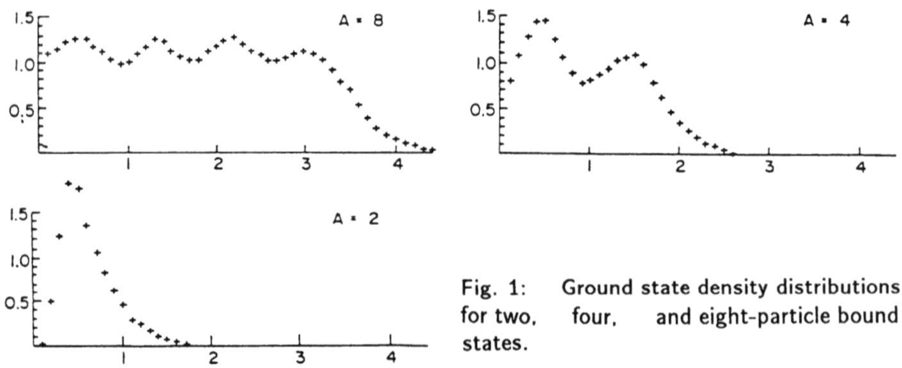

Fig. 1: Ground state density distributions for two, four, and eight-particle bound states.

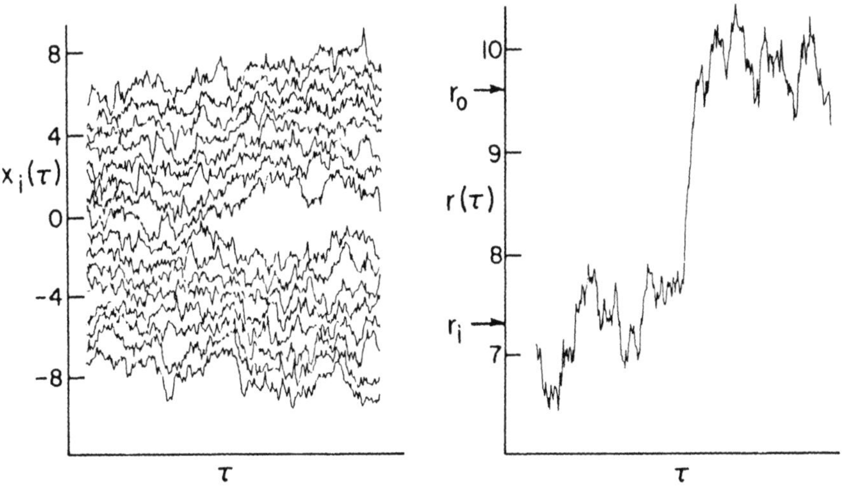

Fig. 2: Typical configurations of the 16-particle system in the one-instanton sector. The left-hand panel shows the coordinates of each of the 16 particles on 400 time slices and the right-hand panel shows the corresponding trajectory of the collective coordinate.

collective variable $r = \frac{1}{8}\sum_{i=9}^{16} x_i - \sum_{i=1}^{8} x_i$. A typical configuration contributing to the matrix element governing the fission lifetime is shown in Fig. 2. Note that the many-body trajectory clearly exhibits separation into two fragments and the corresponding trajectory for the collective coordinate has the same behavior as a single particle in a one-dimensional double well. The details of the fission calculation are described in Refs. [8] and [9].

IV. MESON NUCLEON FIELD THEORY

The simplest nuclear theory containing sub-nuclear degrees of freedom is a theory of the Walecka form containing vector and scalar mesons coupled to nucleons[13] and a one-dimensional version of this theory has been studied in Ref. [14]. Eliminating the time component of the vector field, which is a static mode, and defining E to be the momentum conjugate to the space component of the vector field, V_1, the Hamiltonian may be written

$$\mathcal{H} = \psi^\dagger\left(-i\alpha\frac{\partial}{\partial x} + \beta M\right)\psi + \frac{1}{2}\left(\pi_s^2 + (\phi')^2 + m_s^2\phi^2\right) + \frac{1}{2}\left(E^2 + \frac{(E')^2}{m_v^2} + m_v^2 V_1^2\right)$$
$$+\psi^\dagger\left[-g_v\alpha V_1 - \frac{g_v}{m_v^2}E' - g_s\beta\phi\right]\psi + \frac{1}{2}\,\frac{g_v^2}{m_v^2}\,(\psi^\dagger\psi)^2 \qquad (4.1)$$

where ϕ and π_s are the scalar field and momentum, respectively, and $m_{v,s}$ and $g_{v,s}$ denote masses and coupling constants for the vector and scalar fields. Taking the non-relativistic limit and Fourier transforming to discrete modes in a domain with periodic boundary conditions yields

$$\mathcal{H} = \frac{1}{2M}\sum_{i=1}^{A}\frac{\partial^2}{\partial x_i^2} + \sum_{k=1}^{\infty}\left\{-\frac{1}{2}\left[\frac{\partial^2}{\partial \phi_k^2} - w_k^2\phi_k^2\right] - \frac{1}{2}\left[\frac{\partial^2}{\partial \hat{E}_k^2} - \Omega_k^2\hat{E}_k^2\right]\right.$$
$$\left. - g_s\phi_k\rho_k + \frac{1}{2}\frac{g_v^2}{m_v^2}\rho_k^2 - \sum_{i=1}^{A}\frac{g_v}{m_v}\hat{E}_k\frac{\partial\rho_k}{\partial x_i}\right\} + H_{\text{zero}} , \tag{4.2}$$

where $\hat{E} = \frac{E}{m_v}, w_k^2 = m_s^2 + k^2$, $\Omega_k^2 = m_v^2 + k^2$ and H_z includes the static contribution of all $k = 0$ modes (which is identical to the mean field approximation) as well as the zero-point contribution $-\frac{1}{2}\sum_{k=1}^{\infty}(w_k + \Omega_k)$.

The appropriate variables for the stochastic evolution are the A fermion coordinates $\{r_i\}$ and the Fourier components ϕ_k and E_k of the scalar field and momentum conjugate to the vector field. For static nucleons, the meson modes are just decoupled displaced harmonic oscillators

$$\left(-\frac{1}{2}\frac{\partial^2}{\partial_k^2} + \frac{1}{2}w_k^2\phi_k^2 - g_s\rho_k\phi_k\right)\Psi_s(\phi_k) = \epsilon_k^s\Psi_s(\phi_k) ,$$

$$\left(-\frac{1}{2}\frac{\partial^2}{\partial E_k^2} + \frac{1}{2}\Omega_k^2E_k^2 - \frac{g_v}{m_v}E_k\frac{\partial\rho_k}{\partial x_i} + \frac{1}{2}\left(\frac{g_v}{m_v}\right)^2\rho_k^2\right)\Psi_v(E_k) = \epsilon^v\Psi_v(\phi_k) \tag{4.3}$$

with solutions $\Phi_s(\phi_k) = e^{-\frac{1}{2}w_k[\phi_k - x_k]^2}$ and $\Phi_v(E_k) = e^{-\frac{1}{2}\Omega_k[E_k-\varsigma_k]^2}$, where $\chi_k = \frac{g_s\rho_k}{w_k^2}, \xi = \frac{g_v}{\Omega_k^2 m_v}\sum_i\frac{\partial\rho_k}{\partial x_i}, \epsilon_k^s = -\frac{1}{2}g_s^2\frac{\rho_k^2}{w_k^2}$, and $\epsilon_k^v = \frac{1}{2}g_v^2\frac{\rho_k^2}{\Omega_k^2}$.

For the static two-body problem, the interaction energy yields the familiar static potential with scalar attraction and vector repulsion:

$$\sum_k\left(\frac{-g_s^2}{k^2+m_s^2} + \frac{g_v^2}{k^2+m_v^2}\right)\rho_k^2 \to U(x) = \left[\frac{-g_s^2}{m_s^2}e^{-m_s|x|} + \frac{g_v^2}{m_v^2}e^{-m_v|x|}\right]$$
$$+ \text{periodic images.} \tag{4.4}$$

The physical difference between the full meson field theory and a static potential theory based on the two-body potential $U(x)$ in Equation (4.4) is the non-adiabatic coupling between nucleons and mesons. Physically, for sufficiently high frequencies, the meson motion is sufficiently fast relative to the nucleon motion that the static solutions, Equation (4.3), become exact. The sum over dynamic meson modes may then be truncated by summing the static contributions for all modes above a cutoff $k_{\max}$ as follows:

$$\mathcal{H} \to \sum_{k=1}^{k_{\max}}\left[\mathcal{H}_k - \left(-\frac{g_s^2}{w_k^2} + \frac{g_v^2}{\Omega_k^2}\right)\right] + \sum_{i<j}U(x_i - x_j) . \tag{4.5}$$

The obvious trial wave function to use for the nuclear coordinates and dynamic meson modes is the product of a nucleon Slater determinant and displaced meson oscillator functions

$$\Psi_{\text{Trial}} = \psi_{SD}(x_1\ldots x_n)\prod_k e^{-\frac{1}{2}w_k[\phi_k - \chi_k(x_1\ldots x_n)]^2 - \frac{1}{2}\Omega_k[E_k - \varsigma_k(x_1\ldots x_n)]^2} . \tag{4.6}$$

To the extent to which the low-lying mesons modes are adiabatic, all the meson physics is incorporated in the displaced oscillator functions, and the stochastic evolution is only required to deal with the small non-adiabatic effects. The trial function Equation (4.6) constitutes a form of Born-Oppenheimer approximation in which the dynamic part of the meson wave function is determined by the instantaneous nucleon coordinates and provides a convenient physical description of the role of mesons in nuclei.

The principal result of stochastic solutions of the meson-nucleus field theory is that dynamic contributions of mesons are very small. Dynamic modes were cut off above $3\ fm^{-1}$, justified *a posteriori* by the fact that lower modes were essentially adiabatic as well. In the adiabatic limit, the meson contribution to the nuclear self-energy is $188.4\ MeV$ and this was shifted only by $3.9\ MeV$ and $2.7\ MeV$ from the effect of dynamic scalar and vector mesons, respectively.

The nuclear matter saturation curve for this theory is shown in Fig. 3. The stochastic solution of the many-fermion problem with the static potential of Equation (4.4) is shown by the solid dots, and the effect of adding dynamic scalar or vector modes, shown by the crosses and pluses, respectively, is less than 5%. The parameters of the model were defined such that the mean field approximation, shown by the solid curve, saturated at the usual energy and density, so the vector potential is evidently much weaker than desired by a more realistic model.

No effects of meson dynamics were observable in the one-body density distributions for finite nuclei. Binding energies of finite nuclei displayed roughly the same sensitivity to meson dynamics as nuclear matter, with the total effect of dynamic scalar and vector mesons on the binding energy of the $A = 12$ system being 5%. The nucleon-nucleon correlation function showed no effect of meson dynamics at the one percent level, and the meson correlation function $< \phi_k^2 >$ differed by less than one percent from the adiabatic value $\chi_k^2 + \frac{1}{2w_k}$ for all modes considered.

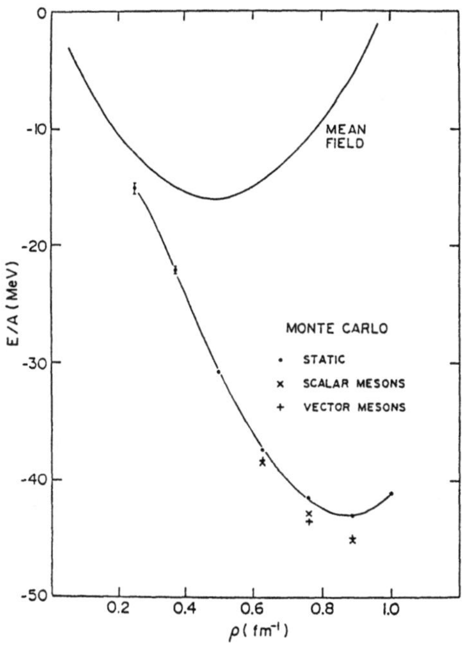

Fig. 3: Nuclear matter saturation curve for meson-nucleon field theory.

In the context of meson theory, these results suggest that the appropriate way to treat subnuclear degrees of freedom is to first solve the nuclear many-body problem with the static interaction derived in the adiabatic limit, and then treat the small dynamic meson effects perturbatively. The physical Born-Oppenheimer argument is independent of dimension and should apply to more realistic models in three dimensions. Extension of the present method to relativistic fermions in one dimension is practical, and for very light nuclei, stochastic calculations in three dimensions may be possible.

V. LATTICE GAUGE THEORY

Ultimately, any treatment of subnuclear degrees of freedom will require the solution of QCD. Although a variety of stochastic approaches are presently utilized to solve lattice gauge theories, the practical and physical appeal of using a trial wave function to guide the stochastic evolution motivated use of the method described in Section II. As a first application, we have studied $U(1)$ lattice gauge theory in 3 spatial dimensions.[11] This case is both simple enough to display the essence of the method and rich enough, because of the presence of a phase transition, to pose a non-trivial test of the use of a trial function.

The dynamical variables are periodic variables $A_\mu(\vec{n})$ defined on the links of a three dimensional lattice with μ denoting the direction of the link and $\vec{n}$ denoting the position of the end of the link. The Hamiltonian may be written

$$H = \sum_{\mu,\vec{n}} \frac{1}{2} \frac{\partial^2}{\partial A_\mu^2(\vec{n})} - \lambda \sum_{\mu,\vec{n}} \Big(1 - \cos B_\mu(\vec{n})\Big) . \tag{5.1}$$

where $B_\mu(\vec{n})$ is the lattice curl

$$B_\mu(\vec{n}) = \epsilon_{\mu\nu\gamma}[A_\gamma(\vec{n}+\vec{e}_\nu) - A_\gamma(\vec{n})] , \tag{5.2}$$

which is just the sum over all plaquettes of directed links around the plaquette. In terms of the coupling constant g of the lattice theory, $\gamma = \frac{1}{g^4}$. In the weak coupling limit, the first term becomes E^2 and the second term becomes B^2 reproducing continuum QED.

For a periodic angular variable θ, the matrix element of the kinetic evolution operator may be expanded in periodic eigenstates

$$\begin{aligned} \langle\theta'|e^{-\frac{1}{2}\frac{\partial^2}{\partial\theta^2}}|\theta\rangle &= \sum_n \langle\theta'|n\rangle\langle n|e^{-\frac{1}{2}\frac{\partial^2}{\partial\theta^2}}|\theta\rangle \\ &= \sum_n e^{in(\theta'-\theta)} e^{-\epsilon\frac{n^2}{2}} \\ &= \sum_m^n e^{-\frac{1}{2\epsilon}(\theta'-\theta+2\pi m)^2} \underset{\epsilon\to 0}{\longrightarrow} e^{-\frac{1}{2\epsilon}(\theta'-\theta)^2} . \end{aligned} \tag{5.3}$$

In the limit of infinitesimal time step ϵ, the sum of the Gaussian propagators over an infinite set of images obtained in the last line using the Poisson sum formula may be replaced on the interval $|\theta'-\theta| < \pi$ by a single Gaussian. Thus, the evolution of a periodic angular variable becomes identical to that of a coordinate in quantum mechanics, and the lattice gauge theory is equivalent to $3L^3$ variables $A_\mu(n)$ evolving with a potential

$$V(A) = -\lambda \sum_{\text{plaquettes}} \Big(1 - \cos(B[A])\Big) . \tag{5.4}$$

A convenient gauge-invariant trial function is the product over all plaquettes P

$$\Phi(A) = \prod_P f(B_P[A]) , \tag{5.5}$$

where

$$f(B) = e^{-b(\lambda)(1-\cos B) - \frac{1}{2}(a(\lambda)-b(\lambda))\sin^2 B} , \tag{5.6}$$

B_P denotes the lattice curl, Equation (4.3.2), for the plaquette P and $a(\lambda)$ and $b(\lambda)$ are variational parameters to optimize the wave function. In the strong coupling limit, that is $\lambda \to 0$, the ground state is dominated by single plaquette contributions $\psi_0 = \prod_P e^{-\frac{1}{2}[1-\cos B_P]} + 0(\lambda^2)$ so that Equation (4.3.5) becomes exact for $a(\lambda) = b(\lambda) = \frac{1}{2}$. In the weak coupling limit, the ground state is just an ensemble of independent harmonic oscillators $\psi_0 = e^{-\frac{\sqrt{\lambda}}{2} \sum_{PP'} B_p \Gamma_{PP'} B_{P'}}$. It is shown in Ref. [11] that the diagonal terms strongly dominate the off-diagonal terms and are reproduced by $a(\lambda) = 0.428\lambda^{\frac{1}{2}}$ and $b(\lambda) = \frac{4}{3}a(\lambda)$. The trial function smoothly interpolates between these two limits for intermediate λ and gives an economical representation of much of the local structure of the ground state wave function.

The primary result of this calculation is that the stochastic method of Section II is well-suited to lattice gauge theory calculations. Results for the binding energy per plaquette and plaquette energy $\langle 1 - \cos(B) \rangle$ are shown in Fig. 4. The use of a trial function yields high statistical accuracy throughout the region of the phase transition, and the results are consistent with strong and weak coupling expansions in the appropriate domains and with other Monte Carlo results.

Whereas the results thus far are not directly relevant to the study of sub-nuclear degrees of freedom, the method may be extended straightforwardly to non-abelian lattice gauge theories.[15] The principal challenge is to find a trial function which has the appropriate physics in the continuum limit. Note that the present independent plaquette trial function has no physical content in the continuum limit, and one must find a tractable,

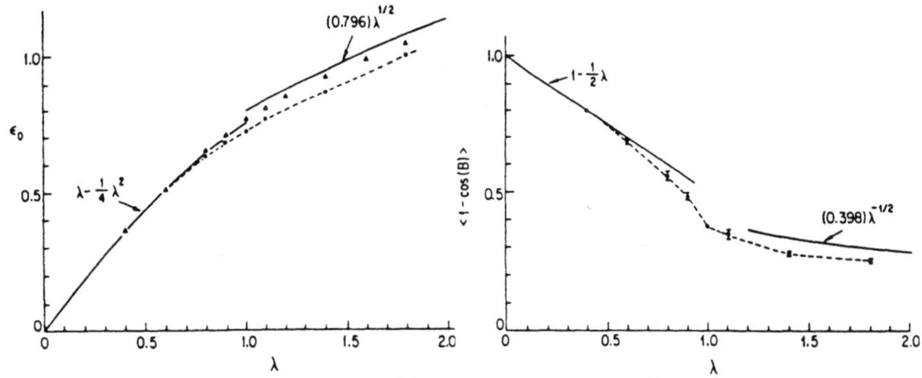

Fig. 4: Monte Carlo results for $U(1)$ lattice gauge theory. The left panel shows the ground state energy per plaquette as a function of λ. The solid lines denote strong and weak coupling expansions and the triangles and dots denote variational and exact Monte Carlo results. The statistical errors are too small to be shown. The right panel shows the plaquette energy as a function of λ. Solid curves again denote the strong and weak coupling expansions. Dots denote perturbative estimates of the plaquette energy and the $\times$ denotes a derivative of the ground state energy with respect to λ.

gauge-invariant wave function which introduces long-range correlations having a physical continuum limit.

VI. CONFINING QUARK MODEL

The confining quark model of Ref. [16] provides a convenient system in which to study how certain aspects of nuclear structure arises from underlying quark degrees of freedom. By studying the model in one spatial dimension, it is possible to obtain exact solutions for uniform matter described both in terms of quark and hadronic degrees of freedom and thus explore the complementarity of the two descriptions.[7]

Physically, the model may be thought of as an adiabatic limit in which for any configuration of $2N$ spinless quarks, the color fields instantaneously adjust themselves to form the lowest energy configuration in which N distinct pairs of quarks are connected by flux tubes. Mathematically, the model is defined by specifying a potential energy $v(r)$ which is an increasing function of r associated with a flux tube of length r and defining the total potential energy of the quarks to be

$$V = \min_{P} \{v(x_{P1} - x_{P2}) + v(x_{P3} - x_{P4}) + \ldots + v(x_{P(N-1)} - x_{PN})\}\,, \qquad (6.1)$$

where the minimum overall permutations selects the lowest energy assignment of pairs. The theory exhibits the desired separability, confinement, and exchange symmetry and is free of Van der Waals interactions. Although finding the optimal pairing in three dimensions entails solution of a non-trivial assignment problem, in one dimension with periodic boundary conditions, the optimal pairing of consecutively labeled quarks either pairs q_{2m} with q_{2m+1} or q_{2m} with q_{2m-1} for all m. Hadrons are composed of two quarks in the simplest version of the model used here and are thus Bosons. The potential $v(x)$ is taken to be quadratic, yielding a quark density in a free nucleon proportional to $x^2 e^{-\frac{x^2}{\sqrt{2}}}$

It is instructive to compare observables calculated in this model with the results obtained from an effective Hamiltonian having only hadronic degrees of freedom. As in traditional nuclear physics, a hadron-hadron potential may be defined to reproduce the exact hadron scattering phase shifts, and the many-hadron problem is solved with this potential. We find that a gross property such as the hadronic matter binding energy as a function of density is well-reproduced by this effective Hamiltonian, whereas the quark-quark correlation function and quark momentum distributions, which would be probed in deep inelastic electron scattering, reflect the underlying quark structure.

The phase shifts for the scattering of a two-quark hadron from a two-quark hadron may be solved stochastically below threshold by calculating the ground state of the four quark system with the appropriate boundary condition on the distance between the centers of mass of the left two quarks and the right two quarks, and the Monte Carlo results are shown in Ref. [7]. Since the phase shifts in this energy range are evidently characterized by a scattering length and an effective range, they are easy to reproduce with a simple local potential $V_N(x) = \frac{52}{\sqrt{2\pi}}\, e^{-\frac{x^2}{2}}$. Of the two effects which occur for overlapping hadrons, diminuation of the potential interaction energy and increased kinetic energy due to the Pauli principle, the later effect dominates and the interaction is purely repulsive.

The binding energy per quark of uniform nuclear matter is shown in Fig. 5. Stochastic solution of the many-quark problem yields the solid dots, and the many-hadron problem defined by Bosons interacting via the phenomenological potential $V_N(x)$ yields the triangular points.

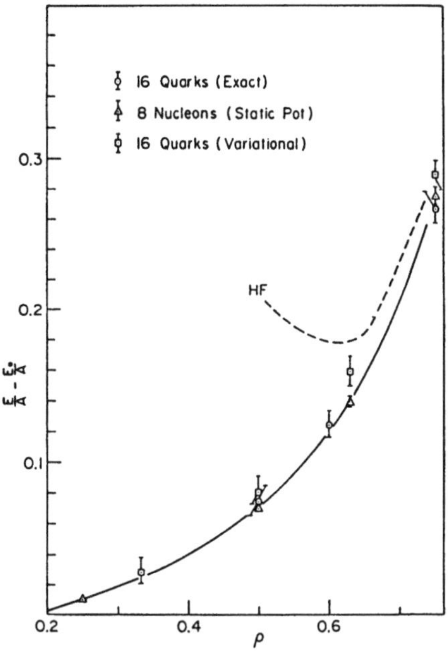

Fig. 5: Monte Carlo calculation of the binding energy per quark of uniform matter calculated with the quark Hamiltonian (circles) and the effective hadron Hamiltonian (triangles).

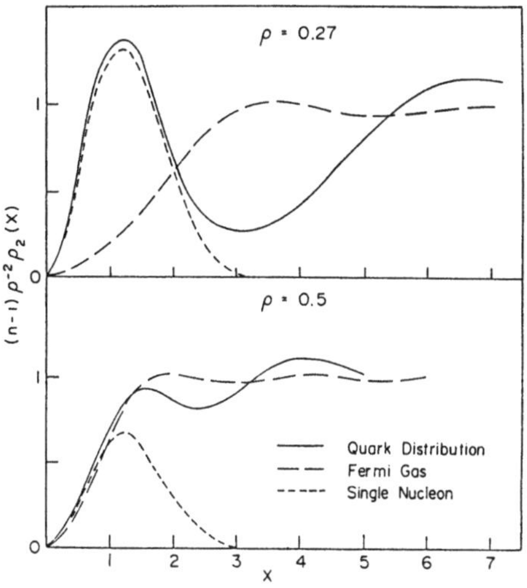

Fig. 6: Quark-quark correlation functions as described in the text.

In order to calibrate densities in the one-dimensional model, it is useful to note that the ratio $\frac{\bar{\rho}_N}{\rho_{NM}} = (1.1)^3$, where $\bar{\rho}_N$ denotes the average density in a proton of uniform density and the observed rms charge radius and $\bar{\rho}_{NM}$ is nuclear matter density. Since the average density of a hadron in the one-dimensional model is $\bar{\rho}_H = 0.5$, a corresponding degree of overlap obtains at a hadronic matter density $\bar{\rho}_{HM} = \frac{0.5}{1.4} \sim 0.45$. Thus, we observe from Fig. 5 that up to densities twice that of hadronic matter, the binding energy yields no signature of quark substructure or, equivalently, of the breakdown of the descriptions in terms of hadronic degrees of freedom.

A microscopic view of the quark behavior is provided by the quark-quark correlation functions shown in Fig. 6 which specify the probability of finding another quark a distance x from a given quark. The normalization is defined such that at density ρ corresponding to n particles in a periodic box of length L, the integral of the correlation function from 0 to L is $L(n-1)$, thus counting the $(n-1)$ remaining particles. The Fermi gas correlation function then approaches 1 in the interior of the box and approaches zero within range $\frac{1}{k_F}$ of 0 and L as shown by the long dashed lines. At very low density, one would expect the correlation function to look like the undistorted ground state nucleon density at short distances and approach a constant corresponding to the low-density nucleon gas at large distances. Hence, the appropriately normalized density distributions for a single hadron are also shown in Fig. 6 by the short dashed curves. As observed in the top portion of the figure, at $\rho = 0.27$, roughly half the density of hadronic matter, the quark distribution exhibits the low density behavior of undistorted hadrons. At $\rho = 0.5$, corresponding to hadronic matter density, however, the nucleon correlations have nearly disappeared and the correlation function is close to that of a Fermi gas. Note that at this density, the binding energy is still well represented by the effective hadron-hadron potention. An analogous transition between the hadron momentum distribution and a Fermi gas distribution is observed in the quark momentum distribution. At hadronic matter density, one observes the enhancement at high momentum, enhancement at low momentum, and depletion at intermediate momentum characteristic of the EMC effect.

This provocative coexistence of bulk properties accurately governed by an effective Hamiltonian containing only hadronic degrees of freedom while leptonic probes reveal definite signatures of quark substructure is presently being investigated in more realistic confining models in three dimensions having nucleons comprised of three quarks.[17]

VII. CONCLUSIONS

Quantum Monte Carlo methods are providing new approaches to a variety of problems involving large numbers of degrees of freedom. I have shown how studies of simple model systems have enabled us to obtain some physical insights into problems in contemporary nuclear physics. With the new developments and ideas emerging from complementary studies of spin systems and condensed matter physics, we can look forward to continued progress in addressing more complicated and realistic problems.

VIII. ACKNOWLEDGEMENTS

It is a pleasure to thank Professor Suzuki and the conference organizers for their kind hospitality. This work was supported in part through funds provided by the U. S. Department of Energy (D.O.E.) under contract #DE-AC02-76ER03069.

IX. REFERENCES

1. M. Suzuki, *Prog. Theor. Phys.* **56**, 1454 (1976).
2. D. M. Ceperley and M. H. Kalos, *Monte Carlo Methods in Statistical Mechanics*, K. Binder, ed., (Springer-Verlag, NY) (1979)
3. S. E. Koonin, *Nuclear Theory 1981*, G. F. Bertsch, ed. World Scientific, (1981).
4. J. W. Negele, *Proceedings of the International Symposium on Time-Dependent Hartree-Fock and Beyond, Lecture Notes in Physics* **171**, K. Geoke and P. G. Reinhardt, eds., Springer-Verlag, NY (1982).
5. J. W. Negele, *Jour. Stat. Phys.* **43**, 991 (1986).
6. J. W. Negele, *Theory of Nuclear Structure and Reactions*, M. Lozano and G. Madurga, eds. (World Scientific:Singapore, 1985) p. 77.
7. C. Horowitz, E. Moniz, and J. W. Negele, *Phys. Rev. D* **31**, 1689 (1985).
8. J. W. Negele, Proceedings of "Path Integrals from mev to MeV", Bielefeld, 1985, J. Klauder, ed.
9. M. Creutz and B. Freedman, *Ann. Phys.* **132**, 427 (1981).
10. J. W. Negele, *Rev. Mod. Phys.* **54**, 913 (1982).
11. S. A. Chin, J. W. Negele and S. E. Koonin, *Ann. Phys.* (NY) **157**, 140 (1984).
12. D. M. Ceperley and B. J. Alder, *Phys. Rev. Lett.* **45**, 566 (1980).
13. J. D. Walecka, *Ann. Phys.* **83**, 491 (1971); S. A. Chin and J. D. Walecka, *Phys. Lett.* **52B**, 24 (1974); S. A. Chin, *Ann. Phys.* **108**, 403 (1977).
14. B. D. Serot, S. E. Koonin and J. W. Negele, *Phys. Rev. C* **28**, 1679 (1983).

15. S. A. Chin, O. S. van Roosmalen, E. A. Umland and S. E. Koonin, *Phys. Rev. D* **31**, 3201 (1985).
16. F. Lenz, J. T. Longergan, E. J. Moniz, R. Rosenfelder, M. Stingl and K. Yazaki, *Ann. Phys.* **170**, 65 (1986).
17. R. M. Panoff, private communication.

Index of Contributors

MIX
Papier aus verantwortungsvollen Quellen
Paper from responsible sources
FSC® C105338

If you have any concerns about our products,
you can contact us on
ProductSafety@springernature.com

In case Publisher is established outside the EU,
the EU authorized representative is:
Springer Nature Customer Service Center GmbH
Europaplatz 3, 69115 Heidelberg, Germany

Printed by Libri Plureos GmbH
in Hamburg, Germany